Digital Video Concepts, Methods, and Metrics

Quality, Compression, Performance, and Power Trade-off Analysis

Shahriar Akramullah

apress
open

Digital Video Concepts, Methods, and Metrics: Quality, Compression, Performance, and Power Trade-off Analysis

Shahriar Akramullah

Managing Director: Welmoed Spahr
Associate Publisher: Jeffrey Pepper
Lead Editors: Steve Weiss (Apress); Patrick Hauke (Intel)
Coordinating Editor: Melissa Maldonado
Cover Designer: Anna Ishchenko

Distributed to the book trade worldwide by Springer Science+Business Media New York, 233 Spring Street, 6th Floor, New York, NY 10013. Phone 1-800-SPRINGER, fax (201) 348-4505, e-mail orders-ny@springer-sbm.com, or visit www.springeronline.com.

For information on translations, please e-mail rights@apress.com, or visit www.apress.com.

About ApressOpen

What Is ApressOpen?

- ApressOpen is an open access book program that publishes high-quality technical and business information.

- ApressOpen eBooks are available for global, free, noncommercial use.

- ApressOpen eBooks are available in PDF, ePub, and Mobi formats.

- The user friendly ApressOpen free eBook license is presented on the copyright page of this book.

Contents at a Glance

Contents

About the Author

 **Shahriar Akramullah** is an expert in the field of video technology. He has worked in the video software industry for over 15 years, including about four years at Apple and six years at various Silicon Valley startups. He received his PhD degree in Electrical Engineering from Hong Kong University of Science and Technology in 1999, where he was also a postdoctoral fellow until 2000. A Commonwealth scholar, Dr. Akramullah has received many awards, including the best paper award in the IEEE (HK section) PG student paper contest in 1995.

Since the days of his graduate studies, he has been performing research and analysis on video compression, quality, and video coding performance on various computing platforms. Recently, he has done a comprehensive tradeoff analysis of visual quality and other factors using GPU-accelerated video coding capabilities of Intel processor graphics drivers. Before enabling new encoding features, his experiments and tradeoff evaluations led to a variety of optimizations in multiple Intel products. On this subject, the author has published several technical papers for various Intel conferences, such as DTTC, SWPC, VPG Tech Summit, and so on. He received a number of Intel awards, including the Intel Software Quality Award in 2013.

Shahriar holds several U.S. patents, and has written many research papers for international journals and conferences. His professional experience spans areas in video compression, communications, storage, editing, and signal processing applications. His general interests include video and image signal processing, parallel and distributed processing, video codec and video delivery algorithms, software and firmware implementation, tradeoff analysis and performance optimization of video codecs on various platforms ranging from Intel supercomputer Paragon to TI DSPs; Sun workstations to Intel servers; IBM PowerPC-based Macbooks to Intel x86-based personal computers; and Equator BSP-15 VLIW processors to Telairity's multicore vector processors. He is currently an architect for Intel's processor graphics software.

In his spare time, Shahriar loves to read religious texts and references, and spend time with his family.

About the Technical Reviewer

Scott Janus is a Principal Engineer at Intel Corporation, where he designs media, display, and security technologies for future computing platforms. He holds several patents. Additionally, he has authored multiple informative and entertaining books in both the technical and fiction domains.

Acknowledgments

All praises and thanks belong to Allah, the Lord of the worlds, the Beneficent, the Merciful. First and foremost, I thank Allah, who enabled me to undertake an endeavor of this extent. It was a great journey for me—quite an experience on an unfamiliar path. This work would not have been complete without the help and support of many individuals, and I am grateful to all of them.

I am indebted to Scott Janus, who reviewed the contents of this book and provided valuable early feedback. After completing the book I began to realize just how much I had learned from him.

Thanks are due to my manager Wael Ashmawi, who not only allowed me all the opportunities to write but also supported me in various meetings with the operations team. I appreciate his active encouragement throughout the book's preparation. I also acknowledge the encouragements I received from Mircea Ouatu-Iascar, Lisa Pearce, and Radha Venkataraman.

Several Intel colleagues helped me in numerous ways. In particular, I would like to express my gratitude to Abhishek Agrawal, Saimanohara Alapati, Naveen Appaji, Vasily Aristarkhov, Vivaik Balasubrawmanian, Aishwarya Bhat, Durgaprasad Bilagi, Liza Blaney, Ilya Brailovskiy, Mark Buxton, Jorge Caviedes, William Cheung, Maxym Dmytrychenko, Alexey Fadeev, Fangwen Fu, Jeff Frizzell, James Holland, Wen-Fu Kao, Dmitry Kozlov, Vidhya Krishnan, Timofei Kulakov, Snow Lam, Penne Lee, Wesley Lee, Ryan Lei, Qing Li, Ping Liu, Shalin Mehta, Niranjan Narvekar, Mustayeen Nayeem, Jenny Nieh, Anthony Pabon, Milind Patel, Sankar Radhakrishnan, Shwetha Ramesh, Raj Rasappan, Sailesh Rathi, Daniel Rhee, Matthew Rozek, Dmitry Ryzhov, Michael Stahl, Travis Schluessler, Makhlouf Serghine, Nikolai Shostak, Subash Sudireddy, Jason Tanner, Balakesan Thevar, James Varga, Vishwas Vedantham, Bei Wang, Ce Wang, Charlie Wang, Wei Xiong, Ximin Zhang, and Bin Zhu. I am thankful for their valuable discussions from which I benefited tremendously. Zachary Plovanic helped me with many scripts to carry out various analyses; I am grateful to him for all his help. I also thank Intel colleagues who helped me with my experiments and analysis over the years—among them are Sherine Abdelhak, Hasan Abuzaina, Lavanya Barathraj, Nipun Barua, Namitha Benjamin, Glen Bertulfo, Michael Chen, Cheehoi Chua, Jeff Fleury, Dan Kong, Eugene Smirnov, Patrick Smith, Raghu Ts, and Karthik Vishwanathan. I may have missed some names, but I appreciate their help nonetheless.

This book is supported by Intel. Patrick Hauke of the Intel Labs deserves recognition for his help throughout its development. I am also thankful to the editing and publishing team, including Carole Berglie, Anna Ishchenko, Dhaneesh Kumar, Melissa Maldonado, James Markham, Jeffrey Pepper, Heinz Weinheimer, Steve Weiss, and the freelance artists who were instrumental in making this book possible.

Finally, I thank my wife Ferdousi, my son Abdullah, and my daughter Nishat for their immense patience, frequent sacrifices, and tireless support, which allowed me to carve out countless hours from my weekends and weeknights for this project.

Preface

End-users of video applications are seeking and receiving increasingly more control over their operations. Uploading video to the Internet is only the beginning of a trend in which a consumer controls video quality by balancing various other factors. Emerging applications will give further control to the end-user, such as for video analysis and classification; open-source applications for private party content generation, editing, and archiving; and applications for cloud asset management. Thus, it is important to understand the concepts and methods for evaluating the various video measures.

Owing to characteristics of the human visual system (HVS), some loss in visual quality is perceptible, but much is not. Taking advantage of this fact, we know that digital video data can be compressed. In fact, in most digital video applications the video data is frequently compressed and coded for easier capture, storage, transmission, and display. Typical compression techniques result in some visual quality loss and require the use of powerful computing systems. The more the compression, the worse is the decline in quality. Therefore, a sweet spot needs to be found in balancing compression with perceptible visual quality, with the simultaneous goal of achieving a high coding speed. Likewise, optimization of power consumption is essential for getting the job done on inexpensive consumer platforms. Here, the goal is to keep the total system cost down while delivering a good user experience.

While tradeoffs can be made now, there needs to be a comprehensive set of engineering principles, strategies, methodologies, and metrics to enable greater understanding of such tradeoffs. This book addresses this shortfall with an explanatory approach, exposing readers to methods of evaluating various coding solutions in terms of their potential gains and losses. Further, it enables the reader to differentiate between two video coding solutions, and thereby better shape a perception of tradeoff potentials. For example, an informed video codec user may consider the requirements of a particular video use and opt to select the opportunities offered by the Intel (R) Quick Sync (TM) Video and its GPU-acceleration capabilities rather than choosing a competing solution.

One approach for such an evaluation of coding solutions, or a comparative study of multiple coding solutions, is to consider ways the video can reach the end-user; doing so, though, means comprehending the requirements of various video applications that dictate visual quality levels. To achieve acceptable visual quality, then, the minimum system capacity must be determined, which impacts performance and power requirements. On available platforms with GPU-acceleration capabilities, optimizations can be done to meet those requirements, which usually require tradeoffs among video measures, obtained by tuning the system and encoding the parameters. This book discusses various video measures and the tradeoff opportunities they offer, providing a solid background for evaluating various coding solutions.

Practical video applications include interactive digital video storage—for example, in Blu-ray discs; local and remote video playback on computer or television screens; video capture and recording; video broadcast over terrestrial, satellite, cable and telecommunication channels; video screencast over wireline or wireless channels to an appropriate display unit; video streaming over a network; cloud-based on-demand video services; video transcoding from one format or resolution to other formats and resolutions for burning to disc or for uploading to the Internet; video conferencing, video e-mail and other visual communication; video editing or video post-production and processing; digital cinema, home television theater, and remote video surveillance; telemedicine; electronic news gathering; and so on. Additionally, many more video applications are emerging, including virtual reality and synthetic video, video composition and analysis, and video classification and retrieval. Users of these existing and new video applications can take advantage of the information in this book to tune the various parameters to suit their needs.

Benefits to the Readers

The topics covered in this book are valuable for a wide range of engineers, codec architects, application developers, system validators, technical marketers, technical reviewers, and end-users. They are important in multiple industries and for platforms ranging from low-power mobile phones to high-end servers as long as they use video coding solutions, be they on the desktop, on cloud-based platforms, or through Internet streaming.

This book, therefore, is for anyone who wants to master the video coding concepts without sustaining the rigors of the underlying mathematics and signal processing; or who wants to tune a video coding solution for a particular video usage while making optimal use of available computing platforms and resources. Additionally, anyone who wants to assess a newly available video solution or compare and rank different solutions will find the material contained herein to be worthy.

As noted above, no standards exist for tradeoff analysis of methodologies or metrics for video codec evaluation. Existing benchmarks primarily deal with either visual quality or system performance. Rarely do they consider the power consumption of CPU-intensive applications. There are no comprehensive multimedia-centric calibration and benchmarking tools for considering visual quality, encoding speed, and power consumption simultaneously, particularly on GPU-accelerated platforms. However, this book attempts to fill that gap. It will guide all interested parties in avoiding erroneous comparisons and in understanding the true strengths and limitations of various coding solutions.

Furthermore, performance, power, and quality of video are important subjects both in the industry and in academia. To our knowledge, this is the first book to deal with tradeoff measures of video coding.

Owing to space limits, the book covers only the concepts, principles, methods, and metrics of video compression, quality, performance, and power use. If you need clarification of some information, find any errors, or have questions or comments, feel free to contact me at: shahriar.m.akramullah@intel.com.

Organization of the Book

The book comprises nine chapters and an appendix. Chapter 1 introduces some key concepts and various considerations for video encoding. It also presents the reasons for tradeoff analysis and the challenges and opportunities in doing such an analysis. Chapter 2 presents the HVS characteristics and how various digital video compression techniques can exploit the HVS to achieve compression. It also notes the factors influencing and characterizing the compression algorithms. Chapter 3 provides an overview of the most popular international video coding standards. Chapter 4 discusses visual quality issues and factors affecting the human observer's perceptual quality of video. Chapter 5 covers video coding speed and performance, as well as factors influencing performance, and it identifies the coding parameters to be tuned in any attempt to trade performance for visual quality. Chapters 6 and 7 present the power consumption aspects of video applications, mentioning the challenges especially encountered on low-power platforms with limited resources. Chapter 8 discusses the considerations for tradeoff analysis, focusing on three major areas of optimization—namely, performance, power, and visual quality. Specific examples of tradeoffs employing these measures are provided. Chapter 9 summarizes the key points of this book, and proposes some considerations for future work. In the appendix, well-known industry benchmarks and interesting references are listed, and their limitations are indicated.

CHAPTER 1

Introduction

Over the past decade, countless multimedia functionalities have been added to mobile devices. For example, front and back video cameras are common features in today's cellular phones. Further, there has been a race to capture, process, and display ever-higher resolution video, making this an area that vendors emphasize and where they actively seek market differentiation. These multimedia applications need fast processing capabilities, but those capabilities come at the expense of increased power consumption. The battery life of mobile devices has become a crucial factor, whereas any advances in battery capacity only partly address this problem. Therefore, the future's winning designs must include ways to reduce the energy dissipation of the system as a whole. Many factors must be weighed and some tradeoffs must be made.

Granted, high-quality digital imagery and video are significant components of the multimedia offered in today's mobile devices. At the same time, there is high demand for efficient, performance- and power-optimized systems in this resource-constrained environment. Over the past couple of decades, numerous tools and techniques have been developed to address these aspects of digital video while also attempting to achieve the best visual quality possible. To date, though, the intricate interactions among these aspects had not been explored.

In this book, we study the concepts, methods, and metrics of digital video. In addition, we investigate the options for tuning different parameters, with the goal of achieving a wise tradeoff among visual quality, performance, and power consumption. We begin with an introduction to some key concepts of digital video, including visual data compression, noise, quality, performance, and power consumption. We then discuss some video compression considerations and present a few video coding usages and requirements. We also investigate the tradeoff analysis—the metrics for its good use, its challenges and opportunities, and its expected outcomes. Finally, there is an introductory look at some emerging applications. Subsequent chapters in this book will build upon these fundamental topics.

The Key Concepts

This section deals with some of the key concepts discussed in this book, as applicable to perceived visual quality in compressed digital video, especially as presented on contemporary mobile platforms.

Digital Video

The term *video* refers to the visual information captured by a camera, and it usually is applied to a time-varying sequence of pictures. Originating in the early television industry of the 1930s, video cameras were electromechanical for a decade, until all-electronic versions based on cathode ray tubes (CRT) were introduced. The analog tube technologies were then replaced in the 1980s by solid-state sensors, particularly CMOS active pixel sensors, which enabled the use of digital video.

Early video cameras captured analog video signals as a one-dimensional, time-varying signal according to a pre-defined scanning convention. These signals would be transmitted using analog amplitude modulation, and they were stored on analog video tapes using video cassette recorders or on analog laser discs using optical technology. The analog signals were not amenable to compression; they were regularly converted to digital formats for compression and processing in the digital domain.

Recently, use of all-digital workflow encompassing digital video signals from capture to consumption has become widespread, particularly because of the following characteristics:

- It is easy to record, store, recover, transmit, and receive, or to process and manipulate, video that's in digital format; it's virtually without error, so digital video can be considered just another data type for today's computing systems.

- Unlike analog video signals, digital video signals can be compressed and subsequently decompressed. Storage and transmission are much easier in compressed format compared to uncompressed format.

- With the availability of inexpensive integrated circuits, high-speed communication networks, rapid-access dense storage media, advanced architecture of computing devices, and high-efficiency video compression techniques, it is now possible to handle digital video at desired data rates for a variety of applications on numerous platforms that range from mobile handsets to networked servers and workstations.

Owing to a high interest in digital video, especially on mobile computing platforms, it has had a significant impact on human activities; this will almost certainly continue to be felt in the future, extending to the entire area of information technology.

Video Data Compression

It takes a massive quantity of data to represent digital video signals. Some sort of data compression is necessary for practical storage and transmission of the data for a plethora of applications. Data compression can be *lossless*, so that the same data is retrieved upon decompression. It can also be *lossy*, whereby only an approximation of the original signal is recovered after decompression. Fortunately, the characteristic of video data is such that a certain amount of loss can be tolerated, with the resulting video signal perceived without objection by the human visual system. Nevertheless, all video signal-processing methods and techniques make every effort to achieve the best visual quality possible, given their system constraints.

Note that video data compression typically involves coding of the video data; the coded representation is generally transmitted or stored, and it is decoded when a decompressed version is presented to the viewer. Thus, it is common to use the terms *compression/decompression* and *encoding/decoding* interchangeably. Some professional video applications may use uncompressed video in coded form, but this is relatively rare.

A *codec* is composed of an encoder and a decoder. Video encoders are much more complex than video decoders are. They typically require a great many more signal-processing operations; therefore, designing efficient video encoders is of primary importance. Although the video coding standards specify the bitstream syntax and semantics for the decoders, the encoder design is mostly open.

Chapter 2 has a detailed discussion of video data compression, while the important data compression algorithms and standards can be found in Chapter 3.

Noise Reduction

Although compression and processing are necessary for digital video, such processing may introduce undesired effects, which are commonly termed *distortions* or *noise*. They are also known as *visual artifacts*. As noise affects the fidelity of the user's received signal, or equivalently the visual quality perceived by the end user, the video signal processing seeks to minimize the noise. This applies to both analog and digital processing, including the process of video compression.

In digital video, typically we encounter many different types of noise. These include noise from the sensors and the video capture devices, from the compression process, from transmission over lossy channels, and so on. There is a detailed discussion of various types of noise in Chapter 4.

Visual Quality

Visual quality is a measure of perceived visual deterioration in the output video compared to the original signal, which has resulted from lossy video compression techniques. This is basically a measure of the *quality of experience* (QoE) of the viewer. Ideally, there should be minimal loss to achieve the highest visual quality possible within the coding system.

Determining the visual quality is important for analysis and decision-making purposes. The results are used in the specification of system requirements, comparison and ranking of competing video services and applications, tradeoffs with other video measures, and so on.

Note that because of compression, the artifacts found in digital video are fundamentally different from those in analog systems. The amount and visibility of the distortions in video depend on the contents of that video. Consequently, the measurement and evaluation of artifacts, and the resulting visual quality, differ greatly from the traditional analog quality assessment and control mechanisms. (The latter, ironically, used signal parameters that could be closely correlated with perceived visual quality.)

Given the nature of digital video artifacts, the best method of visual quality assessment and reliable ranking is subjective viewing experiments. However, subjective methods are complex, cumbersome, time-consuming, and expensive. In addition, they are not suitable for automated environments.

An alternative, then, is to use simple error measures such as the *mean squared error* (MSE) or the *peak signal to noise ratio* (PSNR). Strictly speaking, PSNR is only a measure of the signal fidelity, not the visual quality, as it compares the output signal to the input signal and so does not necessarily represent perceived visual quality. However, it is the most popular metric for visual quality used in the industry and in academia. Details on this use are provided in Chapter 4.

Performance

Video coding *performance* generally refers to the speed of the video coding process: the higher the speed, the better the performance. In this context, *performance optimization* refers to achieving a fast video encoding speed.

In general, the performance of a computing task depends on the capabilities of the processor, particularly the *central processing unit* (CPU) and the *graphics processing unit* (GPU) frequencies up to a limit. In addition, the capacity and speed of the main memory, auxiliary cache memory, and the disk input and output (I/O), as well as the cache hit ratio, scheduling of the tasks, and so on, are among various system considerations for performance optimization.

Video data and video coding tasks are especially amenable to parallel processing, which is a good way to improve processing speed. It is also an optimal way to keep the available processing units busy for as long as necessary to complete the tasks, thereby maximizing resource utilization. In addition, there are many other performance-optimization techniques for video coding, including tuning of encoding parameters. All these techniques are discussed in detail in Chapter 5.

Power Consumption

A mobile device is expected to serve as the platform for computing, communication, productivity, navigation, entertainment, and education. Further, devices that are implantable to human body, that capture intrabody images or videos, render to the brain, or securely transmit to external monitors using biometric keys may become available in the future. The interesting question for such new and future uses would be how these devices can be supplied with power. In short, leaps of innovation are necessary in this area. However, even while we await such breakthroughs in power supply, know that some externally wearable devices are already complementing today's mobile devices.

Power management and optimization are the primary concerns for all these existing and new devices and platforms, where the goal is to prolong battery life. However, many applications are particularly power-hungry, either by their very nature or because of special needs, such as on-the-fly binary translation.

Power—or equivalently, energy—consumption thus is a major concern. Power optimization aims to reduce energy consumption and thereby extend battery life. High-speed video coding and processing present further challenges to power optimization. Therefore, we need to understand the power management and optimization considerations, methods, and tools; this is covered in Chapters 6 and 7.

Video Compression Considerations

A major drawback in the processing, storage, and transmission of digital video is the huge amount of data needed to represent the video signal. Simple scanning and binary coding of the camera voltage variations would produce billions of bits per second, which without compression would result in prohibitively expensive storage or transmission devices. A typical high-definition video (three color planes per picture, a resolution of 1920×1080 pixels per plane, 8 bits per pixel, at a 30 pictures per second rate) necessitates a data rate of approximately 1.5 billion bits per second. A typical transmission channel capable of handling about 5 Mbps would require a 300:1 compression ratio. Obviously, lossy techniques can accommodate such high compression, but the resulting reconstructed video will suffer some loss in visual quality.

However, video compression techniques aim at providing the best possible visual quality at a specified data rate. Depending on the requirements of the applications, available channel bandwidth or storage capacity, and the video characteristics, a variety of data rates are used, ranging from 33.6 kbps video calls in an old-style public switched telephone network to ~20 Mbps in a typical HDTV rebroadcast system.

Varying Uses

In some video applications, video signals are captured, processed, transmitted, and displayed in an on-line manner. Real-time constraints for video signal processing and communication are necessary for these applications. The applications use an end-to-end real-time workflow and include, for example, video chat and video conferencing, streaming, live broadcast, remote wireless display, distant medical diagnosis and surgical procedures, and so on.

A second category of applications involve recorded video in an off-line manner. In these, video signals are recorded to a storage device for archiving, analysis, or further processing. After being used for many years, the main storage medium for the recorded video is shifted from analog video tapes to digital *DV* or *Betacam* tapes, optical discs, hard disks, or flash memory. Apart from archiving, stored video is used for off-line processing and analysis purposes in television and film production, in surveillance and monitoring, and in security and investigation areas. These uses may benefit from video signal processing as fast as possible; thus, there is a need to speed up video compression and decompression processes.

Conflicting Requirements

The conflicting requirements of video compression on modern mobile platforms pose challenges for a range of people, from system architects to end users of video applications. Compressed data is easy to handle, but visual quality loss typically occurs with compression. A good video coding solution must produce videos without too much loss of quality.

Furthermore, some video applications benefit from high-speed video coding. This generally implies a high computation requirement, resulting in high energy consumption. However, mobile devices are typically resource constrained and battery life is usually the biggest concern. Some video applications may sacrifice visual quality in favor of saving energy.

These conflicting needs and purposes have to be balanced. As we shall see in the coming chapters, video coding parameters can be tuned and balanced to obtain such results.

Hardware vs. Software Implementations

Video compression systems can be implemented using dedicated application-specific integrated circuits (ASICs), field-programmable gate arrays (FPGAs), GPU-based hardware acceleration, or purely CPU-based software.

The ASICs are customized for a particular use and are usually optimized to perform specific tasks; they cannot be used for purposes other than what they are designed for. Although they are fast, robust against error, yield consistent, predictable, and offer stable performance, they are inflexible, implement a single algorithm, are not programmable or easily modifiable, and can quickly become obsolete. Modern ASICs often include entire microprocessors, memory blocks including read-only memory (ROM), random-access memory (RAM), flash memory, and other large building blocks. Such an ASIC is often termed a system-on-chip (SoC).

FPGAs consist of programmable logic blocks and programmable interconnects. They are much more flexible than ASICs; the same FPGA can be used in many different applications. Typical uses include building prototypes from standard parts. For smaller designs or lower production volumes, FPGAs may be more cost-effective than an ASIC design. However, FPGAs are usually not optimized for performance, and the performance usually does not scale with the growing problem size.

Purely CPU-based software implementations are the most flexible, as they run on general-purpose processors. They are usually portable to various platforms. Although several performance-enhancement approaches exist for the software-based implementations, they often fail to achieve a desired performance level, as hand-tuning of various parameters and maintenance of low-level codes become formidable tasks. However, it is easy to tune various encoding parameters in software implementations, often in multiple passes. Therefore, by tuning the various parameters and number of passes, software implementations can provide the best possible visual quality for a given amount of compression.

GPU-based hardware acceleration typically provides a middle ground. In these solutions, there are a set of programmable execution units and a few performance- and power-optimized fixed-function hardware units. While some complex algorithms may take advantage of parallel processing using the execution units, the fixed-function units provide fast processing. It is also possible to reuse some fixed-function units with updated parameters based on certain feedback information, thereby achieving multiple passes for those specific units. Therefore, these solutions exhibit flexibility and scalability while also being optimized for performance and power consumption. The tuning of available parameters can ensure high visual quality at a given bit rate.

Tradeoff Analysis

Tradeoff analysis is the study of the cost-effectiveness of different alternatives to determine where benefits outweigh costs. In video coding, a tradeoff analysis looks into the effect of tuning various encoding parameters on the achievable compression, performance, power savings, and visual quality in consideration of the application requirements, platform constraints, and video complexity.

Note that the tuning of video coding parameters affects performance as well as visual quality, so a good video coding solution balances performance optimization with achievable visual quality. In Chapter 8, a case study illustrates this tradeoff between performance and quality.

It is worthwhile to note that, while achieving high encoding speed is desirable, it may not always be possible on platforms with different restrictions. In particular, achieving power savings is often the priority on modern computing platforms. Therefore, a typical tradeoff between performance and power optimization is considered in a case study examined in Chapter 8.

Benchmarks and Standards

The benchmarks typically used today for ranking video coding solutions do not consider all aspects of video. Additionally, industry-standard benchmarks for methodology and metrics specific to tradeoff analysis do not exist. This standards gap leaves the user guessing about which video coding parameters will yield satisfactory outputs for particular video applications. By explaining the concepts, methods, and metrics involved, this book helps readers understand the effects of video coding parameters on the video measures.

Challenges and Opportunities

Several challenges and opportunities in the area of digital video techniques have served as the motivating factors for tradeoff analysis.

- The demand for compressed digital video is increasing. With the desire to achieve ever-higher resolution, greater bit depth, higher dynamic range, and better quality video, the associated computational complexity is snowballing. These developments present a challenge for the algorithms and architectures of video coding systems, which need to be optimized and tuned for higher compression but better quality than standard algorithms and architectures.

- Several international video coding standards are now available to address a variety of video applications. Some of these standards evolved from previous standards, were tweaked with new coding features and tools, and are targeted toward achieving better compression efficiency.

- Low-power computing devices, particularly in the mobile environment, are increasingly the chosen platforms for video applications. However, they remain restrictive in terms of system capabilities, a situation that presents optimization challenges. Nonetheless, tradeoffs are possible to accommodate goals such as preserving battery life.

- Some video applications benefit from increased processing speed. Efficient utilization of resources, resource specialization, and tuning of video parameters can help achieve faster processing speed, often without compromising visual quality.

- The desire to obtain the best possible visual quality on any given platform requires careful control of coding parameters and wise choice among many alternatives. Yet there exists a void where such tools and measures should exist.

- Tuning of video coding parameters can influence various video measures, and desired tradeoffs can be made by such tuning. To be able to balance the gain in one video measure with the loss in another requires knowledge of coding parameters and how they influence each other and the various video measures. However, there is no unified approach to the considerations and analyses of the available tradeoff opportunities. A systematic and in-depth study of this subject is necessary.

- A tradeoff analysis can expose the strengths and weaknesses of a video coding solution and can rank different solutions.

The Outcomes of Tradeoff Analysis

Tradeoff analysis is useful in many real-life video coding scenarios and applications. Such analysis can show the value of a certain encoding feature so that it is easy to make a decision whether to add or remove that feature under the specific application requirements and within the system restrictions. Tradeoff analysis is useful in assessing the strengths and weaknesses of a video encoder, tuning the parameters to achieve optimized encoders, comparing two encoding solutions based on the tradeoffs they involve, or ranking multiple encoding solutions based on a set of criteria.

It also helps a user make decisions about whether to enable some optional encoding features under various constraints and application requirements. Furthermore, a user can make informed product choices by considering the results of the tradeoff analysis.

Emerging Video Applications

Compute performance has increased to a level where computers are no longer used solely for scientific and business purposes. We have a colossal amount of compute capabilities at our disposal, enabling unprecedented uses and applications. We are revolutionizing human interfaces, using vision, voice, touch, gesture, and context. Many new applications are either already available or are emerging for our mobile devices, including perceptual computing, such as 3-D image and video capture and depth-based processing; voice, gesture, and face recognition; and virtual-reality-based education and entertainment.

These applications are appearing in a range of devices and may include synthetic and/or natural video. Because of the fast pace of change in platform capabilities, and the innovative nature of these emerging applications, it is quite difficult to set a strategy on handling the video components of such applications, especially from an optimization point of view. However, by understanding the basic concepts, methods, and metrics of various video measures, we'll be able to apply them to future applications.

Summary

This chapter discussed some key concepts related to digital video, compression, noise, quality, performance, and power consumption. It presented various video coding considerations, including usages, requirements, and different aspects of hardware and software implementations. There was also a discussion of tradeoff analysis and the motivations, challenges, and opportunities that the field of video is facing in the future. This chapter has set the stage for the discussions that follow in subsequent chapters.

Digital Video Compression Techniques

Digital video plays a central role in today's communication, information consumption, entertainment and educational approaches, and has enormous economic and sociocultural impacts on everyday life. In the first decade of the 21st century, the profound dominance of video as an information medium on modern life—from digital television to Skype, DVD to Blu-ray, and YouTube to Netflix–has been well established. Owing to the enormous amount of data required to represent digital video, it is necessary to compress the video data for practical transmission and communication, storage, and streaming applications.

In this chapter we start with a brief discussion of the limits of digital networks and the extent of compression required for digital video transmission. This sets the stage for further discussions on compression. It is followed by a discussion of the human visual system (HVS) and the compression opportunities allowed by the HVS. Then we explain the terminologies, data structures, and concepts commonly used in digital video compression.

We discuss various redundancy reduction and entropy coding techniques that form the core of the compression methods. This is followed by overviews of various compression techniques and their respective advantages and limitations. We briefly introduce the rate-distortion curve both as the measure of compression efficiency and as a way to compare two encoding solutions. Finally, there's a discussion of the factors influencing and characterizing the compression algorithms before a brief summary concludes the chapter.

Network Limits and Compression

Before the advent of the *Integrated Services Digital Network* (ISDN), the *Plain Old Telephone Service* (POTS) was the commonly available network, primarily to be used for voice-grade telephone services based on analog signal transmission. However, the ubiquity of the telephone networks meant that the design of new and innovative communication services such as facsimile (fax) and modem were initially inclined toward using these available analog networks. The introduction of ISDN enabled both voice and video communication to engage digital networks as well, but the standardization delay in *Broadband ISDN* (B-ISDN) allowed packet-based local area networks such as the *Ethernet* to become more popular. Today, a number of network protocols support transmission of images or videos using wire line or wireless technologies, having different bandwidth and data-rate capabilities, as listed in Table 2-1.

Table 2-1. *Various Network Protocols and Their Supported Bit Rates*

Network	Bit Rate
Plain Old Telephone Service (POTS) on conventional low-speed twisted-pair copper wiring	2.4 kbps (ITU* V.27†), 14.4 kbps (V.17), 28.8 kpbs (V.34), 33.6 kbps (V.34bis), etc.
Digital Signal 0 (DS 0), the basic granularity of circuit switched telephone exchange	64 kbps
Integrated Services Digital Network (ISDN)	64 kbps (Basic Rate Interface), 144 kbps (Narrow band ISDN)
Digital Signal 1 (DS 1), aka T-1 or E-1	1.5 – 2 Mbps (Primary Rate Interface)
Ethernet Local Area Network	10 Mbps
Broadband ISDN	100 – 200 Mbps
Gigabit Ethernet	1 Gbps

* *International Telecommunications Union.*
† *The ITU V-series international standards specify the recommendations for vocabulary and related subjects for radiocommuncation.*

In the 1990s, transmission of raw digital video data over POTS or ISDN was unproductive and very expensive due to the sheer data rate required. Note that the raw data rate for the ITU-R 601 formats[1] is ~165 Mbps (million bits per second), beyond the networks' capabilities. In order to partially address the data-rate issue, the 15th specialist group (SGXV) of the *CCITT*[2] defined the *Common Image Format* (CIF) to have common picture parameter values independent of the picture rate. While the format specifies many picture rates (24 Hz, 25 Hz, 30 Hz, 50 Hz, and 60 Hz), with a resolution of 352 × 288 at 30 Hz, the required data rate was brought down to approximately 37 Mbps, which would typically fit into a basic *Digital Signal 0* (DS0) circuit, and would be practical for transmission.

---

[1]The specification was originally known as CCIR-601. The standard body CCIR a.k.a. International Radio Consultative Committee (Comité Consultatif International pour la Radio) was formed in 1927, and was superceded in 1992 by the ITU Recommendations Sector (ITU-R).
[2]CCITT (International Consultative Committee for Telephone and Telegraph) is a committee of the ITU, currently known as the ITU Telecommunication Standardization Sector (ITU-T).

With increased compute capabilities, video encoding and processing operations became more manageable over the years. These capabilities fueled the growing demand of ever higher video resolutions and data rates to accommodate diverse video applications with better-quality goals. One after another, the ITU-R Recommendations BT.601,[3] BT.709,[4] and BT.2020[5] appeared to support video formats with increasingly higher resolutions. Over the years these recommendations evolved. For example, the recommendation BT.709, aimed at high-definition television (HDTV), started with defining parameters for the early days of analog high-definition television implementation, as captured in Part 1 of the specification. However, these parameters are no longer in use, so Part 2 of the specification contains HDTV system parameters with square pixel common image format.

Meanwhile, the network capabilities also grew, making it possible to address the needs of today's industries. Additionally, compression methods and techniques became more refined.

The Human Visual System

The *human visual system* (HVS) is part of the human nervous system, which is managed by the brain. The electrochemical communication between the nervous system and the brain is carried out by about 100 billion nerve cells, called *neurons*. Neurons either generate pulses or inhibit existing pulses, and result in a variety of phenomena ranging from *Mach bands*, band-pass characteristic of the visual frequency response, to the edge-detection mechanism of the eye. Study of the enormously complex nervous system is manageable because there are only two types of signals in the nervous system: one for long distances and the other for short distances. These signals are the same for all neurons, regardless of the information they carry, whether visual, audible, tactile, or other.

Understanding how the HVS works is important for the following reasons:

- It explains how accurately a viewer perceives what is being presented for viewing.

- It helps understand the composition of visual signals in terms of their physical quantities, such as luminance and spatial frequencies, and helps develop measures of signal fidelity.

---

[3]ITU-R. See *ITU-R Recommendation BT. 601-5: Studio encoding parameters of digital television for standard 4:3 and widescreen 16:9 aspect ratios* (Geneva, Switzerland: International Telecommunications Union, 1995).

[4]ITU-R. See *ITU-R Recommendation BT.709-5: Parameter values for the HDTV standards for production and international programme exchange* (Geneva, Switzerland: International Telecommunications Union, 2002).

[5]ITU-R. See *ITU-R Recommendation BT.2020: Parameter values for ultra-high definition television systems for production and international programme exchange* (Geneva, Switzerland: International Telecommunications Union, 2012).

- It helps represent the perceived information by various attributes, such as brightness, color, contrast, motion, edges, and shapes. It also helps determine the sensitivity of the HVS to these attributes.

- It helps exploit the apparent imperfection of the HVS to give an impression of faithful perception of the object being viewed. An example of such exploitation is color television. When it was discovered that the HVS is less sensitive to loss of color information, it became easy to reduce the transmission bandwidth of color television by chroma subsampling.

The major components of the HVS include the *eye,* the *visual pathways* to the brain, and part of the brain called the *visual cortex.* The eye captures light and converts it to signals understandable by the nervous system. These signals are then transmitted and processed along the visual pathways.

So, the eye is the sensor of visual signals. It is an optical system, where an image of the outside world is projected onto the *retina*, located at the back of the eye. Light entering the retina goes through several layers of neurons until it reaches the light-sensitive *photoreceptors*, which are specialized neurons that convert incident light energy into neural signals.

There are two types of photoreceptors: *rods* and *cones*. Rods are sensitive to low light levels; they are unable to distinguish color and are predominant in the periphery. They are also responsible for *peripheral* vision and they help in motion and shape detection. As signals from many rods converge onto a single neuron, sensitivity at the periphery is high, but the resolution is low. Cones, on the other hand, are sensitive to higher light levels of long, medium, and short wavelengths. They form the basis of color perception. Cone cells are mostly concentrated in the center region of the retina, called the *fovea*. They are responsible for *central* or *foveal* vision, which is relatively weak in the dark. Several neurons encode the signal from each cone, resulting in high resolution but low sensitivity.

The number of the rods, about 100 million, is higher by more than an order of magnitude compared to the number of cones, which is about 6.5 million. As a result, the HVS is more sensitive to motion and structure, but it is less sensitive to loss in color information. Furthermore, motion sensitivity is stronger than texture sensitivity; for example, a camouflaged still animal is difficult to perceive compared to a moving one. However, texture sensitivity is stronger than disparity; for example, 3D depth resolution does not need to be so accurate for perception.

Even if the retina perfectly detects light, that capacity may not be fully utilized or the brain may not be consciously aware of such detection, as the visual signal is carried by the optic nerves from the retina to various processing centers in the brain. The *visual cortex*, located in the back of the cerebral hemispheres, is responsible for all high-level aspects of vision.

Apart from the primary visual cortex, which makes up the largest part of the HVS, the visual signal reaches to about 20 other cortical areas, but not much is known about their functions. Different cells in the visual cortex have different specializations, and they are sensitive to different stimuli, such as particular colors, orientations of patterns, frequencies, velocities, and so on.

Simple cells behave in a predictable fashion in response to particular spatial frequency, orientation, and phase, and serve as an oriented band-pass filter. Complex cells, the most common cells in the primary visual cortex, are also orientation-selective, but unlike simple cells, they can respond to a properly oriented stimulus anywhere in their *receptive field*. Some complex cells are direction-selective and some are sensitive to certain sizes, corners, curvatures, or sudden breaks in lines.

The HVS is capable of adapting to a broad range of light intensities or *luminance*, allowing us to differentiate luminance variations relative to surrounding luminance at almost any light level. The actual luminance of an object does not depend on the luminance of the surrounding objects. However, the perceived luminance, or the *brightness* of an object, depends on the surrounding luminance. Therefore, two objects with the same luminance may have different perceived brightnesses in different surroundings. *Contrast* is the measure of such relative luminance variation. Equal logarithmic increments in luminance are perceived as equal differences in contrast. The HVS can detect contrast changes as low as 1 percent.[6]

The HVS Models

The fact that visual perception employs more than 80 percent of the neurons in human brain points to the enormous complexity of this process. Despite numerous research efforts in this area, the entire process is not well understood. Models of the HVS are generally used to simplify the complex biological processes entailing visualization and perception. As the HVS is composed of nonlinear spatial frequency channels, it can be modeled using nonlinear models. For easier analysis, one approach is to develop a linear model as a first approximation, ignoring the nonlinearities. This approximate model is then refined and extended to include the nonlinearities. The characteristics of such an example HVS model[7] include the following.

The First Approximation Model

This model considers the HVS to be linear, isotropic, and time- and space-invariant. The linearity means that if the intensity of the light radiated from an object is increased, the magnitude of the response of the HVS should increase proportionally. *Isotropic* implies invariance to direction. Although, in practice, the HVS is anisotropic and its response to a rotated contrast grating depends on the frequency of the grating, as well as the angle of orientation, the simplified model ignores this nonlinearity. The spatio-temporal invariance is difficult to modify, as the HVS is not homogeneous. However, the spatial invariance assumption partially holds near the optic axis and the foveal region. Temporal responses are complex and are not generally considered in simple models.

In the first approximation model, the contrast sensitivity as a function of spatial frequency represents the *optical transfer function* (OTF) of the HVS. The magnitude of the OTF is called the *modulation transfer function* (MTF), as shown in Figure 2-1.

---

[6]S. Winkler, *Digital Video Quality: Vision Models and Metrics* (Hoboken, NJ: John Wiley, 2005).
[7]C. F. Hall and E. L. Hall, "A Nonlinear Model for the Spatial Characteristics of the Human Visual System," *IEEE Transactions on Systems, Man, and Cybernetics* 7, no. 3 (1977): 161–69.

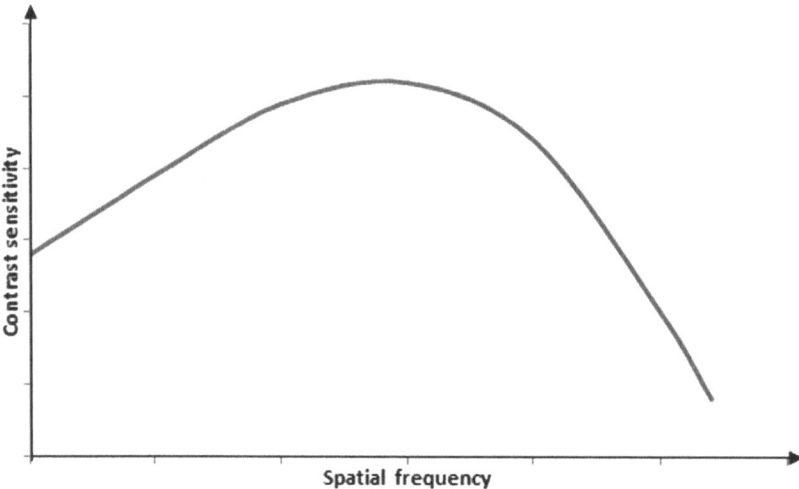

Figure 2-1. *A typical MTF plot*

The curve representing the thresholds of visibility at various spatial frequencies has an inverted U-shape, while its magnitude varies with the viewing distance and viewing angle. The shape of the curve suggests that the HVS is most sensitive to mid-frequencies and less sensitive to high frequencies, showing band-pass characteristics.

The MTF can thus be represented by a band-pass filter. It can be modeled more accurately as a combination of a low-pass and a high-pass filter. The low-pass filter corresponds to the optics of the eye. The lens of the eye is not perfect, even for persons with no weakness of vision. This imperfection results in *spherical aberration*, appearing as a blur in the focal plane. Such blur can be modeled as a two-dimensional low-pass filter. The pupil's diameter varies between 2 and 9 mm. This aperture can also be modeled as a low-pass filter with high cut-off frequency corresponding to 2 mm, while the frequency decreases with the enlargement of the pupil's diameter.

On the other hand, the high-pass filter accounts for the following phenomenon. The post-retinal neural signal at a given location may be inhibited by some of the laterally located photoreceptors. This is known as *lateral inhibition*, which leads to the *Mach band* effect, where visible bands appear near the transition regions of a smooth ramp of light intensity. This is a high-frequency change from one region of constant luminance to another, and is modeled by the high-pass portion of the filter.

Refined Model Including Nonlinearity

The linear model has the advantage that, by using the Fourier transform techniques for analysis, the system response can be determined for any input stimulus as long as the MTF is known. However, the linear model is insufficient for the HVS as it ignores important nonlinearities in the system. For example, it is known that light stimulating the receptor causes a potential difference across the membrane of a receptor cell, and this potential mediates the frequency of nerve impulses. It has also been determined that this frequency is a logarithmic function of light intensity (Weber-Fechner law). Such logarithmic function can approximate the nonlinearity of the HVS. However, some experimental results indicate a nonlinear distortion of signals at high, but not low, spatial frequencies.

These results are inconsistent with a model where logarithmic nonlinearity is followed by linear independent frequency channels. Therefore, the model most consistent with the HVS is the one that simply places the low-pass filter in front of the logarithmic nonlinearity, as shown in Figure 2-2. This model can also be extended for spatial vision of color, in which a transformation from spectral energy space to tri-stimulus space is added between the low-pass filter and the logarithmic function, and the low-pass filter is replaced with three independent filters, one for each band.

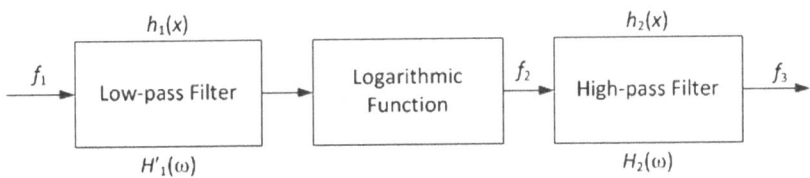

Figure 2-2. *A nonlinear model for spatial characteristics of the HVS*

The Model Implications

The low-pass, nonlinearity, high-pass structure is not limited to spatial response, or even to spectral-spatial response. It was also found that this basic structure is valid for modeling the temporal response of the HVS. A fundamental premise of this model is that the HVS uses low spatial frequencies as features. As a result of the low-pass filter, rapid discrete changes appear as continuous changes. This is consistent with the appearance of discrete time-varying video frames as continuous-time video to give the perception of smooth motion.

This model also suggests that the HVS is analogous to a variable bandwidth filter, which is controlled by the contrast of the input image. As input contrast increases, the bandwidth of the system decreases. Therefore, limiting the bandwidth is desirable to maximize the signal-to-noise ratio. Since noise typically contains high spatial frequencies, it is reasonable to limit this end of the system transfer function. However, in practical video signals, high-frequency details are also very important. Therefore, with this model, noise filtering can only be achieved at the expense of *blurring* the high-frequency details, and an appropriate tradeoff is necessary to obtain optimum system response.

The Model Applications

In image recognition systems, a correlation may be performed between low spatial-frequency filtered images and stored prototypes of the primary receptive area for vision, where this model can act as a pre-processor. For example, in recognition and analysis of complex scenes with variable contrast information, when a human observer directs his attention to various subsections of the complex scene, an automated system based on this model could compute average local contrast of the subsection and adjust filter parameters accordingly. Furthermore, in case of image and video coding, this model can also act as a pre-processor to appropriately reflect the noise-filtering effects, prior to coding only the relevant information. Similarly, it can also be used for bandwidth reduction and efficient storage systems as pre-processors.

A block diagram of the HVS model is shown in Figure 2-3, where parts related to the lens, the retina, and the visual cortex, are indicated.

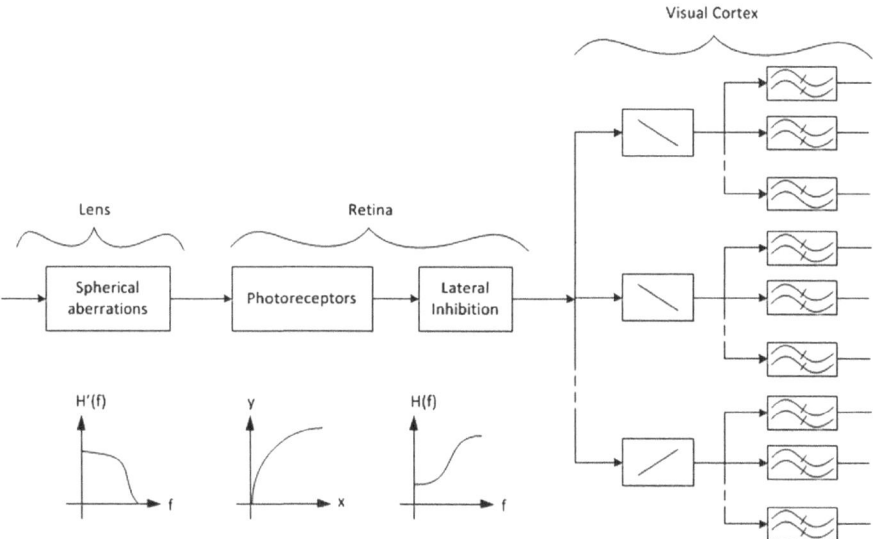

Figure 2-3. *A block diagram of the HVS*

In Figure 2-3, the first block is a spatial, isotropic, low-pass filter. It represents the spherical aberration of the lens, the effect of the pupil, and the frequency limitation by the finite number of photoreceptors. It is followed by the nonlinear characteristic of the photoreceptors, represented by a logarithmic curve. At the level of the retina, this nonlinear transformation is followed by an isotropic high-pass filter corresponding to the lateral inhibition phenomenon. Finally, there is a directional filter bank that represents the processing performed by the cells of the visual cortex. The bars in the boxes indicate the directional filters. This is followed by another filter bank, represented by the double waves, for detecting the intensity of the stimulus. It is worth mentioning that the overall system is shift-variant because of the decrease in resolution away from the fovea.[8]

---

[8]M. Kunt, A. Ikonomopoulos, and M. Kocher, "Second -Generation Image-Coding Techniques," *Proceedings of the IEEE* 73, no. 4 (April 1985): 549–74.

Expoliting the HVS

By taking advantage of the characteristics of the HVS, and by tuning the parameters of the HVS model, tradeoffs can be made between visual quality loss and video data compression. In particular, the following benefits may be accrued.

- By limiting the bandwidth, the visual signal may be sampled in spatial or temporal dimensions at a frequency equal to twice the bandwidth, satisfying the Nyquist criteria of sampling, without loss of visual quality.

- The sensitivity of the HVS is decreased during rapid large-scale scene change and intense motion of objects, resulting in *temporal or motion masking*. In such cases the visibility thresholds are elevated due to temporal discontinuities in intensity. This can be exploited to achieve more efficient compression, without producing noticeable artifacts.

- Texture information can be compressed more than motion information with negligible loss of visual quality. As discussed later in this chapter, several lossy compression algorithms allow quantization and resulting quality loss of texture information, while encoding the motion information losslessly.

- Owing to low sensitivity of the HVS to the loss of color information, chroma subsampling is a feasible technique to reduce data rate without significantly impacting the visual quality.

- Compression of brightness and contrast information can be achieved by discarding high-frequency information. This would impair the visual quality and introduce artifacts, but parameters of the amount of loss are controllable.

- The HVS is sensitive to structural distortion. Therefore, measuring such distortions, especially for highly structured data such as image or video, would give a criterion to assess whether the amount of distortion is *acceptable* to human viewers. Although acceptability is subjective and not universal, structural distortion metrics can be used as an objective evaluation criterion.

- The HVS allows humans to pay more attention to interesting parts of a complex image and less attention to other parts. Therefore, it is possible to apply different amount of compression on different parts of an image, thereby achieving a higher overall compression ratio. For example, more bits can be spent on the foreground objects of an image compared to the background, without substantial quality impact.

An Overview of Compression Techniques

A high-definition uncompressed video data stream requires about 2 billion bits per second of data bandwidth. Owing to the large amount of data necessary to represent digital video, it is desirable that such video signals are easy to compress and decompress, to allow practical storage or transmission. The term *data compression* refers to the reduction in the number of bits required to store or convey data—including numeric, text, audio, speech, image, and video—by exploiting statistical properties of the data. Fortunately, video data is highly compressible owing to its strong vertical, horizontal, and temporal correlation and its redundancy.

Transform and prediction techniques can effectively exploit the available correlation, and information coding techniques can take advantage of the statistical structures present in video data. These techniques can be lossless, so that the reverse operation (decompression) reproduces an exact replica of the input. In addition, however, lossy techniques are commonly used in video data compression, exploiting the characteristics of the HVS, which is less sensitive to some color losses and some special types of noises.

Video compression and decompression are also known as video *encoding* and *decoding*, respectively, as information coding principles are used in the compression and decompression processes, and the compressed data is presented in a coded bit stream format.

Data Structures and Concepts

Digital video signal is generally characterized as a form of computer data. Sensors of video signals usually output three color signals–red, green and blue (*RGB*)—that are individually converted to digital forms and are stored as arrays of picture elements (*pixels*), without the need of the blanking or sync pulses that were necessary for analog video signals. A two-dimensional array of these pixels, distributed horizontally and vertically, is called an *image* or a *bitmap*, and represents a *frame* of video. A time-dependent collection of frames represents the full video signal. There are five parameters[9] associated with a bitmap: the starting address in memory, the number of pixels per line, the pitch value, the number of lines per frame, and the number of bits per pixel. In the following discussion, the terms *frame* and *image* are used interchangeably.

Signals and Sampling

The conversion of a continuous analog signal to a discrete digital signal, commonly known as the analog-to-digital (A/D) conversion, is done by taking samples of the analog signal at appropriate intervals in a process known as *sampling*. Thus $x(n)$ is called the sampled version of the analog signal $x_a(t)$ if $x(n) = x_a(nT)$ for some $T > 0$, where T is known as the *sampling period* and $2\pi/T$ is known as the *sampling frequency* or the *sampling rate*. Figure 2-4 shows a spatial domain representation of $x_a(t)$ and corresponding $x(n)$.

---

[9]A. Tekalp, *Digital Video Processing* (Englewood Cliff: Prentice-Hall PTR, 1995).

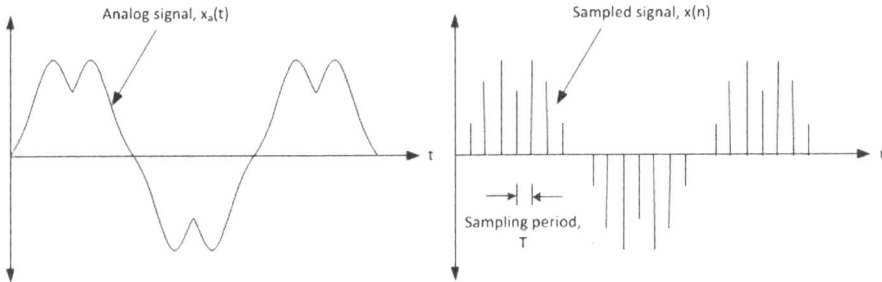

Figure 2-4. *Spatial domain representation of an analog signal and its sampled version*

The frequency-domain representation of the signal is obtained by using the Fourier transform, which gives the analog frequency response $X_a(j\Omega)$ replicated at uniform intervals $2\pi/T$, while the amplitudes are reduced by a factor of T. Figure 2-5 shows the concept.

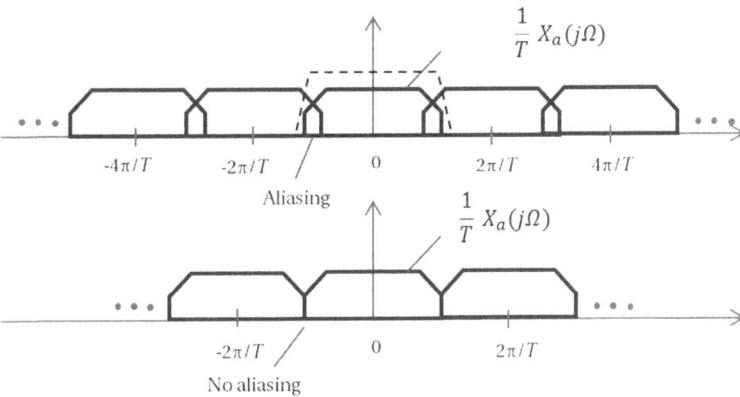

Figure 2-5. *Fourier transform of a sampled analog bandlimited signal*

If there is overlap between the shifted versions of $X_a(j\Omega)$, *aliasing* occurs because there are remnants of the neighboring copies in an extracted signal. However, when there is no aliasing, the signal $x_a(t)$ can be recovered from its sampled version $x(n)$ by retaining only one copy.[10] Thus if the signal is band-limited within a frequency band $-\pi/T$ to π/T, a sampling rate of $2\pi/T$ or more guarantees an alias-free sampled signal, where no actual information is lost due to sampling. This is called the *Nyquist sampling rate*, named after Harry Nyquist, who in 1928 proposed the above sampling theorem. Claude Shannon proved this theorem in 1949, so it is also popularly known as Nyquist-Shannon sampling theorem.

The theorem applies to single- and multi-dimensional signals. Obviously, compression of the signal can be achieved by using fewer samples, but in the case of sampling frequency less than twice the bandwidth of the signal, annoying *aliasing artifacts* will be visible.

---

[10]P. Vaidyanathan, *Multirate Systems and Filter Banks* (Englewood Cliffs: Prentice Hall PTR, 1993).

Common Terms and Notions

There are a few terms to know that are frequently used in digital video. The *aspect ratio* of a geometric shape is the ratio between its sizes in different dimensions. For example, the aspect ratio of an image is defined as the ratio of its width to its height. The *display aspect ratio* (DAR) is the width to height ratio of computer displays, where common ratios are 4:3 and 16:9 (*widescreen*). An aspect ratio for the pixels within an image is also defined. The most commonly used *pixel aspect ratio* (PAR) is 1:1 (square); other ratios, such as 12:11 or 16:11, are no longer popular. The term *storage aspect ratio* (SAR) is used to describe the relationship between the DAR and the PAR such that SAR × PAR = DAR.

Historically, the role of pixel aspect ratio in the video industry has been very important. As digital display technology, digital broadcast technology, and digital video compression technology evolved, using the pixel aspect ratio has been the most popular way to address the resulting video frame differences. However, today, all three technologies use square pixels predominantly.

As other colors can be obtained from a linear combination of primary colors such as red, green and blue in RGB *color model*, or cyan, magenta, yellow, and black in CMYK model, these colors represent the basic components of a *color space* spanning all colors. A complete subset of colors within a given color space is called a *color gamut*. Standard RGB (sRGB) is the most frequently used color space for computers. International Telecommunications Union (ITU) has recommended color primaries for standard definition (SD), high-definition (HD) and ultra-high-definition (UHD) televisions. These recommendations are included in internationally recognized digital studio standards defined by ITU-R recommendation BT.601,[11] BT.709, and BT.2020, respectively. The sRGB uses the ITU-R BT.709 color primaries.

Luma is the brightness of an image, and is also known as the *black-and-white* information of the image. Although there are subtle differences between *luminance* as used in color science and *luma* as used in video engineering, often in the video discussions these terms are used interchangeably. In fact, *luminance* refers to a linear combination of red, green, and blue color representing the intensity or power emitted per unit area of light, while *luma* refers to a nonlinear combination of $R' G' B'$, the nonlinear function being known as the *gamma function* ($y = x^\gamma$, $\gamma = 0.45$). The primes are used to indicate nonlinearity. The gamma function is needed to compensate for properties of perceived vision, so as to perceptually evenly distribute the noise across the tone scale from black to white, and to use more bits to represent the color information that is more sensitive to human eyes. For details, see Poynton.[12]

Luma is often described along with *chroma*, which is the *color* information. As human vision has finer sensitivity to luma rather than chroma, chroma information is often subsampled without noticeable visual degradation, allowing lower resolution processing and storage of chroma. In component video, the three color components are

---

[11]It was originally known as CCIR-601, which defined C_B and C_R components. The standard body CCIR, a.k.a. International Radio Consultative Committee (Comité Consultatif International pour la Radio), was formed in 1927, and was superceded in 1992 by the International Telecommunications Union, Recommendations Sector (ITU-R).

[12]C. Poynton, *Digital Video and HDTV: Algorithms and Interfaces* (Burlington, MA: Morgan Kaufmann, 2003).

transmitted separately.[13] Instead of sending $R'G'B'$ directly, three derived components are sent—namely the luma (Y') and two color difference signals ($B' - Y'$) and ($R' - Y'$).

While in analog video, these color difference signals are represented by U and V, respectively, in digital video, they are known as C_B and C_R components, respectively. In fact, U and V apply to analog video only, but are commonly, albeit inappropriately, used in digital video as well. The term *chroma* represents the color difference signals themselves; this term should not be confused with *chromaticity*, which represents the characteristics of the color signals.

In particular, *chromaticity* refers to an objective measure of the quality of color information only, not accounting for the luminance quality. Chromaticity is characterized by the *hue* and the *saturation*. The hue of a color signal is its "redness," "greenness," and so on. The hue is measured as degrees in a color wheel from a single hue. The saturation or colorfulness of a color signal is the degree of its difference from gray.

Figure 2-6 depicts the chromaticity diagram for the ITU-R recommendation BT.709 and BT.2020, showing the location of the red, green, blue, and white colors. Owing to the differences shown in this diagram, digital video signal represented in BT.2020 color primaries cannot be directly presented to a display that is designed according to BT.709; a conversion to the appropriate color primaries would be necessary in order to faithfully reproduce the actual colors.

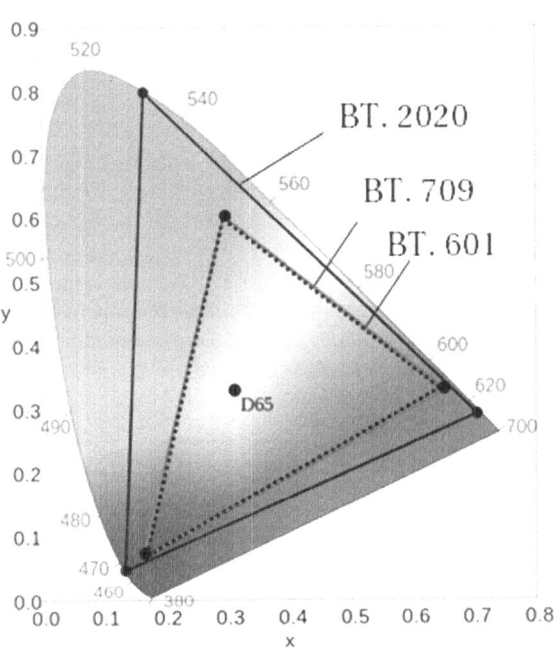

Figure 2-6. *ITU-R Recommendation BT.601, BT.709 and BT.2020 chromaticity diagram and location of primary colors. The point D65 shows the white point. (Courtesy of Wikipedia)*

---

[13]Poynton, *Digital Video.*

In order to convert $R'G'B'$ samples to corresponding $Y'C_BC_R$ samples, in general, the following formulas are used:

$$Y' = K_r R' + K_g G' + K_b B'$$
$$C_B = \frac{B' - Y'}{2(1 - K_b)}$$
$$C_R = \frac{R' - Y'}{2(1 - K_r)}$$

(Eq. 2-1)

Each of the ITU-R recommendations mentioned previously uses the values of constants K_r, K_g, and K_b, as shown in Table 2-2, although the constant names are not defined as such in the specifications.

Table 2-2. *Constants of R'G'B' Coefficients to Form Luma and Chroma Components*

Standard	K_r	K_g	K_b
BT.2020	0.2627	0.6780	0.0593
BT.709	0.2126	0.7152	0.0722
BT.601	0.2990	0.5870	0.1140

It is notable that all of these ITU-R recommendations also define a visible range between black and white for the allowed bit depths. For example, according to BT.2020, for 10-bit the luma ranges from 64 to 940; the ranges 0 to 3 and 1020 to 1023 are used for timing reference, while the ranges 4 to 63 and 941 to 1019 provide foot- and headroom, respectively, to accommodate transient black and white signals that may result from overshoots of filters. Similarly, BT.601 and BT.709 define the active range of luma between 16 and 235 for 8-bit video. In the case of 4:2:2 video, values 0 and 255 are reserved for synchorization and are forbidded from the visible picture area. Values 1 to 15 and 236 to 254 provide the relevant foot- and headroom. Table 2-3 gives the signal formats and conversion formula used in these recommendations.

Table 2-3. *Signal Formats and Conversion Formula in ITU-R Digital Video Studio Standards*

Standard	Parameter	Formula
BT.601	Derivation of luminance signal E_Y'	$E_Y' = 0.299E_R' + 0.587E_G' + 0.114E_B'$
	Derivation of color-difference signal	$E_{CB}' = \dfrac{E_B' - E_Y'}{1.772}$ $E_{CR}' = \dfrac{E_R' - E_Y'}{1.402}$
	Quantization of RGB, luminance and color-difference signals	$D_R' = INT\left[\left(219E_R' + 16\right) \cdot 2^{n-8}\right]$ $D_G' = INT\left[\left(219E_G' + 16\right) \cdot 2^{n-8}\right]$ $D_B' = INT\left[\left(219E_B' + 16\right) \cdot 2^{n-8}\right]$ $D_Y' = INT\left[\left(219E_Y' + 16\right) \cdot 2^{n-8}\right]$ $D_{CB}' = INT\left[\left(224E_{CB}' + 128\right) \cdot 2^{n-8}\right]$ $D_{CR}' = INT\left[\left(224E_{CR}' + 128\right) \cdot 2^{n-8}\right]$
	Derivation of luminance and color-difference signals via quantized RGB signals	$D_Y' = INT\left[0.2126D_R' + 0.7152D_G' + 0.0722D_B'\right]$ $D_{CB}' = INT\left[\left(-\dfrac{0.299}{1.772}D_R' - \dfrac{0.587}{1.772}D_G' + \dfrac{0.886}{1.772}D_B'\right) \cdot \dfrac{224}{219} + 2^{n-1}\right]$ $D_{CR}' = INT\left[\left(\dfrac{0.701}{1.402}D_R' - \dfrac{0.587}{1.402}D_G' - \dfrac{0.114}{1.402}D_B'\right) \cdot \dfrac{224}{219} + 2^{n-1}\right]$

(continued)

25

Table 2-3. (*continued*)

Standard	Parameter	Formula
BT.709	Derivation of luminance signal E'_Y	$E'_Y = 0.2126 E'_R + 0.7152 E'_G + 0.0722 E'_B$
	Derivation of color-difference signal	$E'_{CB} = \dfrac{E'_B - E'_Y}{1.8556}$ $E'_{CR} = \dfrac{E'_R - E'_Y}{1.5748}$
	Quantization of RGB, luminance and color-difference signals	$D'_R = INT\left[\left(219 E'_R + 16\right) \cdot 2^{n-8}\right]$ $D'_G = INT\left[\left(219 E'_G + 16\right) \cdot 2^{n-8}\right]$ $D'_B = INT\left[\left(219 E'_B + 16\right) \cdot 2^{n-8}\right]$ $D'_Y = INT\left[\left(219 E'_Y + 16\right) \cdot 2^{n-8}\right]$ $D'_{CB} = INT\left[\left(224 E'_{CB} + 128\right) \cdot 2^{n-8}\right]$ $D'_{CR} = INT\left[\left(224 E'_{CR} + 128\right) \cdot 2^{n-8}\right]$
	Derivation of luminance and color-difference signals via quantized RGB signals	$D'_Y = INT\left[0.2126 D'_R + 0.7152 D'_G + 0.0722 D'_B\right]$ $D'_{CB} = INT\left[\left(-\dfrac{0.2126}{1.8556} D'_R - \dfrac{0.7152}{1.8556} D'_G + \dfrac{0.9278}{1.8556} D'_B\right) \cdot \dfrac{224}{219} + 2^{n-1}\right]$ $D'_{CR} = INT\left[\left(\dfrac{0.7874}{1.5748} D'_R - \dfrac{0.7152}{1.5748} D'_G - \dfrac{0.0722}{1.5748} D'_B\right) \cdot \dfrac{224}{219} + 2^{n-1}\right]$

(*continued*)

Table 2-3. *(continued)*

Standard	Parameter	Formula
BT.2020	Derivation of luminance signal Y'	$Y' = 0.2627R' + 0.678G' + 0.0593B'$
	Derivation of color-difference signal	$C'_B = \dfrac{B' - Y'}{1.8814}$ $C'_R = \dfrac{R' - Y'}{1.4746}$
	Quantization of RGB, luminance and color-difference signals	$D'_R = INT\left[\left(219R' + 16\right) \cdot 2^{n-8}\right]$ $D'_G = INT\left[\left(219G' + 16\right) \cdot 2^{n-8}\right]$ $D'_B = INT\left[\left(219B' + 16\right) \cdot 2^{n-8}\right]$ $D'_Y = INT\left[\left(219Y' + 16\right) \cdot 2^{n-8}\right]$ $D'_{CB} = INT\left[\left(224C'_B + 128\right) \cdot 2^{n-8}\right]$ $D'_{CR} = INT\left[\left(224C'_R + 128\right) \cdot 2^{n-8}\right]$
	Derivation of luminance and color-difference signals via quantized RGB signals	$D'_Y = INT\left[0.2627D'_R + 0.6780D'_G + 0.0593D'_B\right]$ $D'_{CB} = INT\left[\left(-\dfrac{0.2627}{1.8814}D'_R - \dfrac{0.6780}{1.8814}D'_G + \dfrac{0.9407}{1.8814}D'_B\right) \cdot \dfrac{224}{219} + 2^{n-1}\right]$ $D'_{CR} = INT\left[\left(\dfrac{0.7373}{1.4746}D'_R - \dfrac{0.6780}{1.4746}D'_G - \dfrac{0.0593}{1.4746}D'_B\right) \cdot \dfrac{224}{219} + 2^{n-1}\right]$

Note: Here, E_k is the original analog signal, D_k is the coded digital signal, n is the number of bits in the quantized signal, and $INT[\cdot]$ is rounding to nearest integer.

In addition to the signal formats, the recommendations also specify the opto-electronic conversion parameters and the picture characteristics. Table 2-4 shows some of these parameters.

Table 2-4. *Important Parameters in ITU-R Digital Video Studio Standards*

Standard	Parameter	Value
BT. 601	Chromaticity co-ordinates (x, y)	60 field/s: R: (0.63, 0.34), G: (0.31, 0.595), B: (0.155, 0.07) 50 field/s: R: (0.64, 0.33), G: (0.29, 0.6), B: (0.15, 0.06)
	Display aspect ratio	13.5 MHz sampling frequency: 4:3 and 16:9 18 MHz sampling frequency: 16:9
	Resolution	4:4:4, 13.5 MHz sampling frequency: 60 field/s: 858 × 720 50 field/s: 864 × 720 4:4:4, 18 MHz sampling frequency: 60 field/s: 1144 × 960 50 field/s: 1152 × 960 4:2:2 systems have appropriate chroma subsampling.
	Picture rates	60 field/s, 50 field/s
	Scan mode	Interlaced
	Coding format	Uniformly quantized PCM, 8 (optionally 10) bits per sample
BT. 709	Chromaticity co-ordinates (x, y)	R: (0.64, 0.33), G: (0.3, 0.6), B: (0.15, 0.06)
	Display aspect ratio	16:9
	Resolution	1920×1080
	Picture rates	60p, 50p, 30p, 25p, 24p, 60i, 50i, 30psf, 25psf, 24psf
	Scan modes	Progressive (*p*), interlaced (*i*), progressive capture but segmented frame transmission (*psf*)
	Coding format	Linear 8 or 10 bits per component
BT. 2020	Chromaticity co-ordinates (x, y)	R: (0.708, 0.292), G: (0.17, 0.797), B: (0.131, 0.046)
	Display aspect ratio	16:9
	Resolution	3840 × 2160, 7680 × 4320
	Picture rates	120, 60, 60/1.001, 50, 30, 30/1.001, 25, 24, 24/1.001
	Scan mode	Progressive
	Coding format	10 or 12 bits per component

Chroma Subsampling

As mentioned earlier, the HVS is less sensitive to color information compared to its sensitivity to brightness information. Taking advantage of this fact, technicians developed methods to reduce the chroma information without significant loss in visual quality. Chroma subsampling is a common data-rate reduction technique and is used in both analog and digital video encoding schemes. Besides video, it is also used, for example, in popular single-image coding algorithms, as defined by the Joint Photographic Experts Group (JPEG), a joint committee between the International Standards Organization (ISO) and the ITU-T.

Exploiting the high correlation in color information and the characteristics of the HVS, chroma subsampling reduces the overall data bandwidth. For example, a 2:1 chroma subsampling of a rectangular image in the horizontal direction results in only two-thirds of the bandwidth required for the image with full color resolution. However, such saving in data bandwidth is achieved with little perceptible visual quality loss at normal viewing distances.

4:4:4 to 4:2:0

Typically, images are captured in the $R'G'B'$ color space, and are converted to the $Y'UV$ color space (or for digital video $Y'C_BC_R$; in the discussion we use $Y'UV$ and $Y'C_BC_R$ interchangeably for simplicity) using the conversion matrices described earlier. The resulting $Y'UV$ image is a full-resolution image with a 4:4:4 sampling ratio of the Y', U and V components, respectively. This means that for every four samples of Y' (luma), there are four samples of U and four samples of V chroma information present in the image.

The ratios are usually defined for a 4×2 sample region, for which there are four 4×2 luma samples. In the ratio $4:a:b$, a and b are determined based on the number of chroma samples in the top and bottom row of the 4×2 sample region. Accordingly, a 4:4:4 image has full horizontal and vertical chroma resolution, a 4:2:2 image has a half-horizontal and full vertical resolution, and a 4:2:0 image has half resolutions in both horizontal and vertical dimensions.

The 4:2:0 is different from 4:1:1 in that in 4:1:1, one sample is present in each row of the 4×2 region, while in 4:2:0, two samples are present in the top row, but none in the bottom row. An example of the common chroma formats (4:4:4, 4:2:2 and 4:2:0) is shown in Figure 2-7.

4:4:4

Number of chroma samples in a
4×2 sample region: top row = 4,
bottom row = 4

4:2:2

Number of chroma samples in a
4×2 sample region: top row = 2,
bottom row = 2

4:2:0

Number of chroma samples in a
4×2 sample region: top row = 2,
bottom row = 0

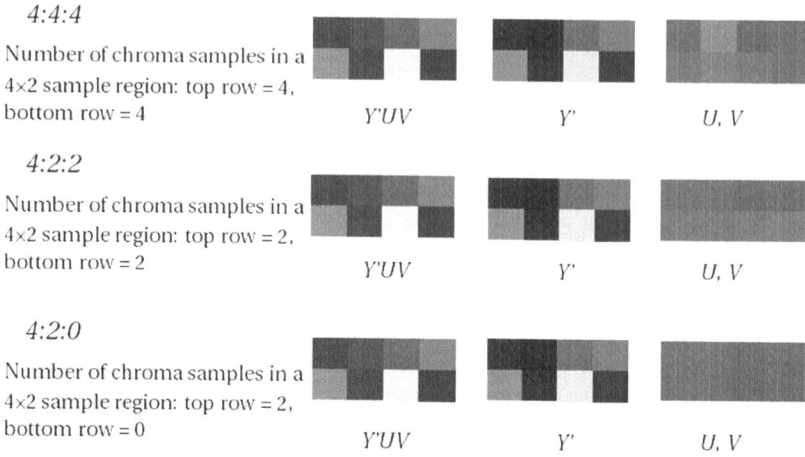

                       Y'UV   *Y'*   *U, V*

Figure 2-7. *Explanation of 4:a:b subsamples*

A *subsampling* is also known as *downsampling*, or *sampling rate compression*. If the input signal is not bandlimited in a certain way, subsampling results in aliasing and information loss, and the operation is not reversible. To avoid aliasing, a low pass filter is used before subsampling in most appplications, thus ensuring the signal to be bandlimited.

The 4:2:0 images are used in most international standards, as this format provides sufficient color resolution for an acceptable perceptual quality, exploiting the high correlation between color components. Therefore, often a camera-captured $R'G'B'$ image is converted to $Y'UV$ 4:2:0 format for compression and processing. In order to convert a 4:4:4 image to a 4:2:0 image, typically a two-step approach is taken. First, the 4:4:4 image is converted to a 4:2:2 image via filtering and subsampling horizontally; then, the resulting image is converted to a 4:2:0 format via vertical filtering and subsampling. Example filters are shown in Figure 2-8.

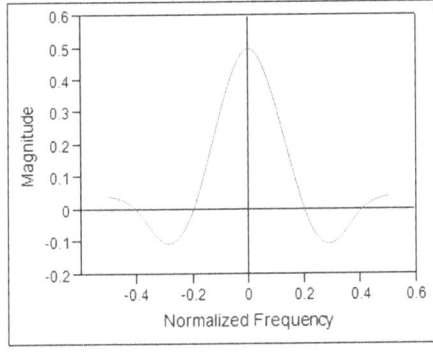

Frequency response of a 2:1 horizontal
subsampling filter

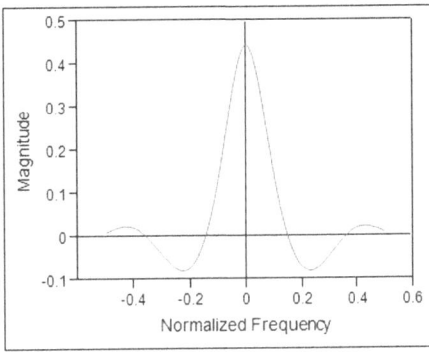

Frequency response of a 2:1 vertical
subsampling filter

Figure 2-8. *Typical symmetric finite impulse response (FIR) filters used for 2:1 subsampling*

The filter coefficients for the Figure 2-8 finite impulse response (FIR) filters are given in Table 2-5. In this example, while the horizontal filter has zero phase difference, the vertical filter has a phase shift of 0.5 sample interval.

Table 2-5. *FIR Filter Coefficients of a 2:1 Horizontal and a 2:1 Vertical Filter, Typically Used in 4:4:4 to 4:2:0 Conversion*

Filter Coefficients

Horiz.	0.0430	0.0000	-0.1016	0.0000	0.3105	0.5000	0.3105	0.0000	-0.1016	0.0000	0.0430
Vert.	0.0098	0.0215	-0.0410	-0.0723	0.1367	0.4453	0.1367	-0.0723	-0.0410	0.0215	0.0098
Norm. Freq.	-0.5	-0.4	-0.3	-0.2	-0.1	0	0.1	0.2	0.3	0.4	0.5

Reduction of Redundancy

Digital video signal contains a lot of similar and correlated information between neighboring pixels and neighboring frames, making it an ideal candidate for compression by removing or reducing the redundancy. We have already discussed chroma subsampling and the fact that very little visual difference is seen because of such subsampling. In that sense, the full resolution of chroma is redundant information, and by doing the subsampling, a reduction in data rate—that is, data compression—is achieved. In addition, there are other forms of redundancy present in a digital video signal.

Spatial Redundancy

The digitization process ends up using a large number of bits to represent an image or a video frame. However, the number of bits necessary to represent the information content of a frame may be substantially less, due to redundancy. Redundancy is defined as 1 minus the ratio of the minimum number of bits needed to represent an image to the actual number of bits used to represent it. This typically ranges from 46 percent for images with a lot of spatial details, such as a scene of foliage, to 74 percent[14] for low-detail images, such as a picture of a face. Compression techniques aim to reduce the number of bits required to represent a frame by removing or reducing the available redundancy.

Spatial redundancy is the consequence of the correlation in horizontal and the vertical spatial dimensions between neighboring pixel values within the same picture or frame of video (also known as *intra-picture* correlation). Neighboring pixels in a video frame are often very similar to each other, especially when the frame is divided into the luma and the chroma components. A frame can be divided into smaller blocks of pixels to take advantage of such pixel correlations, as the correlation is usually high within a block. In other words, within a small area of the frame, the rate of change in a spatial dimension is usually low. This implies that, in a frequency-domain representation of the video frame, most of the energy is often concentrated in the low-frequency region, and high-frequency edges are relatively rare. Figure 2-9 shows an example of spatial redundancy present in a video frame.

Original frame

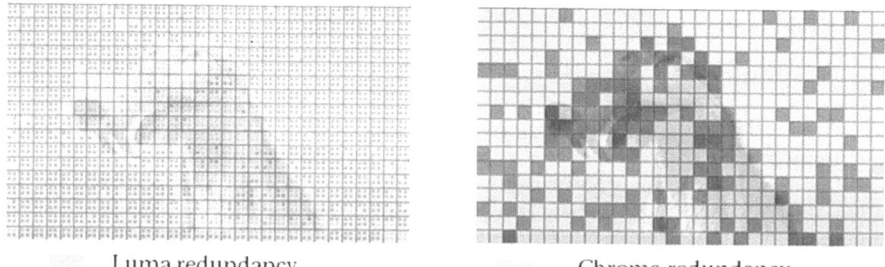

Luma redundancy   Chroma redundancy

Figure 2-9. *An example of spatial redundancy in an image or a video frame*

---

[14]M. Rabbani and P. Jones, *Digital Image Compression Techniques* (Bellingham, WA: SPIE Optical Engineering Press, 1991).

The redundancy present in a frame depends on several parameters. For example, the sampling rate, the number of quantization levels, and the presence of source or sensor noise can all affect the achievable compression. Higher sampling rates, low quantization levels, and low noise mean higher pixel-to-pixel correlation and higher exploitable spatial redundancy.

Temporal Redundancy

Temporal redundancy is due to the correlation between different pictures or frames in a video (also known as *inter-picture* correlation). There is a significant amount of temporal redundancy present in digital videos. A video is frequently shown at a frame rate of more than 15 *frames per second* (fps) in order for a human observer to perceive a smooth, continuous motion; this requires neighboring frames to be very similar to each other. One such example is shown in Figure 2-10. It may be noted that a reduced frame rate would result in data compression, but that would be at the expense of perceptible *flickering artifact*.

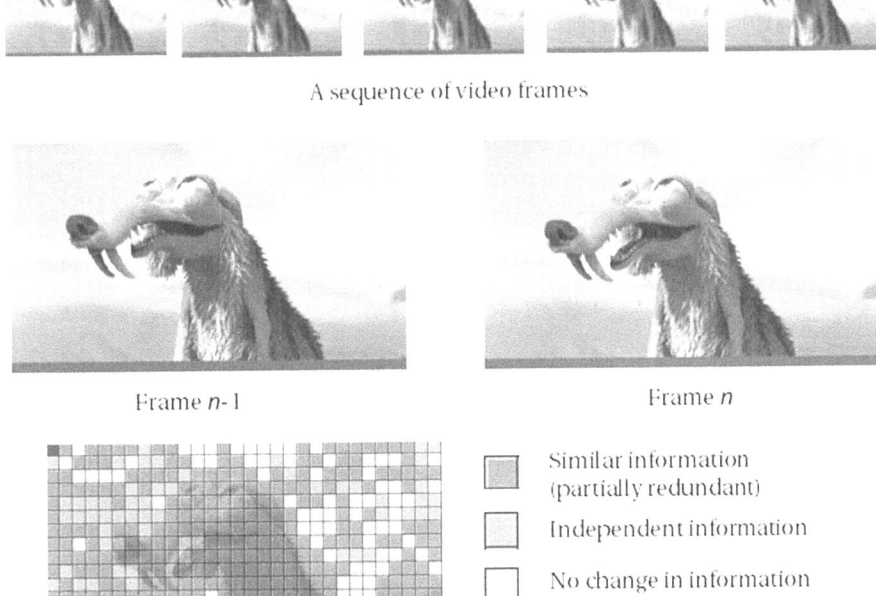

A sequence of video frames

Frame *n*-1
Frame *n*

Similar information
(partially redundant)

Independent information

No change in information
(fully redundant)

Figure 2-10. An example of temporal redundancy among video frames. Neighboring video frames are quite similar to each other

Thus, a frame can be represented in terms of a neighboring reference frame and the difference information between these frames. Because an independent frame is reconstructed at the receiving end of a transmission system, it is not necessary for a dependent frame to be transmitted. Only the difference information is sufficient for the successful reconstruction of a dependent frame using a prediction from an already received reference frame. Due to temporal redundancy, such difference signals are often quite small. Only the difference signal can be coded and sent to the receiving end, while the receiver can combine the difference signal with the predicted signal already available and obtain a frame of video, thereby achieving very high amount of compression. Figure 2-11 shows an example of how temporal redundancy is exploited.

Reconstructed frame   Predicted frame from   Difference signal
information already received   (coded & transmitted)

Figure 2-11. *Prediction and reconstruction process exploiting temporal redundancy*

The difference signal is often motion-compensated to minimize the amount of information in it, making it amenable to a higher compression compared to an uncompensated difference signal. Figure 2-12 shows an example of reduction of information using motion compensation from one video frame to another.

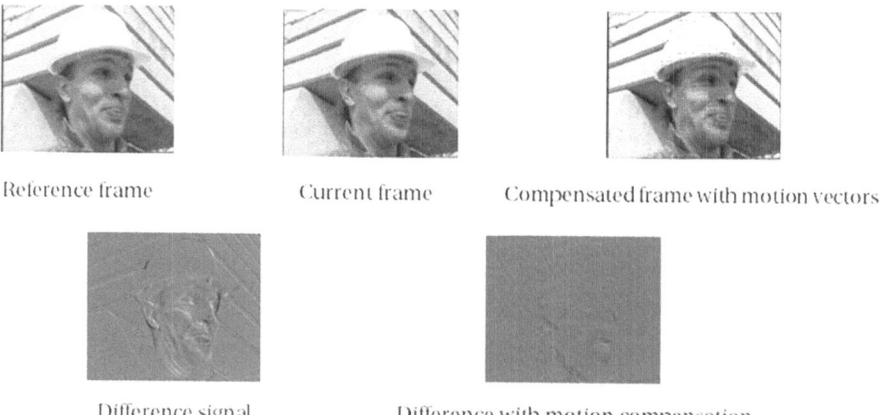

Reference frame   Current frame   Compensated frame with motion vectors

Difference signal   Difference with motion compensation

Figure 2-12. *An example of reduction of informataion via motion compensation*

The prediction and reconstruction process is lossless. However, it is easy to understand that the better the prediction, the less information remains in the difference signal, resulting in a higher compression. Therefore, every new generation of international video coding standards has attempted to improve upon the prediction process of the previous generation.

Statistical Redundancy

In information theory, redundancy is the number of bits used to transmit a signal minus the number of bits of actual information in the signal, normalized to the number of bits used to transmit the signal. The goal of data compression is to reduce or eliminate unwanted redundancy. Video signals characteristically have various types of redundancies, including spatial and temporal redundancies, as discussed above. In addition, video signals contain statistical redundancy in its digital representation; that is, there are usually extra bits that can be eliminated before transmission.

For example, a region in a binary image (e.g., a fax image or a video frame) can be viewed as a string of 0s and 1s, the 0s representing the white pixels and 1s representing the black pixels. These strings, where the same bit occurs in a series or *run* of consecutive data elements, can be represented using run-length codes; these codes the address of each string of 1s (or 0s) followed by the length of that string. For example, 1110 0000 0000 0000 0000 0011 can be coded using three codes (1,3), (0,19), and (1,2), representing 3 1s, 19 0s, and 2 1s. Assuming only two symbols, 0 and 1, are present, the string can also be coded using two codes (0,3) and (22,2), representing the length of 1s at locations 0 and 22.

Variations on the run-length are also possible. The idea is this: instead of the original data elements, only the number of consecutive data elements is coded and stored, thereby achieving significant data compression. Run-length coding is a lossless data compression technique and is effectively used in compressing quantized coefficients, which contains runs of 0s and 1s, especially after discarding high-frequency information.

According to Shannon's source coding theorem, the maximum achievable compression by exploiting statistical redundancy is given as:

$$C = \frac{average\ bit\ rate\ of\ the\ original\ signal\ (B)}{average\ bit\ rate\ of\ the\ encoded\ data\ (H)}$$

Here, H is the entropy of the source signal in bits per symbol. Although this theoretical limit is achievable by designing a coding scheme, such as *vector quantization* or *block coding*, for practical video frames—for instance, video frames of size 1920×1080 pixels with 24 bits per pixel—the codebook size can be prohibitively large.[15] Therefore, international standards instead often use entropy coding methods to get arbitrarily close to the theoretical limit.

---

[15]A. K. Jain, *Fundamentals of Digital Image Processing* (Englewood Cliffs: Prentice-Hall International, 1989).

Entropy Coding

Consider a set of quantized coefficients that can be represented using B bits per pixel. If the quantized coefficients are not uniformly distributed, then their entropy will be less than B bits per pixel. Now, consider a block of M pixels. Given that each bit can be one of two values, we have a total number of $L = 2^{MB}$ different pixel blocks.

For a given set of data, let us assign the probability of a particular block i occurring as p_i, where $i = 0, 1, 2, \cdots, L - 1$. Entropy coding is a lossless coding scheme, where the goal is to encode this pixel block using $-\log_2 p_i$ bits, so that the average bit rate is equal to the entropy of the M pixel block: $H = \sum_i p_i(-\log_2 p_i)$. This gives a variable length code for each block of M pixels, with smaller code lengths assigned to highly probable pixel blocks. In most video-coding algorithms, quantized coefficients are usually run-length coded, while the resulting data undergo entropy coding for further reduction of statistical redundancy.

For a given block size, a technique called *Huffman coding* is the most efficient and popular variable-length encoding method, which asymptotically approaches Shannon's limit of maximum achievable compression. Other notable and popular entropy coding techniques are *arithmetic coding* and *Golomb-Rice coding*.

Golomb-Rice coding is especially useful when the approximate entropy characteristics are known—for example, when small values occur more frequently than large values in the input stream. Using sample-to-sample prediction, the Golomb-Rice coding scheme produces output rates within 0.25 bits per pixel of the one-dimensional difference entropy for entropy values ranging from 0 to 8 bits per pixel, without needing to store any code words. Golomb-Rice coding is essentially an optimal run-length code. To compare, we discuss now the Huffman coding and the arithmetic coding.

Huffman Coding

Huffman coding is the most popular lossless entropy coding algorithm; it was developed by David Huffman in 1952. It uses a variable-length code table to encode a source symbol, while the table is derived based on the estimated probability of occurrence for each possible value of the source symbol. Huffman coding represents each source symbol in such a way that the most frequent source symbol is assigned the shortest code and the least frequent source symbol is assigned the longest code. It results in a prefix code, so that a bit string representing a source symbol is never a prefix of the bit string representing another source symbol, thereby making it uniquely decodable.

To understand how Huffman coding works, let us consider a set of four source symbols $\{a_0, a_1, a_2, a_3\}$ with probabilities $\{0.47, 0.29, 0.23, 0.01\}$, respectively. First, a binary tree is generated from left to right, taking the two least probable symbols and combining them into a new equivalent symbol with a probability equal to the sum of the probablities of the two symbols. In our example, therefore, we take a_2 and a_3 and form a new symbol b_2 with a probability $0.23 + 0.01 = 0.24$. The process is repeated until there is only one symbol left.

The binary tree is then traversed backwards, from right to left, and codes are assigned to different branches. In this example, codeword 0 (one bit) is assigned to symbol a_0, as this is the most probable symbol in the source alphabet, leaving codeword

1 for c_1. This codeword is the prefix for all its branches, ensuring unique decodeability. At the next branch level, codeword 10 (two bits) is assigned to the next probable symbol a_1, while 11 goes to b_2 and as a prefix to its branches. Thus, a_2 and a_3 receive codewords 110 and 111 (three bits each), respectively. Figure 2-13 shows the process and the final Huffman codes.

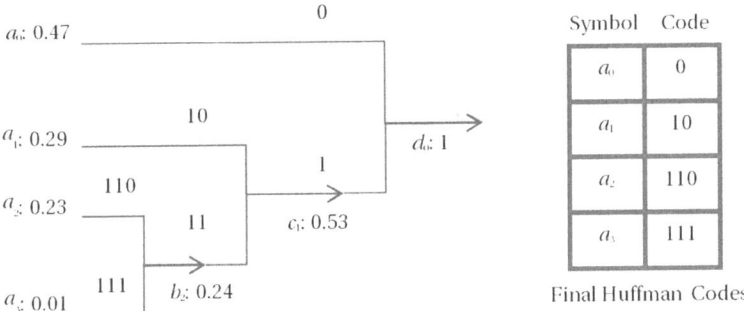

Symbol	Code
a_0	0
a_1	10
a_2	110
a_3	111

Final Huffman Codes

Figure 2-13. *Huffman coding example*

While these four symbols could have been assigned fixed length codes of 00, 01, 10, and 11 using two bits per symbol, given that the probability distribution is non-uniform and the entropy of these symbols is only 1.584 bits per symbol, there is room for improvement. If these codes are used, 1.77 bits per symbol will be needed instead of two bits per symbol. Although this is still 0.186 bits per symbol apart from the theoretical minimum of 1.584 bits per symbol, it still provides approximately 12 percent compression compared to fixed-length code. In general, the larger the difference in probabilities between the most and the least probable symbols, the larger the coding gain Huffman coding would provide. Huffman coding is optimal when the probability of each input symbol is the inverse of a power of 2.

Arithmetic Coding

Arithmetic coding is a lossless entropy coding technique. Arithmetic coding differs from Huffman coding in that, rather than separating the input into component symbols and replacing each with a code, arithmetic coding encodes the entire message into a single fractional number between 0.0 and 1.0. When the probability distribution is unknown, not independent and not identically distributed, arithmetic coding may offer better compression capability than Huffman coding, as it can combine an arbitrary number of symbols for more efficient coding and is usually adaptable to the actual input statistics. It is also useful when the probability of one of the events is much larger than ½. Arithmetic coding gives optimal compression, but it is often complex and may require dedicated hardware engines for fast and practical execution.

In order to describe how arithmetic coding[16] works, let us consider an example of three events (e.g., three letters in a text): the first event is either a_1 or b_1, the second is either a_2 or b_2, and the third is either a_3 or b_3. For simplicity, we choose between only two events at each step, although the algorithm works for multi-events as well. Let the input text be $b_1a_2b_3$, with probabilities as given in Figure 2-14.

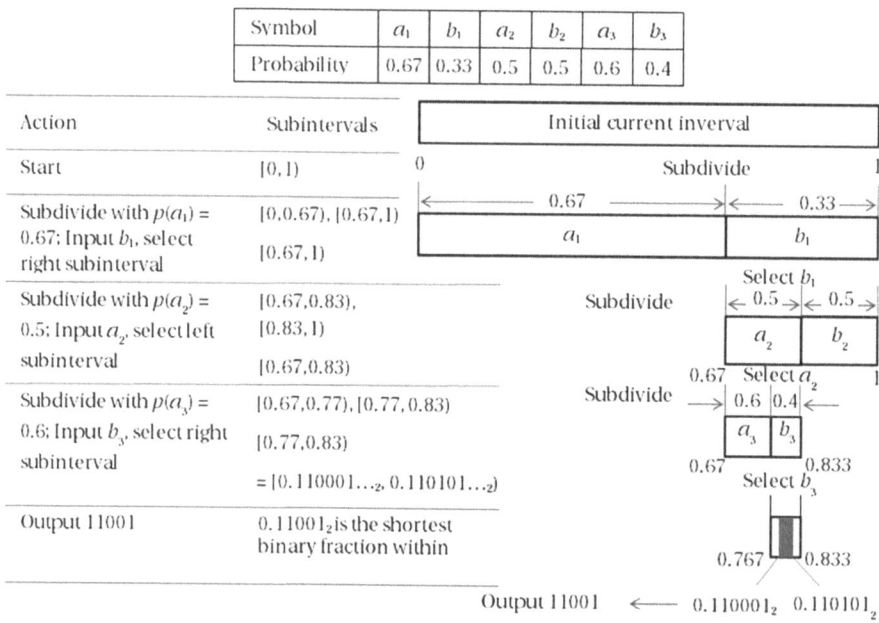

Figure 2-14. *Example of arithmetic coding*

Compression Techniques: Cost-benefit Analysis

In this section we discuss several commonly used video-compression techniques and analyze their merits and demerits in the context of typical usages.

Transform Coding Techniques

As mentioned earlier, pixels in a block are similar to each other and have spatial redundancy. But a block of pixel data does not have much statistical redundancy and is not readily suitable for variable-length coding. The decorrelated representation in the transform domain has more statistical redundancy and is more amenable to compression using variable-length codes.

---

[16]P. Howard and J. Vitter, "Arithmetic Coding for Data Compression," *Proceedings of the IEEE* 82, no. 6 (1994): 857–65.

In transform coding, typically a video frame of size $N \times M$ is subdivided into smaller $n \times n$ blocks, and a reversible linear transform is applied on these blocks. The transform usually has a set of complete orthonormal discrete-basis functions, while its goal is to decorrelate the original signal and to redistribute the signal energy among a small set of transform coefficients. Thus, many coefficients with low signal energy can be discarded through the quantization process prior to coding the remaining few coefficients. A block diagram of transform coding is shown in Figure 2-15.

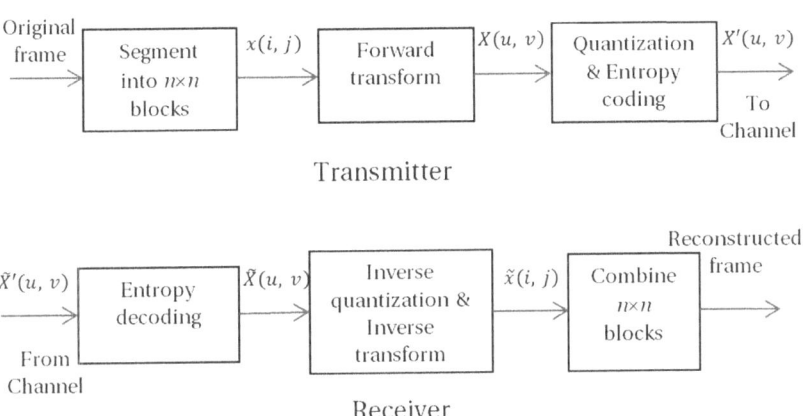

Figure 2-15. *A block diagram of transform coding in a transmission system*

Discrete Cosine Transform

A *discrete cosine transform* (DCT) expresses a finite sequence of discrete data points in terms of a sum of cosine functions with different frequencies and amplitudes. The DCT is a linear, invertible, lossless transform that can very effectively decorrelate the redundancy present in a block of pixels. In fact, the DCT is the most efficient, practical transform available for this purpose and it approaches the theoretically optimum *Karhunen-Loève transform* (KLT), as very few cosine functions are needed to approximate a typical signal. For this reason, the DCT is widely used in video and audio compression techniques.

There are four representations of the DCT, of which DCT-II[17] is the most common form:

$$X(k) = \sum_{n=0}^{N-1} x(n) \cos\left[\frac{\pi}{N}\left(n + \frac{1}{2}\right)k\right] \qquad k = 0, \ldots, N-1.$$

---

[17]K. R. Rao and P. Yip, *Discrete Cosine Transform: Algorithms, Advantages, Aapplications* (New York: Academic Press, 1990).

Here, X_k is the transformed DCT coefficient, and x_n is the input signal. This one-dimensional DCT can be separately used vertically and horizontally, one after the other, to obtain a two-dimensional DCT. For image and video compression, the DCT is most popularly performed on 8×8 blocks of pixels. The 8×8 two-dimensional DCT can be expressed as follows:

$$X(u,v) = \frac{1}{4}\alpha(u)\alpha(v)\sum_{m=0}^{7}\sum_{n=0}^{7}x(m,n)\cos\left[\frac{(2m+1)u\pi}{16}\right]\cos\left[\frac{(2n+1)v\pi}{16}\right]$$

Here, u and v are the horizontal and vertical spatial frequencies, $0 \le u, v < 8$; $a(k)$ is a normalizing factor equal to $1/\sqrt{2}$ for $k = 0$, and equal to 1 otherwise; $x(m, n)$ is the pixel value at spatial location (m,n); and $X(u, v)$ is the DCT coefficient at frequency coordinates (u,v).

The DCT converts an 8×8 block of input values to a linear combination of the 64 two-dimensional DCT *basis* functions, which are represented in 64 different patterns, as shown in Figure 2-16.

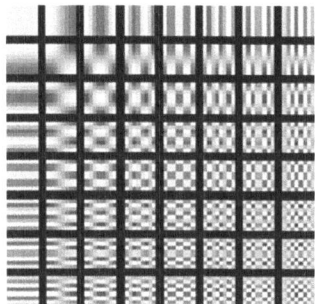

Figure 2-16. *The 64 two-dimensional DCT basis functions for an 8×8 input block*

Although the transform is lossless, owing to limitations in arithmetic precision of a computing system, it may introduce inaccuracies so that the same, exact input may not be obtained upon an inverse operation. In order to handle such inaccuracies, standard committees often take measures such as defining the IEEE standard 1180, which is described later in this chapter.

A signal flow diagram of an eight-point DCT (and inverse DCT) is shown in Figure 2-17, representing a one-dimensional DCT, where the input data set is $(u_0, \ldots, u_7)$, the output data set is $(v_0, \ldots, v_7)$, and $(f_0, \ldots, f_7)$ are the cosine function-based multiplication factors for the intermediate results. There are many fast algorithms and implementations of the DCT available in the literature, as nearly all international standards adopt the DCT as the transform of choice to reduce spatial redundancy.

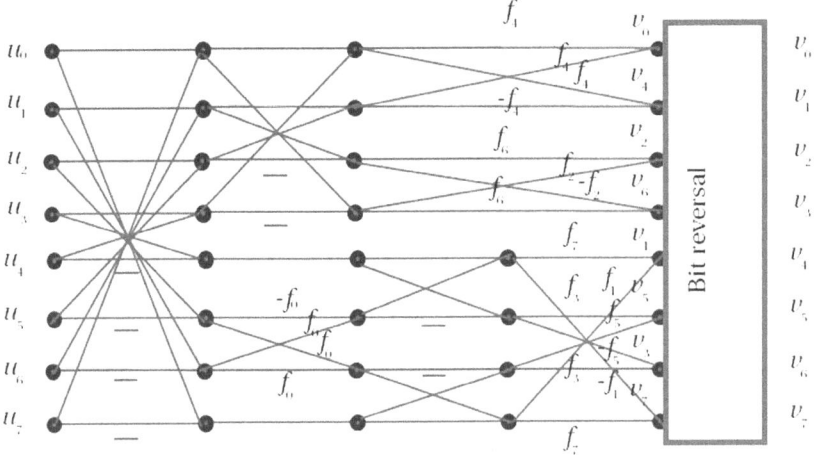

Here, $f_0 = \cos(\pi/4)$, and $f_k = \frac{1}{2}\cos(k\pi/16)$, $k = 1, ..., 7$.

Figure 2-17. *Signal flow graph of eight-point DCT, left to right (and IDCT from right to left)*

The DCT can easily be implemented using hardware or software. An optimized software implementation can take advantage of *single instruction multiple data* (SIMD) parallel constructs available in multimedia instruction sets such as MMX or SSE.[18] Furthermore, there are dedicated hardware engines available in Intel integrated graphics processor based codec solutions.

An example of a block of pixels and its DCT-transformed coefficients is depicted in Figure 2-18.

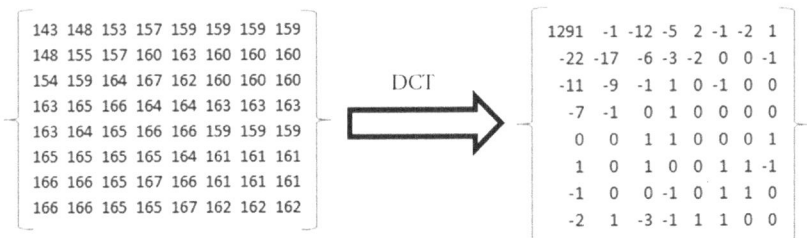

Figure 2-18. *A block of pixels and its DCT transformed version*

[18]S. Akramullah, I. Ahmad, and M. Liou, "Optimization of H.263 Video Encoding Using a Single Processor Computer: Performance Tradeoffs and Benchmarking," *IEEE Transactions on Circuits and Systems for Video Technology* 11, no. 8 (2001): 901–15.

Quantization

As the DCT is characteristically lossless, it does not provide compression by itself; it merely decorrelates the input data. However, the DCT is usually followed by a quantization process, which truncates the high-frequency information of the transformed data block, exploiting the spatial redundancy present in frames of video.

A quantizer is a staircase function that maps a continuous input signal or a discrete input signal with many values, into a smaller, finite number of output levels. If x is a real scalar random variable with $p(x)$ being its probability density function, a quantizer maps x into a discrete variable $\tilde{x} \in \{r_i, i=0,\ldots,N-1\}$, where each level r_i is known as a *reconstruction level*. The values of x that map to a particular x^* are defined by a set of *decision levels* $\{d_i, i = 0, \ldots, N-1\}$. According to the quantization rule, if x lies in the interval $(d_i, d_{i+1}]$, it is mapped—that is, quantized to r_i—which also lies in the same interval. Quantizers are designed to optimize the r_i and d_i for a given $p(x)$ and a given optimization criterion.

Figure 2-19 shows an example eight-level nonlinear quantizer. In this example, any value of x between (-255, 16] is mapped to -20, similarly any value between (-16, -8] is mapped to -11, any value between (-8, -4] is mapped to -6, and so on. This quantization process results in only eight nonlinear reconstruction levels for any input value between (-255, 255).

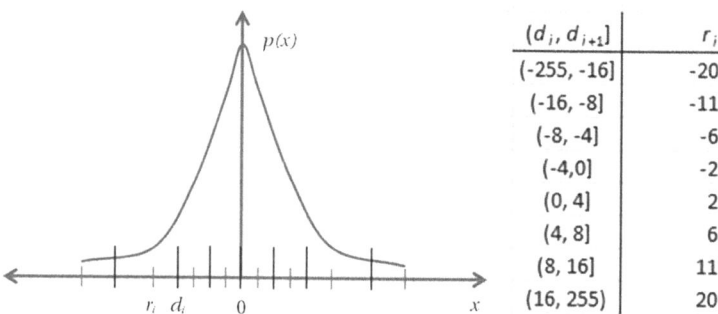

$(d_i, d_{i+1}]$	r_i
(-255, -16]	-20
(-16, -8]	-11
(-8, -4]	-6
(-4, 0]	-2
(0, 4]	2
(4, 8]	6
(8, 16]	11
(16, 255)	20

Figure 2-19. *An example eight-level nonlinear quantizer*

After quantization, an 8×8 transform data block typically reduces from 64 coefficients to from 5 to 10 coefficients, and approximately 6- to 12-fold data compression is usually achieved. However, note that quantization is a lossy process where the discarded high-frequency information cannot be regained upon performing the inverse operation. Although the high-frequency information is frequently negligible, that is not always the case. Thus, the transform and quantization process usually introduces a quality loss, which is commonly known as the *quantization noise*. All international standards define the transform and quantization process in detail and require conformance to the defined process.

In the case of a two-dimensional signal, such as an image or a video frame where quantization is usually performed on blocks of pixels, contouring effect is produced at the block boundaries because the blocks are transformed and quantized independently; as a result, the block boundaries become visible. This is commonly known as the *blocking or blocky artifact*. Although a coarser quantization level would yield a greater data compression, it is worthwhile to note that the coarser the quantization level for a signal, the more blocking artifact will be introduced.

Walsh-Hadamard and Other Transforms

The *Walsh-Hadamard transform* (WHT) is a linear, orthogonal, and symmetric transform that usually operate on 2^m real numbers. It has only modest decorrelation capability, but it is a popular transform owing to its simplicity. The WHT basis functions consist of values of either +1 or -1, and can be obtained from the rows of orthonormal Hadamard matrices. Orthonormal Hadamard matrices can be constructed recursively from the smallest 2×2 matrix of the same kind, which is a size 2 discrete Fourier transform (DFT), as follows:

$$H_2 = \frac{1}{\sqrt{2}}\begin{bmatrix} 1 & 1 \\ 1 & -1 \end{bmatrix}, \: and \: H_{2n} = \frac{1}{\sqrt{2}}\begin{bmatrix} H_n & H_n \\ H_n & -H_n \end{bmatrix}$$

There are fast algorithms available for computation of the Hadamard transform, making it suitable for many applications, including data compression, signal processing, and data encryption algorithms. In video-compression algorithms, it is typically used in the form of *sum of absolute transform differences* (SATD), which is a video-quality metric used to determine if a block of pixel matches another block of pixel.

There are other less frequently used transforms found in various video-compression schemes. Notable among them is the *discrete wavelet transform* (DWT), the simplest form of which is called the *Haar transform* (HT). The HT is an invertible, linear transform based on the Haar matrix. It can be thought of as a sampling process in which rows of the Haar matrix act as samples of finer and finer resolution. It provides a simple approach to analyzing the local aspects of a signal, as opposed to non-localized WHT, and is very effective in algoithms such as subband coding. An example of a 4×4 Haar matrix is this:

$$H_4 = \begin{bmatrix} 1 & 1 & 1 & 1 \\ 1 & 1 & -1 & -1 \\ 1 & -1 & 0 & 0 \\ 0 & 0 & 1 & -1 \end{bmatrix}$$

Predictive Coding Techniques

Prediction is an important coding technique. In the receiving end of a transmission system, if the decoder can somehow predict the signal, even with errors, it can reconstruct an approximate version of the input signal. However, if the error is known or transmitted to the decoder, the reconstruction will be a more faithful replica of the original signal. Predictive coding takes advantage of this principle. Predictive coding can be lossy or lossless. Here are some predictive techniques.

Lossless Predictive Coding

By exploiting the spatial redundancy, a pixel can be predicted from its neighbor. As the difference between the neighbors is usually small, it is more efficient to encode the difference rather than the actual pixel. This approach is called *differential pulse code modulation* (DPCM) technique. In DPCM, the most probable estimates are stored, and the difference between the actual pixel x and its most likely prediction x' is formed. This difference, $e = x - x'$, is called the *error signal*, which is typically entropy coded using variable-length codes.

To get a better estimate, the prediction can be formed as a linear combination of a few previous pixels. As the decoder already decodes the previous pixels, it can use these to predict the current pixel to obtain x', and upon receiving the error signal e, the decoder can perform $e + x'$ to obtain the true pixel value. Figure 2-20 shows the concept.

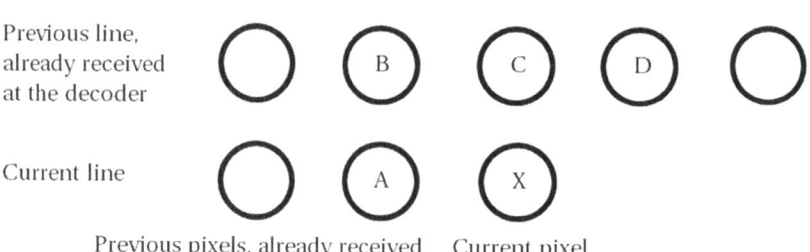

Figure 2-20. *An example of two lines of pixels showing DPCM predictor configuration*

In the Figure 2-20 example, the current pixel X can be predicted from a linear combination of the previously decoded pixels; for example, depending on the correlation, $X = 0.75A - 0.25B + 0.5C$ can be a good predicted value for X. The error image usually has a reduced variance and much less spatial correlation compared to the original image. Therefore, in DPCM, the error image is coded using a variable-length code such as the Huffman code or the arithmetic code. This approach yields the desired lossless compression.

There are some applications—for instance, in medical imaging—that benefit from combining lossless and lossy predictions, mainly to achieve a shorter transmission time. However, these applications may tolerate only small quality degradation and very high quality reconstructions are expected. In such cases, a low bit-rate version of the image is first constructed by using an efficient lossy compression algorithm. Then, a residual image is generated by taking the difference between the lossy version and the original image, which is followed by a lossless coding of the residual image.

Lossy Predictive Coding

In order to accommodate a reduced bit rate, some visual quality loss is allowed in lossy coding, while greater compression may be achieved by allowing more quality degradation. Figure 2-21 shows the general concept of lossy coding, where the original image is decomposed and/or transformed to frequency domain, the frequency-domain information is quantized, and the remaining information is coded using entropy coding.

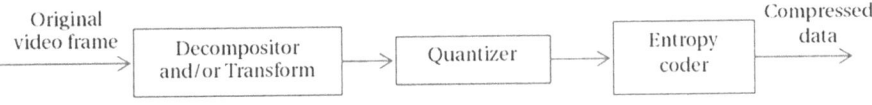

Figure 2-21. *A lossy coding scheme*

The decomposition and transformation reduce the dynamic range of the signal and also decorrelate the signal, resulting in a form that can be coded more efficiently. This step is usually reversible and lossless. However, in the next step, quantization is performed, which introduces information loss and consequently quality degradation but achieves compression as well. The entropy coding is again a lossless process, but it provides some compression by exploiting statistical redundancy.

Lossy DPCM

In predictive coding, as mentioned earlier in connection with lossless DPCM, a prediction or estimate is formed based on a reference, then an error signal is generated and coded. However, DPCM schemes can be used in lossy coding as well, resulting in lossy predictive coding.

The reference used for the prediction can be the original signal; however, the decoder at the receiving end of the transmission channel would only have the partially reconstructed signal based on the bits received so far. Note that the received signal is reconstructed from a quantized version of the signal and contains quantization noise. Therefore, typically there is a difference between the reconstructed and the original signal.

In order to ensure identical prediction at both ends of the transmission channel, the encoder also needs to form its predictions based on the reconstructed values. To achieve this, the quantizer is included in the prediction loop, as shown in Figure 2-22, which essentially incorporates the decoder within the encoding structure.

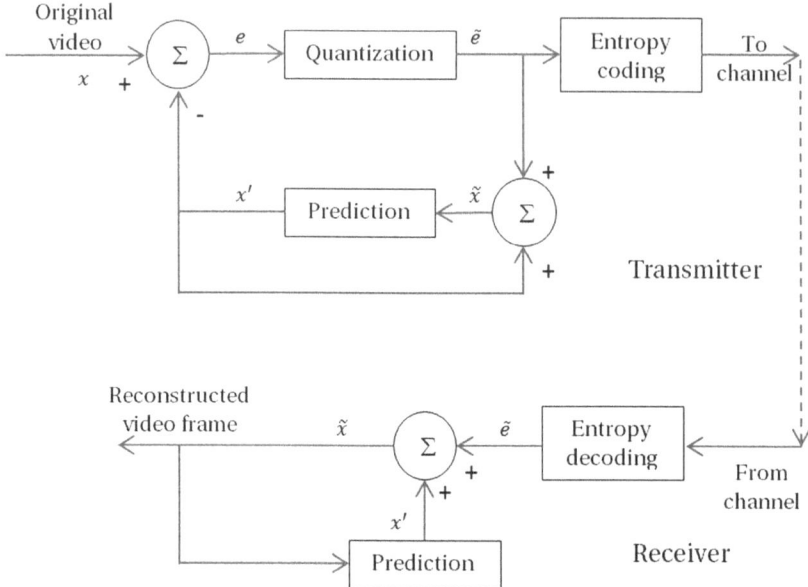

Figure 2-22. *A block diagram for lossy DPCM*

Temporal Prediction

In addition to the prediction from a neighboring pixel exploiting spatial redundancy, prediction may be formed from neighboring frames, exploiting temporal redundancy. Since neighboring frames are similar except for the small movement of objects from one frame to another, the difference signal can be captured and the residual frame can be compensated for the motion.

When the frame is divided in blocks, each block may move to a different location in the next frame. So *motion vectors* are usually defined for each block to indicate the amount of movement in horizontal and vertical dimensions. The motion vectors are integers and expressed as $mv(x, y)$; however, motion vectors from a subsampled residual frame can be combined with those from the original resolution of the residual frame such that the subsampled motion vectors are expressed as fractions. Figure 2-23 illustrates this concept, where the final motion vector is 12.5 pixels away horizontally in the current frame from the original location $(0, 0)$ in the reference frame; using a half-pixel (or half-pel) precision of motion vectors.

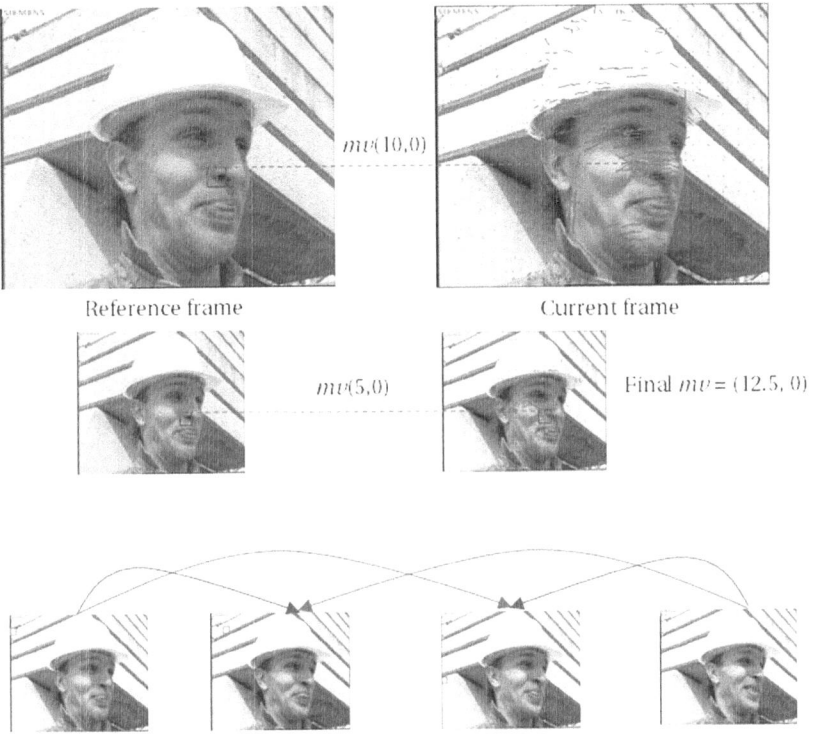

Figure 2-23. Temporal prediction examples

In order to perform motion compensation, motion vectors are determined using a process called *motion estimation*, which is typically done for a 16×16 pixel block or picture partitions of other shapes. The motion-estimation process defines a search window in the reference frame where a search is performed for the best maching block relative to the current block in the current frame. The search window is usually formed around the co-located (0, 0) position, which has the same horizontal and vertical co-ordinates in the reference frame compared to the current block in the current frame. However, in some algorithms, the search windows may be formed around a predicted motion vector candidate as well. A matching criterion, typically a distortion metric, is defined to determine the best match.

This method of *block-matching motion estimation* is different from a *pel-recursive* motion estimation, which involves matching all the pixels of the frame in a recursive manner. Figure 2-24 illustrates an example of a block-matching motion estimation.

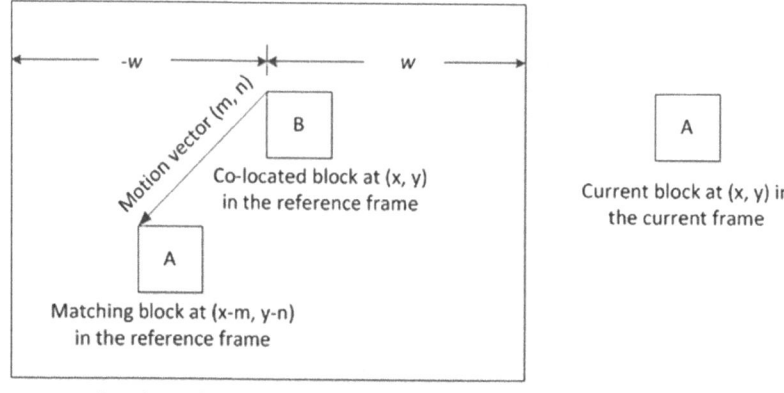

```
int GetSAD4x4(int src[4][4], int ref[4][4]) {
        int i, j, sad = 0;

        for (j=0; j<4; j++)
                for (i=0; i<4; i++)
                        sad += abs(src[i+j*4] - ref[i+j*4]);
        return sad;
}
```

Distortion calculation example

Figure 2-24. *Block-matching motion estimation*

There are a large number of motion-estimation algorithms available in the literature. Block-matching techniques attempt to minimize the distortion metric and try to find a global minimum distance between the two blocks within the search area. Typical distortion metrics are *mean absolute difference* (MAD), *sum of absolute difference* (SAD), and *sum of absolute transform difference* (SATD) involving Haar transform, having different computational complexities and matching capabilities. The motion estimation is a computationally intensive process—so much so that the encoding speed is largely determined by this process. Therefore, the choice of the distortion metric is important in lossy predictive video coding.

Additional Coding Techniques

There are additional popular coding algorithms, including vector quantization and subband coding. These algorithms are also well known owing to their individual, special characteristics.

Vector Quantization

In *vector quantization* (VQ), a frame of video is decomposed into an n-dimensional vector. For example, the $Y'C_BC_R$ components may form a three-dimensional vector, or each column of a frame may be used as elements of the vectors forming a w-dimensional vector, where w is the width of the frame. Each image vector X is compared to several *codevectors* Y_i, $i = 1, \ldots, N$, which are taken from a previously generated *codebook*.

Based on a minimum distortion criterion, such as the *mean square error* (MSE), the comparison results in a best match between X and Y_k, the kth codevector. The index k is transmitted using $\log_2 N$ bits. At the receiving end, a copy of the codebook is already available, where the decoder simply looks up the index k from the codebook to reproduce Y_k. Figure 2-25 shows the VQ block diagram.

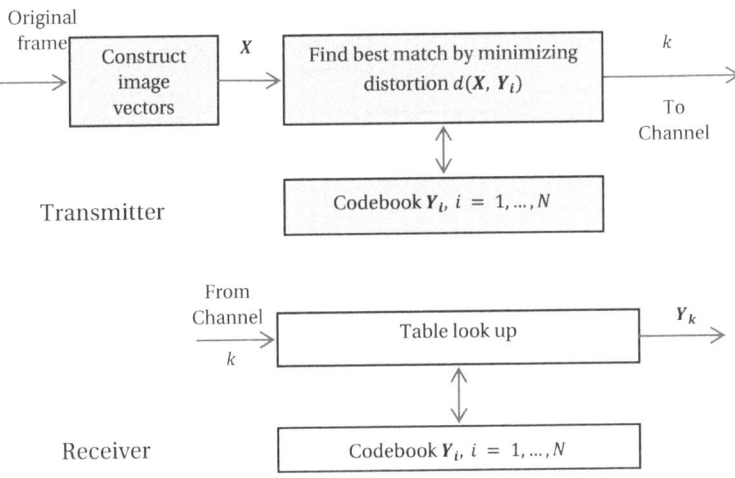

Figure 2-25. *A block diagram for vector quantization scheme*

Compression is achieved because a codebook with relatively few codevectors is used compared to the number of possible codevectors. Although theoretically VQ can achieve compression efficiency[19] close to the rate-distortion bound, in practice an unreasonably large value of n is needed. However, with modest dimensions, sensible compression efficiency can still be achieved, using smart training algorithms. A detailed discussion of VQ can be found in Rabbani and Jones.[20]

Subband Coding

In subband coding (SBC) technique, an image is filtered to create a set of images called *subbands*, each with limited spatial frequencies. As each subband has reduced bandwidth, a subsampled version of the original image is used for each subband.

---

[19]*Compression efficiency* refers to the bit rate used for a certain distortion or video quality.
[20]Rabbani and Jones, *Digital Image Compression*.

The process of filtering and subsampling is known as the *analysis stage*. The subbands are then encoded using one or more encoders, possibly using different encode parameters. This allows the coding error to be distributed among different subbands so that a visually optimal reconstruction can be achieved by performing a corresponding upsampling, filtering, and subsequent combining of the subbands. This manner of reconstruction is known as the *synthesis stage*.

Subband decomposition by itself does not provide any compression. However, subbands can be coded more efficiently compared to the original image, thereby providing an overall compression benefit. Figure 2-26 shows a block diagram of the scheme. Many coding techniques may be used for coding of different subbands, including DWT, Haar transform, DPCM, and VQ. An elaborate discussion on SBC can be found in Rabbani and Jones.[21]

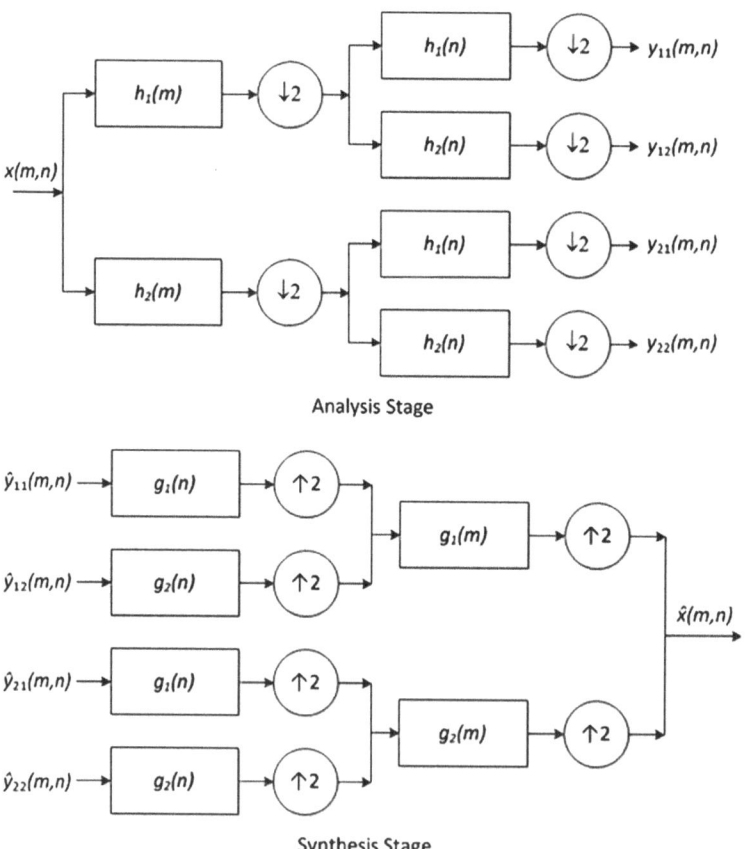

Analysis Stage

Synthesis Stage

Figure 2-26. *A block diagram of a two-dimensional subband coding scheme*

---

[21]Rabbani and Jones, *Digital Image Compression.*

Rate-Distortion Theory

The source entropy defines the minimum number of bits necessary to encode an image or video frame. However, this only applies to lossless encoding. In practice, owing to the characteristics of the human visual system, some irreversible visual quality loss can be tolerated. The extent of loss can be controlled by adjusting encode parameters such as quantization levels.

In order to determine an acceptable amount of visual degradation for a given number of bits, a branch of information theory called the *rate-distortion theory* was developed. The theory establishes theoretical bounds on compression efficiency for lossy data compression, according to a fidelity criterion, by defining a *rate-distortion function* $R(D)$ for various distortion measures and source models. The function has the following properties:

- For a given distortion D, a coding scheme exists for which a rate $R(D)$ is obtained with a distortion D.

- For any coding scheme, $R(D)$ represents the minimum rate for a given distortion D.

- $R(D)$ is a convex cup $\cup$, and continuous function of D.

Figure 2-27 shows a typical rate-distortion function. For distortion-free or visually lossless compression, the minimum rate required is the value of R at $D = 0$, which may be equal to or less than the source entropy, depending on the distortion measure.

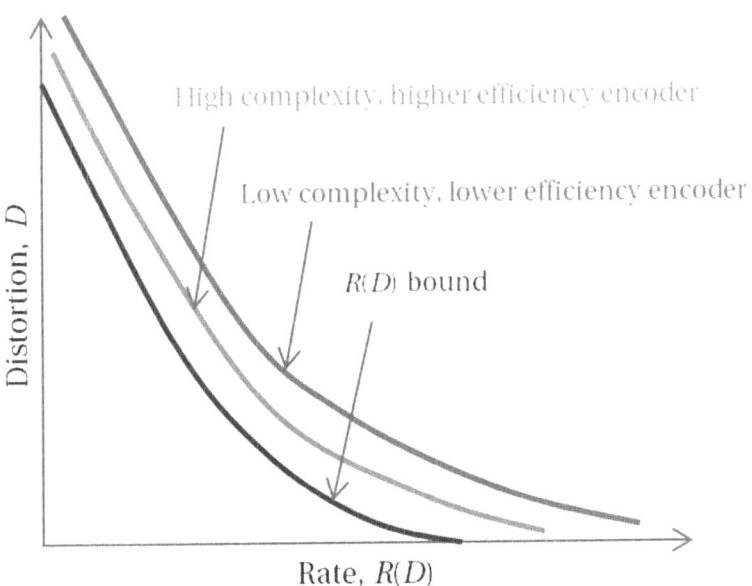

Figure 2-27. *An example of a rate-distortion curve, and compression efficiencies of typical encoders*

The $R(D)$ bound depends on the source model and distortion measures. Usually encoders can achieve compression efficiency closer to $R(D)$ at the expense of higher complexity; a better encoder uses a lower rate and tolerates a lower distortion, but may have higher complexity compared to another encoder. To determine compression efficiency relative to $R(D)$, a two-dimensional Gauss-Markov source image model with unity correlation coefficient is often used as a reference. However, for natural image and video, finding good source model and suitable distortion criteria that correlate well with the human visual system is a topic of active research.

Lossy Compression Aspects

There are several factors that influence and distinguish compression algorithms. These factors should be carefully considered while tuning or choosing a compression algorithm for a particular usage model. Among these factors are:

- **Sensitivity to input frame types:** Compression algorithms may have different compression efficiencies based on input frame characteristics, such as dynamic range, camera noise, amount of pixel to pixel correlation, resolution, and so on.

- **Target bit rate:** Owing to limited bandwidth availability, some applications may need to adhere to a certain bit rate, but would sacrifice visual quality if needed. Compression algorithms usually have different sweet spots in the rate-distortion curve, and target bit rates outside its sweet spots would result in poor visual quality. Some algorithms may not be able to operate below a certain bit rate; for example, the AVC algorithm for HD resolution may need to use more than 1.5 Mbps for any meaningful visual quality), regardless of the spatio-temporal complexity. This bit rate corresponds to approximately 500 times compression for a 1920×1080 resolution at 30 fps video.

- **Constant bit rate vs. constant quality:** Some algorithms are more suitable for transmission without buffering, as they operate with a constant bit rate. However, they may need to maintain the bit rate at the expense of visual quality for complex scenes. As video complexity varies from scene to scene, the constant bit rate requirement will result in a variable reconstruction quality. On the other hand, some algorithms maintain a somewhat constant quality throughout the video by allowing a fixed amout of distortion, or by adjusting levels of quantization based on the scene complexity. In doing so, however, they end up with a variable bit rate, which may require adequate buffering for transmission.

- **Encoder-decoder asymmetry:** Some algorithms, such as vector quantization schemes, use a very complex encoder, while the decoder is implemented with a simple table look-up. Other schemes, such as the MPEG algorithms, need a higher decoder complexity compared to vector quantization, but simplify the encoder. However, MPEG encoders are typically more complex than the decoders, as MPEG encoders also contain complete decoders within them. Depending on the end-user platform, certain schemes may be more suitable than others for a particular application.

- **Complexity and implementation issues:** The computational complexity, memory requirements, and openness to parallel processing are major differentiating factors for hardware or software implementation of compression algorithms. While software-based implementations are more flexible to parameter tuning for highest achievable quality and are amenable to future changes, hardware implementations are usually faster and power-optimized. Appropriate tradeoff is called for depending on end-ser platform and the usage model.

- **Error resilience:** Compressed data is usually vulnerable to channel errors, but the degree of susceptibility varies from one algorithm to another. DCT-based algorithms may lose one or more blocks owing to channel errors, while a simple DPCM algorithm with variable-length codes may be exposed to the loss of an entire frame. Error-correcting codes can compensate for certain errors at the cost of complexity, but often this is cost-prohitive or does not work well in case of burst errors.

- **Artifacts:** Lossy compression algorithms typically produce various artifacts. The type of artifacts and its severity may vary from one algorithm to another, even at the same bit rate. Some artifacts, such as visible block boundaries, jagged edges, ringing artifacts around objects, and the like, may be visually more objectionable than random noise or a softer image. Also, the artifacts are dependent on nature of the content and the viewing condition.

- **Effect of multi-generational coding:** Applications such as video editing may need multiple generations of coding and decoding, where a decoded output is used as the input to the encoder again. The output from the encoder is a second-generation compressed output. Some applications support multiple such generations of compression. Some compression algorithms are not suitable for multi-generational schemes, and often result in poor quality after the second generation of encoding the same frame.

- **System compatibility:** Not all standards are available on all
  systems. Although one of the goals of standardization is to obtain
  use of common format across the industry, some vendors may
  emphasize one compression algorithm over another. Although
  standardization yields commonly acknowledges formats such
  as AVC and HEVC, vendors may choose to promote similar
  algorithm such as VC-1, VP8, or VP9. Overall, this is a larger issue
  encompassing definitions of technologies such as Blu-ray vs.
  HD-DVD. However, compatibility with the targeted eco-system
  is a factor worthy of consideration when choosing a compression
  solution.

Summary

We first discussed typical compression requirements for video transmission with various
networks. This was followed by a discussion of how the characteristics of the human
visual system provide compression opportunities for videos. Then, aiming to familiarize
the reader with various video compression methods and concepts, in particular the most
popular technologies, we presented various ways to perform compression of digital video.
We explained a few technical terms and concepts related to video compression. Then we
presented the various compression techniques targeted to reducing spatial, temporal,
and statistical redundancy that are available in digital video. We briefly described
important video coding techniques, such as transform coding, predictive coding, vector
quantization, and subband coding. These techniques are commonly employed in
presently available video-compression schemes. We then introduced the rate-distortion
curve as the compression efficiency metric, and as a way of comparing two encoding
solutions. Finally, we presented the various factors that influence the compression
algorithms, the understanding of which will facilitate choosing a compression algorithm.

CHAPTER 3

Video Coding Standards

The ubiquity of video has been possible owing to the establishment of a common representation of video signals through *international standards*, with the goal of achieving common formats and interoperability of products from different vendors.

In the late 1980s, the need for standardization of digital video was recognized, and special expert groups from the computer, communications, and video industries came together to formulate practical, low-cost, and easily implementable standards for digital video storage and transmission. To determine these standards, the expert groups reviewed a variety of video data compression techniques, data structures, and algorithms, eventually agreeing upon a few common technologies, which are described in this chapter in some detail.

In the digital video field, many international standards exist to address various industry needs. For example, standard video formats are essential to exchange digital video between various products and applications. Since the amount of data necessary to represent digital video is huge, the data needs to be exchanged in compressed form—and this necessitates video data compression standards. Depending on the industry, standards were aimed at addressing various aspects of end-user applications. For example, display resolutions are standardized in the computer industry, digital studio standards are standardized in the television industry, and network protocols are standardized in the telecommunication industry. As various usages of digital video have emerged to bring these industries ever closer together, standardization efforts have concurrently addressed those cross-industry needs and requirements. In this chapter we discuss the important milestones in international video coding standards, as well as other video coding algorithms popular in the industry.

Overview of International Video Coding Standards

International video coding standards are defined by committees of experts from organizations like the International Standards Organization (ISO) and the International Telecommunications Union (ITU). The goal of this standardization is to have common video formats across the industry, and to achieve interoperability among different vendors and video codec related hardware and software manufacturers.

The standardization of algorithms started with image compression schemes such as JBIG (ITU-T Rec. T82 and ISO/IEC 11544, March 1993) for binary images used in fax and other applications, and the more general JPEG (ITU-T Rec. T81 and ISO/IEC 10918-1), which includes color images as well. The JPEG standardization activities started in 1986, but the standard was ratified in 1992 by ITU-T and in 1994 by ISO. The main standardization activities for video compression algorithms started in the 1980s, with the ITU-T H.261, ratified in 1988, which was the first milestone standard for visual telecommunication. Following that effort, standardization activities increased with the rapid advancement in the television, film, computer, communication, and signal processing fields, and with the advent of new usages requiring contributions from all these diverse industries. These efforts subsequently produced MPEG-1, H.263, MPEG-2, MPEG-4 Part 2, AVC/H.264, and HEVC/H.265 algorithms. In the following sections, we briefly describe the major international standards related to image and video coding.

JPEG

The JPEG is a continuous-tone still image compression standard, designed for applications like desktop publishing, graphic arts, color facsimile, newspaper wirephoto transmission, medical imaging, and the like. The baseline JPEG algorithm uses a DCT-based coding scheme where the input is divided into 8×8 blocks of pixels. Each block undergoes a two-dimensional forward DCT, followed by a uniform quantization. The resulting quantized coefficients are scanned in a zigzag order to form a one-dimensional sequence where high-frequency coefficients, likely to be zero-valued, are placed later in the sequence to facilitate run-length coding. After run-length coding, the resulting symbols undergo more efficient entropy coding.

The first DCT coefficient, often called the DC coefficient as it is a measure of the average value of the pixel block, is differentially coded with respect to the previous block. The run length of the AC coefficients, which have non-zero frequency, are coded using variable-length Huffman codes, assigning shorter codewords for more probable symbols. Figure 3-1 shows the JPEG codec block diagram.

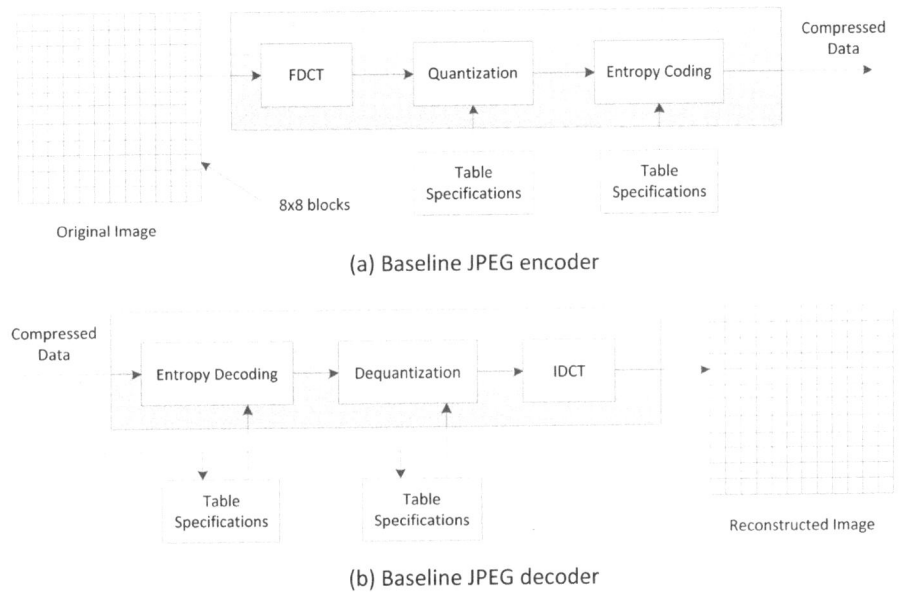

Figure 3-1. *Baseline JPEG codec block diagram*

H.261

H.261 is the first important and practical digital video coding standard adopted by the industry. It is the video coding standard for audiovisual services at $p{\times}64$ kbps, where $p = 1, \ldots, 30$, primarily aimed at providing videophone and video-conferencing services over ISDN, and unifying all aspects of video transmission for such applications in a single standard.[1] The objectives include delivering video at real time, typically at 15–30 fps, with minimum delay (less than 150 ms). Although a successful standard providing industry-wide interoperability, common format, and compression techniques, H.261 is obsolete and is rarely used today.

In H.261, it is mandatory for all codecs to operate at quarter-CIF (QCIF) video format, while the use of CIF is optional. Since the uncompressed video bit rates for CIF and QCIF at 29.97 fps are 26.45 Mbps and 9.12 Mbps, respectively, it is extremely difficult to transport these video signals using an ISDN channel while providing reasonable video quality. To accomplish this goal, H.261 divides a video into a hierarchical block structure comprising *pictures*, *groups of blocks* (GOB), *macroblocks* (MB), and *blocks*. A macroblock consists of four 8×8 luma blocks and two 8×8 chroma blocks; a 3×11 array of macroblocks, in turn, constitutes a GOB. A QCIF picture has three GOBs, while a CIF picture has 12. The hierarchical data structure is shown in Figure 3-2.

---

[1]M. L. Liou, "Overview of the px64 kbit/s Video Coding Standard," *Communications of the ACM* 34, no. 4 (April 1991): 60–63.

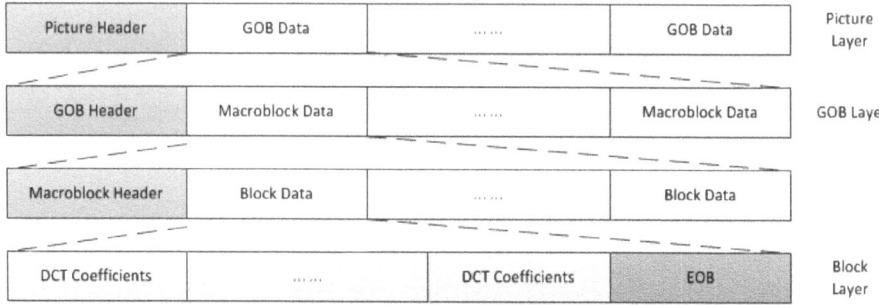

Figure 3-2. *Data structures of the H.261 video multiplex coder*

The H.261 source coding algorithm is a hybrid of intra-frame and inter-frame coding, exploiting spatial and temporal redundancies. Intra-frame coding is similar to baseline JPEG, where block-based 8×8 DCT is performed, and the DCT coefficients are quantized. The quantized coefficients undergo entropy coding using variable-length Huffman codes, which achieves bit-rate reduction using statistical properties of the signal.

Inter-frame coding involves motion-compensated inter-frame prediction and removes the temporal redundancy between pictures. Prediction is performed only in the forward direction; there is no notion of bi-directional prediction. While the motion compensation is performed with integer-pel accuracy, a loop filter can be switched into the encoder to improve picture quality by removing coded high-frequency noise when necessary. Figure 3-3 shows a block diagram of the H.261 source encoder.

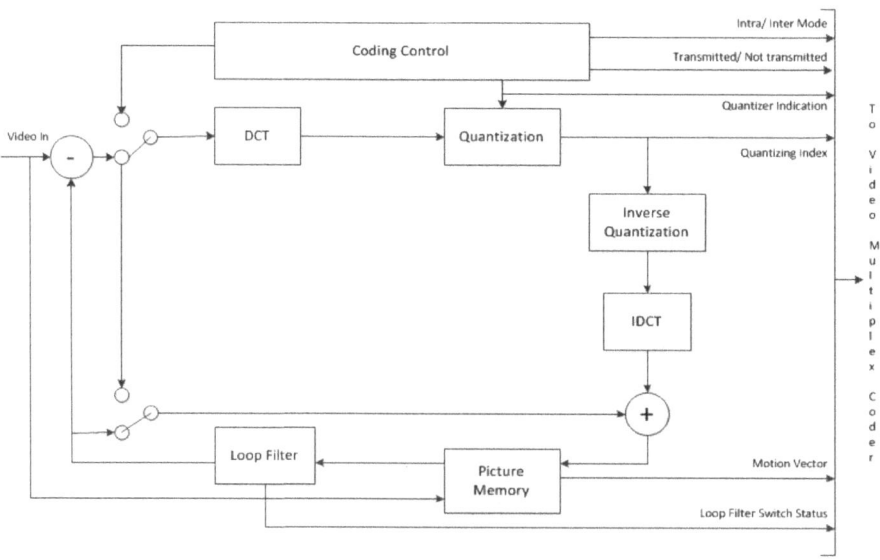

Figure 3-3. *Block diagram of the source encoder of ITU-T H.261*

MPEG-1

In the 1990s, instigated by the market success of compact disc digital audio, CD-ROMs made remarkable inroads into the data storage domain. This prompted the inception of the MPEG-1 standard, targeted and optimized for applications requiring 1.2 to 1.5 Mbps with *video home system* (VHS)-quality video. One of the initial motivations was to fit compressed video into widely available CD-ROMs; however, a surprisingly large number of new applications have emerged to take advantage of the highly compressed video with reasonable video quality provided by the standard algorithm. MPEG-1 remains one of the most successful developments in the history of video coding standards. Arguably, however, the most well-known part of the MPEG-1 standard is the MP3 audio format that it introduced. The intended applications for MPEG-1 include CD-ROM storage, multimedia on computers, and so on. The MPEG-1 standard was ratified as ISO/IEC 11172 in 1991. The standard consists of the following five parts:

1. Systems: Deals with storage and synchronization of video, audio, and other data.

2. Video: Defines standard algorithms for compressed video data.

3. Audio: Defines standard algorithms for compressed audio data.

4. Conformance: Defines tests to check correctness of the implementation of the standard.

5. Reference Software: Software associated with the standard as an example for correct implementation of the encoding and decoding algorithms.

The MEPG-1 bitstream syntax is flexible and consists of six layers, each performing a different logical or signal-processing function. Figure 3-4 depicts various layers arranged in an onion structure.

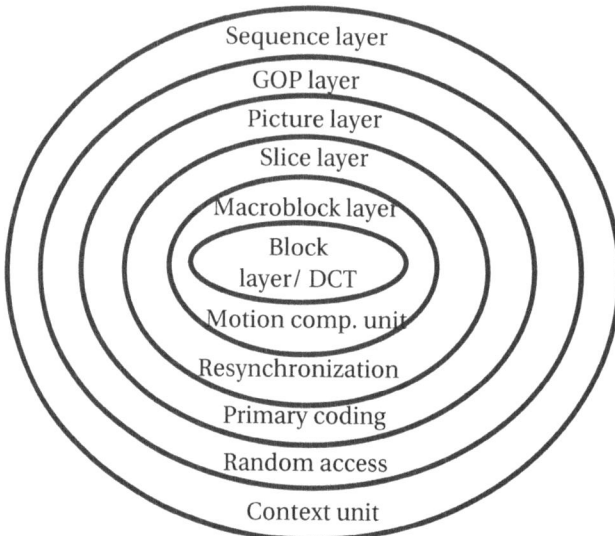

Figure 3-4. *Onion structure of MPEG-1 bitstream syntax*

MPEG-1 is designed for coding progressive video sequences, and the recommended picture size is 360×240 (or 352×288, a.k.a. CIF) at about 1.5 Mbps. However, it is not restricted to this format, and can be applied to higher bit rates and larger image sizes. The intended chroma format is 4:2:0 with 8 bits of pixel depth. The standard mandates real-time decoding and supports features to facilitate interactivity with stored bitstreams. It only specifies syntax for the bitstream and the decoding process, allowing sufficient flexibility for the encoder implementation. Encoders are usually designed to meet specific usage needs, but they are expected to provide sufficient tradeoffs between coding efficiency and complexity.

The main goal of the MPEG-1 video algorithm, as in any other standard, is to achieve the highest possible video quality for a given bit rate. Toward this goal, the MPEG-1 compression approach is similar to that of H.261: it is also a hybrid of intra- and inter-frame redundancy-reduction techniques. For intra-frame coding, the frame is divided into 8×8 pixel blocks, which are transformed to frequency domain using 8×8 DCT, quantized, zigzag scanned, and the run length of the generated bits are coded using variable-length Huffman codes.

Temporal redundancy is reduced by computing a difference signal, namely the *prediction error*, between the original frame and its motion-compensated prediction constructed from a reconstructed reference frame. However, temporal redundancy reduction in MPEG-1 is different from H.261 in a couple of significant ways:

- MPEG-1 permits bi-directional temporal prediction, providing higher compression for a given picture quality than would be attainable using forward-only prediction. For bi-directional prediction, some frames are encoded using either a past or a future frame in display order as the prediction reference. A block of pixels can be predicted from a block in the past reference frame, from a block in the future reference frame, or from the averge of two blocks, one from each reference frame. In bi-directional prediction, higher compression is achieved at the expense of greater encoder complexity and additional coding delay. However, it is still very useful for storage and other off-line applications.

- Further, MPEG-1 introduces half-pel (a.k.a. half-pixel) accuracy for motion compensation and eliminates the loop filter. The half-pel accuracy partly compensates for the benefit provided by the H.261 loop filter in that high-frequency coded noise does not propagate and coding efficiency is not sacrificed.

The *video sequence* layer specifies parameters such as the size of the video frames, frame rate, bit rate, and so on. The *group of pictures* (GOP) layer provides support for random access, fast search, and editing. The first frame of a GOP must be intra-coded (I-frame), where compression is achieved only in the spatial dimension using DCT, quantization, and variable-length coding. The I-frame is followed by an arrangement of forward-predictive coded frames (P-frames) and bi-directionally predictive coded frames (B-frames). I-frames provide ability for random access to the bitstream and for fast search (or VCR-like trick play, such as fast-forward and fast-rewind), as they are coded independently and serve as entry points for further decoding.

The *picture* layer deals with a particular frame and contains information of the frame type (I, P, or B) and the display order of the frame. The bits corresponding to the motion vectors and the quantized DCT coefficients are packages in the *slice* layer, the *macroblock* layer, and the *block* layer. A slice is a contiguous segment of the macroblocks. In the event of a bit error, the slice layer helps resynchronize the bitstream during decoding. The macroblock layer contains the associated motion vector bits and is followed by the block layer, which consists of the coded quantized DCT coefficients. Figure 3-5 shows the MPEG picture structure in coding and display order, which applies to both MPEG-1 and MPEG-2.

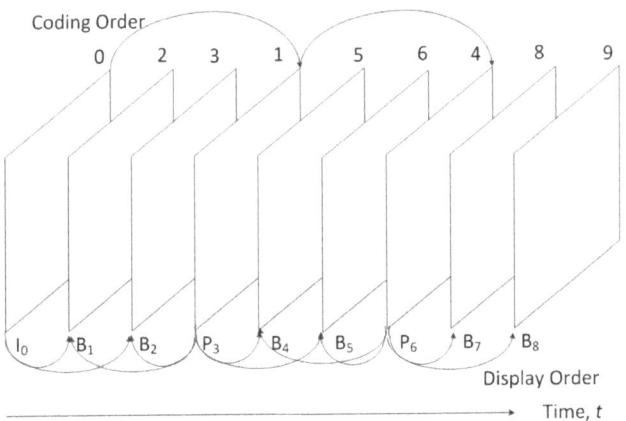

Figure 3-5. *I/P/B frame structure, prediction relationships, coding order, and display order*

MPEG-2

MPEG-2 was defined as the standard for *generic coding of moving pictures and associated audio*. The standard was specified by a joint technical committee of the ISO/IEC and ITU-T, and was ratified in 1993, both as the ISO/IEC international standard 13818 and as the ITU-T Recommendation H.262.

With a view toward resolving the existing issues in MPEG-1, the standardization activity in MPEG-2 focused on the following considerations:

- Extend the number of audio compression channels from 2 channels to 5.1 channels.

- Add standardization support for interlaced video for broadcast applications.

- Provide more standard profiles, beyond the Constrained Parameters Bitstream available in MPEG-1, in order to support higher-resolution video contents.

- Extend support for color sampling from 4:2:0, to include 4:2:2 and 4:4:4.

For MPEG standards, the standards committee addressed video and audio compression, as well as system considerations for multiplexing the compressed audio-visual data. In MPEG-2 applications, the compressed video and audio elementary streams are multiplexed to construct a program stream; several program streams are packetized and combined to form a transport stream before transmission. However, in the following discussion, we will focus on MPEG-2 video compression.

MPEG-2 is targeted for a variety of applications at a bit rate of 2 Mbps or more, with a quality ranging from good-quality NTSC to HDTV. Although widely used as the format of digital television signal for terrestrial, cable, and direct-broadcast satellite TV systems, other typical applications include digital videocassette recorders (VCR), digital video discs (DVD), and the like. As a generic standard supporting a variety of applications generally ranging from 2 Mbps to 40 Mbps, MPEG-2 targets a compression ratio in the range of 30 to 40. To provide application independence, MPEG-2 supports a variety of video formats with resolutions ranging from source input format (SIF) to HDTV. Table 3-1 shows some typical video formats used in MPEG-2 applications.

Table 3-1. *Typical MPEG-2 Paramters*

Format	Resolution	Compressed Bit Rate (Mbps)
SIF	360×240 @ 30 fps	1.2 – 3
ITU-R 601	720×480 @ 30 fps	5 – 10
EDTV	960×480 @ 30 fps	7 – 15
HDTV	1920×1080 @ 30 fps	18 – 40

The aim of MPEG-2 is to provide better picture quality while keeping the provisions for random access to the coded bitstream. However, it is a rather difficult task to accomplish. Owing to the high compression demanded by the target bit rates, good picture quality cannot be achieved by intra-frame coding alone. Contrarily, the random-access requirement is best satisfied with pure intra-frame coding. This dilemma necessitates a delicate balance between the intra- and inter-picture coding. And this leads to the definition of I, P, and B pictures, similar to MPEG-1. I-frames are the least compressed, and contain approximately the full information of the picture in a quantized form in frequency domain, providing robustness against errors. The P-frames are predicted from past I- or P-frames, while the B-frames offer the greatest compression by using past and future I- or P-frames for motion compensation. However, B-frames are the most vulnerable to channel errors.

An MPEG-2 encoder first selects an appropriate spatial resolution for the signal, followed by a block-matching motion estimation to find the displacement of a macroblock (16×16 or 16×8 pixel area) in the current frame relative to a macroblock obtained from a previous or future reference frame, or from their average. The search for the best matching block is based on the *mean absolute difference* (MAD) distortion criterion; the best matching occurs when the accumulated absolute values of the pixel differences for all macroblocks are minimized. The motion estimation process then defines a motion vector representing the displacement of the current block's location from the best matched block's location. To reduce temporal redundancy, motion compensation is used both for causal prediction of the current picture from a previous reference picture and for non-causal, interpolative prediction from past and future reference pictures. The prediction of a picture is constructed based on the motion vectors.

To reduce spatial redundancy, the difference signal—that is, the prediction error—is further compressed using the block transform coding technique that employs the two-dimensional orthonormal 8×8 DCT to remove spatial correlation. The resulting transform coefficients are ordered in an alternating or zigzag scanning pattern before they are quantized in an irreversible process that discards less important information. In MPEG-2, adaptive quantization is used at the macroblock layer, allowing smooth bit-rate control and perceptually uniform video quality. Finally, the motion vectors are combined with the residual quantized coefficients, and are transmitted using variable-length Huffman codes. The Huffman coding tables are pre-determined and optimized for a limited range of compression ratios appropriate for some target applications. Figure 3-6 shows an MPEG-2 video encoding block diagram.

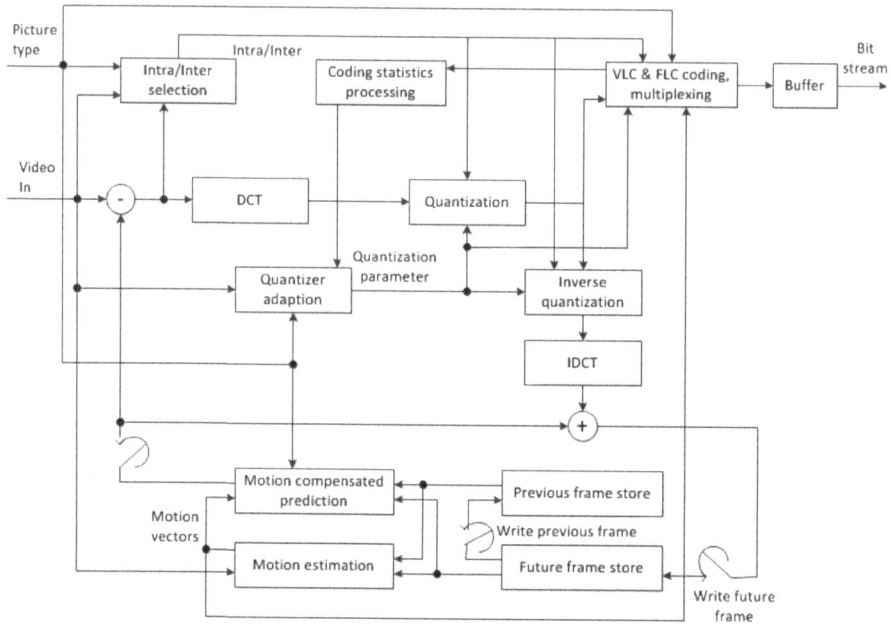

Figure 3-6. *The MPEG-2 video encoding block diagram*

The bitstream syntax in MPEG-2 is divided into subsets known as *profiles*, which specify constraints on the syntax. Profiles are further divided into *levels*, which are sets of constraints imposed on parameters in the bitstream. There are five profiles defined in MPEG-2:

- Main: Aims at the maximum quality of standard definition pictures.

- Simple: Is directed to memory savings by not interpolating pictures.

- SNR scalable: Aims to provide better signal-to-noise ratio on demand by using more than one layer of quantization.

- Spatially scalable: Aims to provide variable resolution on demand by using additional layers of weighted and reconstructed reference pictures.

- High: Intended to support 4:2:2 chroma format and full scalability.

Within each profile, up to four levels are defined:

- Low: Provides compatibility with H.261 or MPEG-1.

- Main: Corresponds to conventional TV.

- High 1440: Roughly corresponds to HDTV, with 1,440 samples per line.

- High: Roughly corresponds to HDTV, with 1,920 samples per line.

The Main profile, Main level (MP @ ML) reflects the initial focus of MPEG-2 with regard to entertainment applications. The permitted profile-level combinations are: Simple profile with Main level, Main profile with all levels, SNR scalable profile with Low and Main levels, Spatially scalable profile with High 1440 level, and High profile with all levels except Low level.

The bitstream syntax can also be divided as follows:

- Non-scalable syntax: A super-set of MPEG-1, featuring extra compression tools for interlaced video signals along with variable bit rate, alternate scan, concealment motion vectors, intra-DCT format, and so on.

- Scalable syntax: A base layer similar to the non-scalable syntax and one or more enhancement layers with the ability to enable the reconstruction of useful video.

The structure of the compressed bitstream is shown in Figure 3-7. The layers are similar to those of MPEG-1. A compressed video sequence starts with a sequence header containing picture resolutions, picture rate, bit rate, and so on. There is a sequence extension header in MPEG-2 containing video format, color primaries, display resolution, and so on. The sequence extension header may be followed by an optional GOP header having the time code, which is subsequently followed by a frame header containing temporal reference, frame type, video buffering verifier (VBV) delay, and so on. The frame header can be succeeded by a picture coding extension containing interlacing, DCT type and quantizer-scale type information, which is usually followed by a slice header to facilitate resynchronization. Inside a slice, several macroblocks are grouped together, where the macroblock address and type, motion vector, coded block pattern, and so on are placed before the actual VLC-coded quantized DCT coefficients for all the blocks in a macroblock. The slices can start at any macroblock location, and they are not restricted to the beginning of macroblock rows.

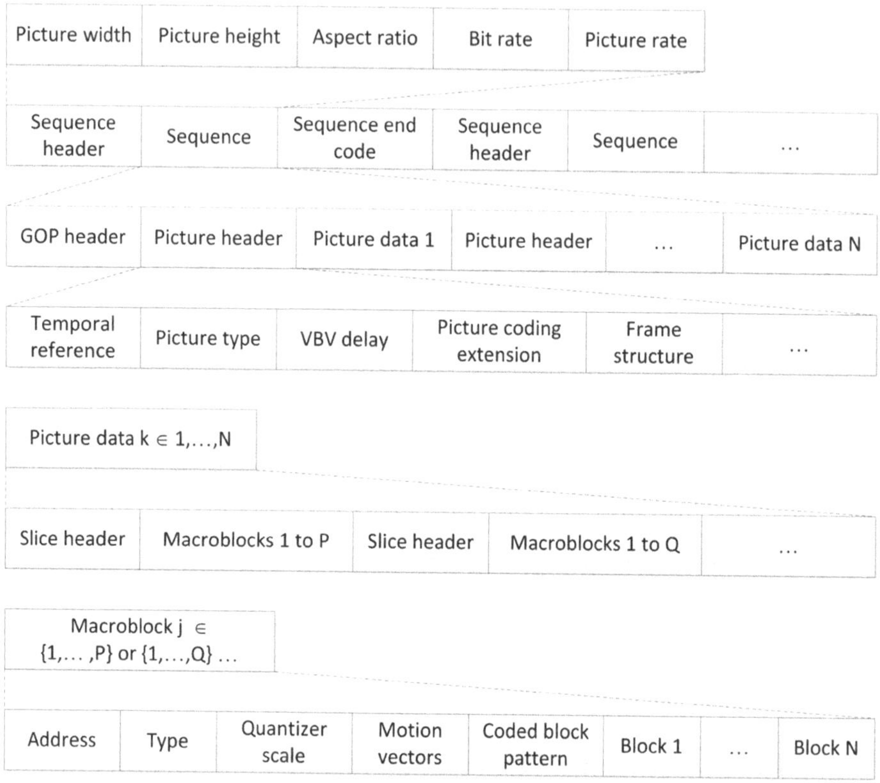

Picture width	Picture height	Aspect ratio	Bit rate	Picture rate

Sequence header	Sequence	Sequence end code	Sequence header	Sequence	...

GOP header	Picture header	Picture data 1	Picture header	...	Picture data N

Temporal reference	Picture type	VBV delay	Picture coding extension	Frame structure	...

Picture data k ∈ 1,...,N

Slice header	Macroblocks 1 to P	Slice header	Macroblocks 1 to Q	...

Macroblock j ∈ {1,... ,P} or {1,...,Q} ...

Address	Type	Quantizer scale	Motion vectors	Coded block pattern	Block 1	...	Block N

Figure 3-7. *Structure of MPEG-2 video bitstream syntax*

Since the MPEG-2 base layer is a super-set of MPEG-1, standard-compliant decoders can decode MPEG-1 bitstreams providing backward compatibility. Furthermore, MPEG-2 is capable of selecting the optimum mode for motion-compensated prediction, such that the current frame or field can be predicted either from the entire reference frame or from the top or bottom field of the reference frame, thereby finding a better relationship of the fields. MPEG-2 also adds the alternate scanning pattern, which suits interlaced video better than the zigzag scanning pattern. Besides, a choice is offered between linear and nonlinear quantization tables, and up to 11 bits DC precision is supported for intra macroblocks. These are improvements on MPEG-1, which does not support nonlinear quantization tables and provides only 8 bits of intra-DC precision. At the same bit rate, MPEG-2 yields better quality than MPEG-1, especially for interlaced video sources. Moreover, MPEG-2 is more flexible for parameter variation at a given bit rate, helping a smoother buffer control. However, these benefits and improvements come at the expense of increased complexity.

H.263

H.263 defined by ITU-T is aimed at low-bit-rate video coding but does not specify a constraint on video bit rate; such constraints are given by the terminal or the network. The objective of H.263 is to provide significantly better picture quality than its predecessor, H.261. Conceptually, H.263 is network independent and can be used for a wide range of applications, but its target applications are visual telephony and multimedia on low-bit-rate networks like PSTN, ISDN, and wireless networks. Some important considerations for H.263 include small overhead, low complexity resulting in low cost, interoperability with existing video communication standards (e.g., H.261, H.320), robustness to channel errors, and quality of service (QoS) parameters. Based on these considerations, an efficient algorithm is developed, which gives manufacturers the flexibility to make tradeoffs between picture quality and complexity. Compared to H.261, it provides the same subjective image quality at less than half the bit rate.

Similar to other standards, H.263 uses inter-picture prediction to reduce temporal redundancy and transform coding of the residual prediction error to reduce spatial redundancy. The transform coding is based on 8×8 DCT. The transformed signal is quantized with a scalar quantizer, and the resulting symbol is variable length coded before transmission. At the decoder, the received signal is inverse quantized and inverse transformed to reconstruct the prediction error signal, which is added to the prediction, thus creating the reconstructed picture. The reconstructed picture is stored in a frame buffer to serve as a reference for the prediction of the next picture. The encoder consists of an embedded decoder where the same decoding operation is performed to ensure the same reconstruction at both the encoder and the decoder.

H.263 supports five standard resolutions: sub-QCIF (128×96), QCIF (176×144), CIF (352×288), 4CIF (704×576), and 16CIF (1408×1152), covering a large range of spatial resolutions. Support for both sub-QCIF and QCIF formats in the decoder is mandatory, and either one of these formats must be supported by the encoder. This requirement is a compromise between high resolution and low cost.

A picture is divided into 16×16 macroblocks, consisting of four 8×8 luma blocks and two spatially aligned 8×8 chroma blocks. One or more macroblocks rows are combined into a *group of blocks* (GOB) to enable quick resynchronization in the event of transmission errors. Compared to H.261, the GOB structure is simplified; GOB headers are optional and may be used based on the tradeoff between error resilience and coding efficiency.

For improved inter-picture prediction, the H.263 decoder has a block motion compensation capability, while its use in the encoder is optional. One motion vector is transmitted per macroblock. Half-pel precision is used for motion compensation, in contrast to H.261, where full-pel precision and a loop filter is used. The motion vectors, together with the transform coefficients, are transmitted after variable-length coding. The bit rate of the coded video may be controlled by preprocessing or by varying the following encoder parameters: quantizer scale size, mode selections, and picture rate.

In addition to the core coding algorithm described above, H.263 includes four negotiable coding options, as mentioned below. The first three options are used to improve inter-picture prediction, while the fourth is related to lossless coding. The coding options increase the complexity of the encoder but improve picture quality, thereby allowing tradeoff between picture quality and complexity.

- **Unrestricted motion vector (UMV) mode**: In the UMV mode, motion vectors are allowed to point outside the coded picture area, enabling a much better prediction, particularly when a reference macroblock is partly located outside the picture area and part of it is not available for prediction. Those unavailable pixels would normally be predicted using the edge pixels instead. However, this mode allows utilization of the complete reference macroblock, producing a gain in quality, especially for the smaller picture formats when there is motion near the picture boundaries. Note that, for the sub-QCIF format, about 50 percent of all the macroblocks are located at or near the boundary.

- **Advanced prediction (AP) mode**: In this optional mode, the *overlapping block motion compensation* (OBMC) is used for luma, resulting in a reduction in blocking artifacts and improvement in subjective quality. For some macroblocks, four 8×8 motion vectors are used instead of a 16×16 vector, providing better prediction at the expense of more bits.

- **PB-frames (PB) mode**: The principal purpose of the PB-frames mode is to increase the frame rate without significantly increasing the bit rate. A PB-frame consists of two pictures coded as one unit. The P-picture is predicted from the last decoded P-picture, and the B-picture is predicted both from the last and from the current P-pictures. Although the names "P-picture" and "B-picture" are adopted from MPEG, B-pictures in H.263 serve an entirely different purpose. The quality of the B-pictures is intentionally kept low, in particular to minimize the overhead of bi-directional prediction, while such overhead is important for low-bit-rate applications. B-pictures use only 15 to 20 percent of the allocated bit rate, but result in better subjective impression of smooth motion.

- **Syntax-based arithmetic coding (SAC) mode**: H.263 is optimized for very low bit rates. As such, it allows the use of optional syntax-based arithmetic coding mode, which replaces the Huffman codes with arithmetic codes for variable-length coding. While Huffman codes must use an integral number of bits, arithmetic coding removes this restriction, thus producing a lossless coding with reduced bit rate.

The video bitstream of H.263 is arranged in a hierarchical structure composed of the following layers: picture layer, group of blocks layer, macroblock layer, and block layer. Each coded picture consists of a picture header followed by coded picture data arranged as group of blocks. Once the transmission of the pictures is completed, an end-of-sequence (EOS) code and, if needed, stuffing bits (ESTUF) are transmitted. There are some optional elements in the bitstream. For example, temporal reference of B-pictures (TRB) and the quantizer parameter (DBQUANT) are only available if the picture type (PTYPE) indicates a B-picture. For P-pictures, a quantizer parameter PQUANT is transmitted.

The GOB layer consists of a GOB header followed by the macroblock data. The first GOB header in each picture is skipped, while for other GOBs, a header is optional and is used based on available bandwidth. Group stuffing (GSTUF) may be necessary for a GOB start code (GBSC). Group number (GN), GOB frame ID (GFID), and GOB quantizer (GQUANT) can be present in the GOB header.

Each macroblock consists of a macroblock header followed by the coded block data. A coded macroblock is indicated by a flag called COD; for P-pictures, all the macroblocks are coded. A macroblock type and coded block pattern for chroma (MCBPC) are present when indicated by COD or when PTYPE indicates an I-picture. A macroblock mode for B-pictures (MODB) is present for non-intra macroblocks for PB-frames. The luma coded block pattern (CBPY), and the codes for the differential quanitizer (DQUANT) and motion vector data (MVD or MVD_{2-4} for advanced prediction), may be present according to MCBPC. The CBP and motion vector data for B-blocks (CBPB and MVDB) are present only if the coding mode is B (MODB). As mentioned before, in the normal mode a macroblock consists of four luma and two chroma blocks; however, in PB-frames mode a macroblock can be thought of as containing 12 blocks. The block structure is made up of intra DC followed by the transform coefficients (TCOEF). For intra macroblocks, intra DC is sent for every P-block in the macroblock. Figure 3-8 shows the structure of various H.263 layers.

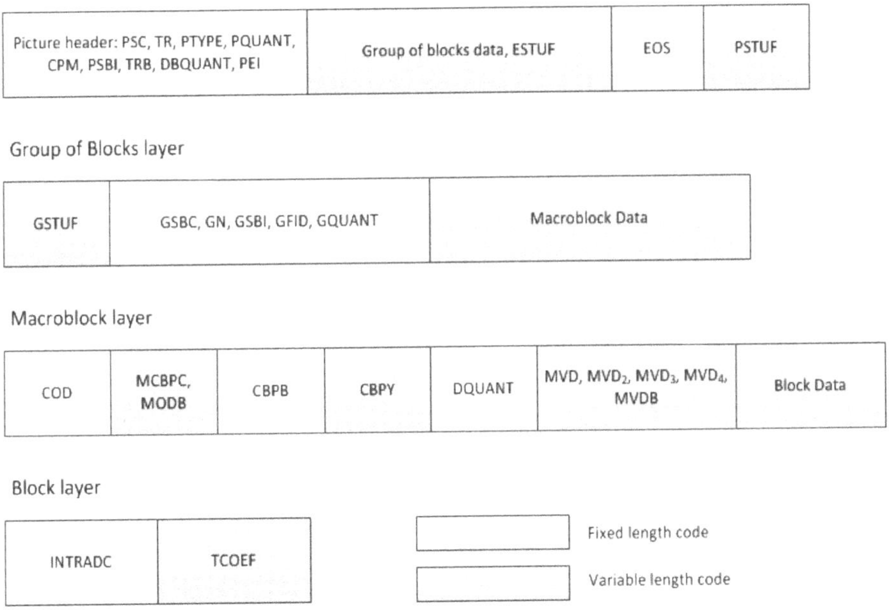

Picture header: PSC, TR, PTYPE, PQUANT, CPM, PSBI, TRB, DBQUANT, PEI	Group of blocks data, ESTUF	EOS	PSTUF

Group of Blocks layer

GSTUF	GSBC, GN, GSBI, GFID, GQUANT	Macroblock Data

Macroblock layer

COD	MCBPC, MODB	CBPB	CBPY	DQUANT	MVD, MVD₂, MVD₃, MVD₄, MVDB	Block Data

Block layer

INTRADC	TCOEF

	Fixed length code
	Variable length code

Figure 3-8. *Structure of various layers in H.263 bitstream*

69

MPEG-4 (Part 2)

MPEG-4, formally the standard ISO/IEC 14496, was ratified by ISO/IEC in March 1999 as the standard for multimedia data representation and coding. In addition to video and audio coding and multiplexing, MPEG-4 addresses coding of various two- or three-dimensional synthetic media and flexible representation of audio-visual scene and composition. As the usage of multimedia developed and diversified, the scope of MPEG-4 was extended from its initial focus on very low bit-rate coding of limited audio-visual materials to encompass new multimedia functionalities.

Unlike pixel-based treatment of video in MPEG-1 or MPEG-2, MPEG-4 supports content-based communication, access, and manipulation of digital audio-visual objects, for real-time or non-real-time interactive or non-interactive applications. MPEG-4 offers extended functionalities and improves upon the coding efficiency provided by previous standards. For instance, it supports variable pixel depth, object-based transmission, and a variety of networks including wireless networks and the Internet. Multimedia authoring and editing capabilities are particularly attractive features of MPEG-4, with the promise of replacing existing word processors. In a sense, H.263 and MPEG-2 are embedded in MPEG-4, ensuring support for applications such as digital TV and videophone, while it is also used for web-based media streaming.

MPEG-4 distinguishes itself from earlier video coding standards in that it introduces object-based representation and coding methodology of real or virtual audio-visual (AV) objects. Each AV object has its local 3D+T coordinate system serving as a handle for the manipulation of time and space. Either the encoder or the end-user can place an AV object in a scene by specifying a co-ordinate transformation from the object's local co-ordinate system into a common, global 3D+T co-ordinate system, known as the scene co-ordinate system. The composition feature of MPEG-4 makes it possible to perform bitstream editing and authoring in compressed domain.

One or more AV objects, including their spatio-temporal relationships, are transmitted from an encoder to a decoder. At the encoder, the AV objects are compressed, error-protected, multiplexed, and transmitted downstream. At the decoder, these objects are demultiplexed, error corrected, decompressed, composited, and presented to an end user. The end user is given an opportunity to interact with the presentation. Interaction information can be used locally or can be transmitted upstream to the encoder.

The transmitted stream can either be a control stream containing connection setup, the profile (subset of encoding tools), and class definition information, or be a data stream containing all other information. Control information is critical, and therefore it must be transmitted over reliable channels; but the data streams can be transmitted over various channels with different quality of service.

Part 2 of the standard deals with video compression. As the need to support various profiles and levels was growing, Part 10 of the standard was introduced to handle such demand, which soon became more important and commonplace in the industry than Part 2. However, MPEG-4 Part 10 can be considered an independent standardization effort as it does not provide backward compatibility with MPEG-4 Part 2. MPEG-4 Part 10, also known as advanced video coding (AVC), is discussed in the next section.

MPEG-4 Part 2 is an object-based hybrid natural and synthetic coding standard. (For simplicity, we will refer to MPEG-4 Part 2 simply as MPEG-4 in the following discussion.) The structure of the MPEG-4 video is hierarchical in nature. At the top layer is a *video session* (VS) composed of one or more *video objects* (VO). A VO may consist of one or more *video object layers* (VOL). Each VOL consists of an ordered time sequence of snapshots, called *video object planes* (VOP). The *group of video object planes* (GOV) layer is an optional layer between the VOL and the VOP layer. The bitstream can have any number of the GOV headers, and the frequency of the GOV header is an encoder issue. Since the GOV header indicates the absolute time, it may be used for random access and error-recovery purposes.

The video encoder is composed of a number of encoders and corresponding decoders, each dedicated to a separate video object. The reconstructed video objects are composited together and presented to the user. The user interaction with the objects such as scaling, dragging, and linking can be handled either in the encoder or in the decoder.

In order to describe arbitrarily shaped VOPs, MPEG-4 defines a VOP by means of a bounding rectangle called a VOP window. The video object is circumscribed by the tightest VOP window, such that a minimum number of image macroblocks are coded. Each VO consists of three main functions: shape coding, motion compensation, and texture coding. In the event of a rectangular VOP, the MPEG-4 encoder structure is similar to that of the MPEG-2 encoder, and shape coding can be skipped. Figure 3-9 shows the structure of a video object encoder.

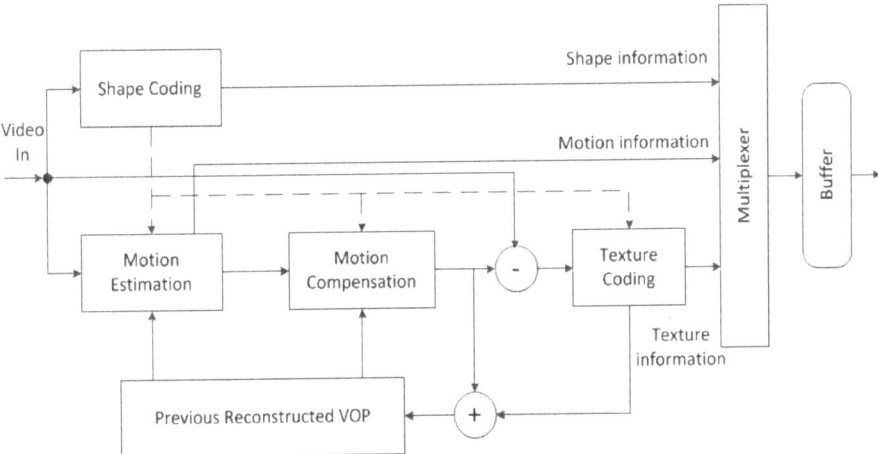

Figure 3-9. *Video object encoder structure in MPEG-4*

The shape information of VOP is referred to as the *alpha plane* in MPEG-4. The alpha plane has the same format as the luma and its data indicates the characteristics of the relevant pixels, whether or not the pixels are within a video object. The shape coder compresses the alpha plane. Binary alpha planes are encoded by modified *content-based arithmetic encoding* (CAE), while gray-scale alpha planes are encoded by motion-compensated DCT, similar to texture coding. The macroblocks that lie completely outside the object (*transperant* macroblocks) are not processed for the motion or texture coding; therefore, no overhead is required to indicate this mode, since this transparency information can be obtained from shape coding.

Motion estimation and compensation are used to reduce temporal redundancies. A padding technique is applied on the reference VOP that allows polygon matching instead of block matching for rectangular images. Padding methods aim at extending arbitrarily shaped image segments to a regular block grid by filling in the missing data corresponding to signal extrapolation such that common block-based coding techniques can be applied. In addition to the basic motion compensation technique, unrestricted motion compensation, advanced prediction mode, and bi-directional motion compensation are supported by MPEG-4 video to obtain a significant improvement in quality at the expense of very little increased complexity.

The intra and residual data after motion compensation of VOPs are coded using a block-based DCT scheme, similar to previous standards. Macroblocks that lie completely inside the VOP are coded using a technique identical to H.263; the region outside the VOP within the *contour* macroblocks (i.e., macroblocks with an object edge) can either be padded for regular DCT transformation or can use shape adaptive DCT (SA-DCT). Transperant blocks are skipped and are not coded in the bitstream.

MPEG-4 supports scalable coding of video objects in spatial and temporal domains, and provides error resilience across various media. Four major tools, namely video packet re-synchronization, data partitioning, header extension code, and reversible VLC, provide loss-resilience properties such as resynchronization, error detection, data recovery, and error concealment.

AVC

The *Advanced Video Coding* (AVC), also known as the ITU-T H.264 standard (ISO/IEC 14496-10), is currently the most common video compression format used in the industry for video recording and distribution. It is also known as MPEG-4 Part 10. The AVC standard was ratified in 2003 by the Joint Video Team (JVT) of the ITU-T Video Coding Experts Group (VCEG) and ISO/IEC Moving Picture Experts Group (MPEG) standardization organizations. One of the reasons the AVC standard is so well known is that it is one of the three compression standards for Blu-ray (the others being MPEG-2 and VC-1), and it is also widely used by Internet streaming applications like YouTube and iTunes, software applications like Flash Player, software frameworks like Silverlight, and various HDTV broadcasts over terrestrial, cable, and satellite channels.

The AVC video coding standard has the same basic functional elements as previous standards MPEG-4 Part 2, MPEG-2, H.263, MPEG-1, and H.261. It uses a lossy predictive, block-based hybrid DPCM coding technique. This involves transform for reduction of spatial correlation, quantization for bit-rate control, motion-compensated prediction for reduction of temporal correlation, and entropy encoding for reduction of statistical correlation. However, with a goal of achieving better coding performance than previous standards, AVC incorporates changes in the details of each functional element by including in-picture prediction, a new 4×4 transform, multiple reference pictures, variable block sizes, and a quarter-pel precision for motion compensation, a deblocking filter, and improved entropy coding. AVC also introduces coding concepts such as generalized *B*-slices, which supports not only bidirectional forward-backward prediction pair but also forward-forward and backward-backward prediction pairs. There are several other tools, including direct modes and weighted prediction, defined by AVC to obtain a very good prediction of the source signal so that the error signal has a minimum energy. These tools help AVC perform significantly better than prior standards for a variety of applications. For example, compared to MPEG-2, AVC typically obtains the same quality at half the bit rate, especially for high-resolution contents coded at high bit rates.

However, the improved coding efficiency comes at the expense of additional complexity to the encoder and decoder. So, to compensate, AVC utilizes some methods to reduce the implementation complexity—for example, multiplier-free integer transform is introduced where multiplication operations for the transform and quantization are combined. Further, to facilitate applications on noisy channel conditions and error-prone environments such as the wireless networks, AVC utilizes some methods to exploit error resilience to network noise. These include *flexible macroblock ordering* (FMO), switched slice, redundant slice methods, and data partitioning.

The coded AVC bitstream has two layers, the *network abstraction layer* (NAL) and *video coding layer* (VCL). The NAL abstracts the VCL data to help transmission on a variety of communication channels or storage media. A NAL unit specifies both byte-stream and packet-based formats. The byte-stream format defines unique start codes for the applications that deliver the NAL unit stream as an ordered stream of bytes or bits, encapsulated in network packets such as MPEG-2 transport streams. Previous standards contained header information about slice, picture, and sequence at the start of each element, where loss of these critical elements in a lossy environment would render the rest of the element data useless. AVC resolves this problem by keeping the sequence and picture parameter settings in the non-VCL NAL units that are transmitted with greater error-protection. The VCL unit contains the core video coded data, consisting of video sequence, picture, slice, and macroblock.

Profile and Level

A profile is a set of features of the coding algorithm that are identified to meet certain requirements of the applications. This means that some features of the coding algorithm are not supported in some profiles. The standard defines 21 sets of capabilities, targeting specific classes of applications.

For non-scalable two-dimensional video applications, the following are the important profiles:

- **Constrained Baseline Profile**: Aimed at low-cost mobile and video communication applications, the Constrained Baseline Profile uses the subset of features that are in common with the Baseline, Main, and High Profiles.

- **Baseline Profile**: This profile is targeted for low-cost applications that require additional error resiliency. As such, on top of the features supported in the Constrained Baseline Profile, it has three features for enhanced robustness. However, in practice, Constrained Baseline Profile is more commonly used than Baseline Profile. The bitstreams for these two profiles share the same profile identifier code value.

- **Extended Profile:** This is intended for video streaming. It has higher compression capability and more robustness than Baseline Profile, and it supports server stream switching.

- **Main Profile:** Main profile is used for standard-definition digital TV broadcasts, but not for HDTV broadcasts, for which High Profile is primarily used.

- **High Profile:** It is the principal profie for HDTV broadcast and for disc storage, such as the Blu-ray Disc storage format.

- **Progressive High Profile:** This profile is similar to High profile, except that it does not support the field coding tools. It is intended for applications and displays using progressive scanned video.

- **High 10 Profile:** Mainly for premium contents with 10-bit per sample decoded picture precision, this profile adds 10-bit precision support to the High Profile.

- **High 4:2:2 Profile:** This profile is aimed at professional applications that use interlaced video. On top of the High 10 Profile, it adds support for the 4:2:2 chroma subsampling format.

- **High 4:4:4 Predictive Profile:** Further to the High 4:2:2 Profile, this profile supports up to 4:4:4 chroma sampling and up to 14 bits per sample precision. It additionally supports lossless region coding and the coding of each picture as three separate color planes.

In addition to the above profiles, the *Scalable Video Coding* (SVC) extension defines five more scalable profiles: Scalable Constrained Baseline, Scalable Baseline, Scalable High, Scalable Constrained High, and Scalable High Intra profiles. Also, the *Multi-View Coding* (MVC) extension adds three more profiles for three-dimensional video—namely Stereo High, Multiview High, and Multiview Depth High profiles. Furthermore, four intra-frame-only profiles are defined for professional editing applications: High 10 Intra, High 4:2:2 Intra, High 4:4:4 Intra, and CAVLC 4:4:4 Intra profiles.

Levels are constraints that specify the degree of decoder performance needed for a profile; for example, a level designates the maximum picture resolution, bit rate, frame rate, and so on that the decoder must adhere to within a profile. Table 3-2 shows some examples of level restrictions; for full description, see the standard specification.[2]

---

[2]*ITU-T Rec. H.264: Advanced Video Coding for Generic Audiovisual Services* (**Geneva, Switzerland**: International Telecommunications Union, 2007).

Table 3-2. Examples of Level Restrictions in AVC

Level	Max Luma samples/ second	Max Macro-blocks/ second	Max Number of Macro-blocks	Max video bit rate, kbps (Baseline, Extended, Main)	Max video bit rate, kbps (High)	Examples (width × height @ frames per second)
1	380,160	1,485	99	64	80	176×144@15
1.1	768,000	3,000	396	192	240	320×240@10
1.2	1,536,000	6,000	396	384	480	352×288@15
1.3	3,041,280	11,880	396	768	960	352×288@30
2	3,041,280	11,880	396	2,000	2,500	352×288@30
2.1	5,068,800	19,800	792	4,000	5,000	352×576@25
2.2	5,184,000	20,250	1,620	4,000	5,000	720×480@15
3	10,368,000	40,500	1,620	10,000	12,500	720×480@30, 720×576@25
3.1	27,648,000	108,000	3,600	14,000	17,500	1280×720@30
3.2	55,296,000	216,000	5,120	20,000	25,000	1280×720@60
4	62,914,560	245,760	8,192	20,000	25,000	1920×1080@30
4.1	62,914,560	245,760	8,192	50,000	62,500	2048×1024@30
4.2	133,693,440	522,240	8,704	50,000	62,500	2048×1080@60
5	150,994,944	589,824	22,080	135,000	168,750	2560×1920@30
5.1	251,658,240	983,040	36,864	240,000	300,000	4096×2048@30
5.2	530,841,600	2,073,600	36,864	240,000	300,000	4096×2160@60

Picture Structure

The video sequence has *frame* pictures or *field* pictures. The pictures usually comprise three sample arrays, one luma and two chroma sample arrays (RGB arrays are supported in High 4:4:4 Profile only). AVC supports either progressive-scan or interlaced-scan, which may be mixed in the same sequence. Baseline Profile is limited to progressive scan.

Pictures are divided into *slices*. A slice is a sequence of a flexible number of macroblocks within a picture. Multiple slices can form *slice groups*; there is macroblock to slice group mapping to determine which slice group includes a particular macroblock. In the 4:2:0 format, each macroblock has one 16×16 luma and two 8×8 chroma sample arrays, while in the 4:2:2 and 4:4:4 formats, the chroma sample arrays are 8 ×16 and 16 ×16, respectively. A picture may be partitioned into 16×16 or smaller partitions with various shapes such as 16×8, 8×16, 8×8, 8×4, 4×8, and 4×4. These partitions are used for prediction purposes. Figure 3-10 shows the different partitions.

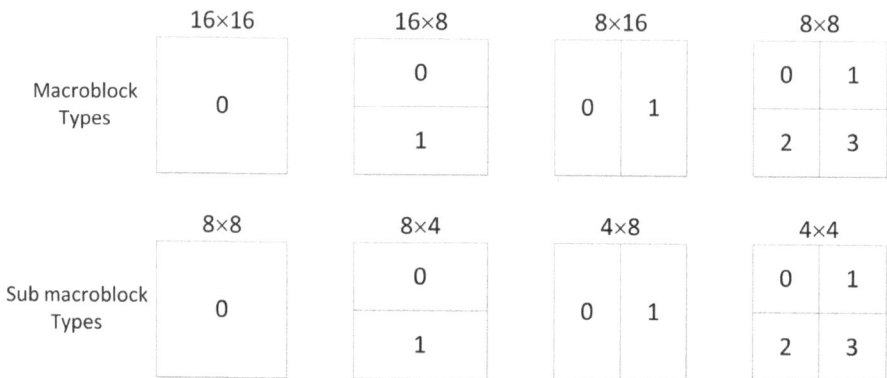

Figure 3-10. *Macroblock and block partitions in AVC*

Coding Algorithm

In the AVC algorithm, the encoder may select between intra and inter coding for various partitions of each picture. *Intra coding* (*I*) provides random access points in the bitstream where decoding can begin and continue correctly. Intra coding uses various spatial prediction modes to reduce spatial redundancy within a picture. In addition, AVC defines *inter* coding that uses *motion vectors* for block-based inter-picture prediction to reduce temporal redundancy. Inter coding are of two types: *predictive* (*P*) and *bi-predictive* (*B*). Inter coding is more efficient as it uses inter prediction of each block of pixels relative to some previously decoded pictures. Prediction is obtained from a *deblocked* version of previously reconstructed pictures that are used as *references* for the prediction. The deblocking filter is used in order to reduce the blocking artifacts at the block boundaries. Motion vectors and intra prediction modes may be specified for a variety of block sizes in the picture. Further compression is achieved by applying a transform to the prediction residual to remove spatial correlation in the block before it is quantized. The intra prediction modes, the motion vectors, and the quantized transform coefficient information are encoded using an entropy code such as *context-adaptive variable length codes* (CAVLC) or *context adaptive binary arithmetic codes* (CABAC). A block diagram of the AVC coding algorithm, showing the encoder and decoder blocks, is presented in Figure 3-11.

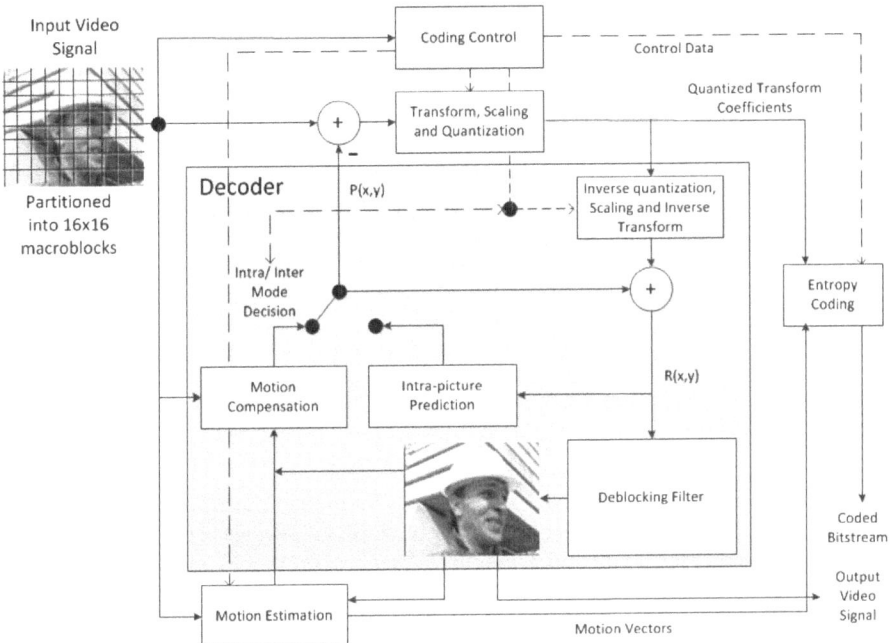

Figure 3-11. *The AVC codec block diagram*

Intra Prediction

In the previous standards, intra-coded macroblocks were coded independently without any reference, and it was necessary to use intra macroblocks whenever a good prediction was not available for a predicted macroblock. As intra macroblocks use more bits than the predicted ones, this was often less efficient for compression. To alleviate this drawback—that is, to reduce the number of bits needed to code an intra picture—AVC introduced intra prediction, whereby a prediction block is formed based on previously reconstructed blocks belonging to the same picture. Fewer bits are needed to code the residual signal between the current and the predicted blocks, compared to the coding the current block itself.

The size of an intra prediction block for the luma samples may be 4×4, 8×8, or 16×16. There are several intra prediction modes, out of which one mode is selected and coded in the bitstream. AVC defines a total of nine intra prediction modes for 4×4 and 8×8 luma blocks, four modes for a 16×16 luma block, and four modes for each chroma block. Figure 3-12 shows an example of intra prediction modes for a 4×4 block. In this example, [$a, b, \ldots, p$] are the predicted samples of the current block, which are predicted from already decoded left and above blocks with samples [$A, B, \ldots, M$]; the arrows show the direction of the prediction, with each direction indicated as a intra prediction mode in the coded bitstream. For mode 0 (vertical), the prediction is formed by extrapolation

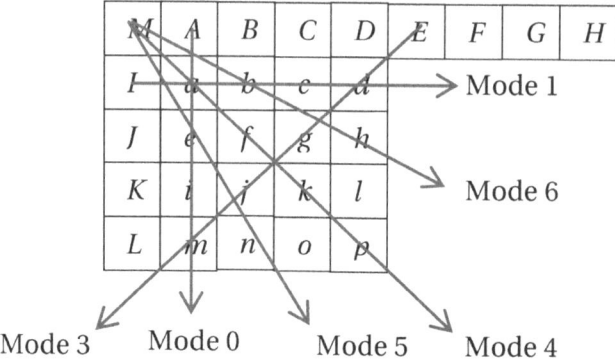

Figure 3-12. *An example of intra prediction modes for a 4×4 block (few modes are shown as examples)*

of the samples above, namely [A, B, C, D]. Similarly for mode 1 (horizontal), left samples [I, J, K, L] are extrapolated. For mode 2 (DC prediction), the average of the above and left samples are used as the prediction. For mode 3 (diagonal down left), mode 4 (diagonal down right), mode 5 (vertical right), mode 6 (horizontal down), mode 7 (vertical left), and mode 8 (horizontal up), the predicted samples are formed from a weighted average of the prediction samples A—M.

Inter Prediction

Inter prediction reduces temporal correlation by using motion estimation and compensation. As mentioned before, AVC partitions the picture into several shapes from 16×16 down to 4×4 for such predictions. The motion compensation results in reduced information in the residual signal, although for the smaller partitions, an overhead of bits is incurred for motion vectors and for signaling the partition type.

Intra prediction can be applied to blocks as small as 4×4 luma samples with up to a quarter-pixel (a.k.a. quarter-pel) motion vector accuracy. Sub-pel motion compensation gives better compression efficiency than using integer-pel alone; while quarter-pel is better than half-pel, it involves more complex computation. For luma, the half-pel samples are generated first and are interpolated from neighboring integer-pel samples using a six-tap *finite impulse response* (FIR) filter with weights (1, -5, 20, 20, -5, 1)/32. With the half-pel samples available, quarter-pel samples are produced using bilinear interpolation between neighboring half- or integer-pel samples. For 4:2:0 chroma, eighth-pel samples correspond to quarter-pel luma, and are obtained from linear interpolation of integer-pel chroma samples. Sub-pel motion vectors are differentially coded relative to predictions from neighboring motion vectors. Figure 3-13 shows the location of sub-pel predictions relative to full-pel.

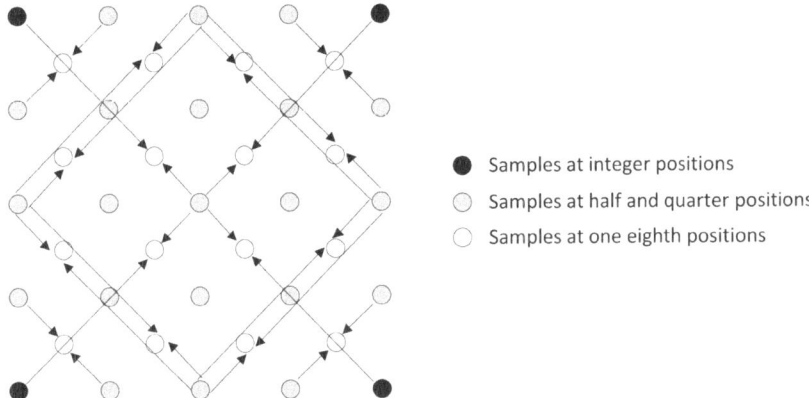

Figure 3-13. *Locations of sub-pel prediction*

For inter prediction, reference pictures can be used from a list of previously reconstructed pictures, which are stored in the picture buffer. The distance of a reference picture from the current picture in display order determines whether it is a *short-term* or a *long-term* reference. Long-term references help increase the motion search range by using multiple decoded pictures. As a limited size of picture buffer is used, some pictures may be marked as *unused for reference*, and may be deleted from the reference list in a controlled manner to keep the memory size practical.

Transform and Quantization

The AVC algorithm uses block-based transform for spatial redundancy removal, as the residual signal from intra or inter prediction is divided into 4×4 or 8×8 (High profile only) blocks, which are converted to transform domain before they are quantized. The use of 4×4 integer transform in AVC results in reduced ringing artifacts compared to those produced by previous standards using fixed 8×8 DCT. Also, multiplications are not necessary at this smaller size. AVC introduced the concept of hierarchical transform structure, in which the DC components of neighboring 4×4 luma transforms are grouped together to form a 4×4 block, which is transformed again using a Hadamard transform for further improvement in compression efficiency.

Both 4×4 and 8×8 transforms in AVC are integer transforms based on DCT. The integer transform, post-scaling, and quantization are grouped together in the encoder, while in the decoder the sequence is inverse quantization, pre-scaling, and inverse integer transform. For a deeper understanding of the process, consider the matrix *H* below.

$$H = \begin{bmatrix} a & a & a & a \\ b & c & -c & -b \\ a & -a & -a & a \\ c & -b & b & -c \end{bmatrix}.$$

79

A 4×4 DCT can be done using this matrix and the formula: $X = HFH^T$, where H^T is the transpose of the matrix H, F is the input 4×4 data block, and X is the resulting 4×4 transformed block. For DCT, the variables a, b, and c are as follows:

$$a = \frac{1}{2}, b = \sqrt{\frac{1}{2}} \cos\left(\frac{\pi}{8}\right), c = \sqrt{\frac{1}{2}} \cos\left(\frac{3\pi}{8}\right).$$

The AVC algorithm simplifies these coefficients with approximations, and still maintains orthogonaity property by using:

$$a = \frac{1}{2}, b = \sqrt{\frac{2}{5}}, c = \frac{1}{2}.$$

Further simplification is made to avoid multiplication by combining the transform with the quantization step, using a scaled transform $X = \hat{H}F\hat{H}^T \otimes SF$, where,

$$\widehat{H} = \begin{bmatrix} 1 & 1 & 1 & 1 \\ 2 & 1 & -1 & -2 \\ 1 & -1 & -1 & 1 \\ 1 & -2 & 2 & -1 \end{bmatrix}, \text{ and } SF = \begin{bmatrix} a^2 & \dfrac{ab}{2} & a^2 & \dfrac{ab}{2} \\ \dfrac{ab}{2} & \dfrac{b^2}{4} & \dfrac{ab}{2} & \dfrac{b^2}{4} \\ a^2 & \dfrac{ab}{2} & a^2 & \dfrac{ab}{2} \\ \dfrac{ab}{2} & \dfrac{b^2}{4} & \dfrac{ab}{2} & \dfrac{b^2}{4} \end{bmatrix},$$

and SF is a 4×4 matrix representing the scaling factors needed for orthonormality, and $\otimes$ represents element-by-element multiplication. The transformed and quantized signal Y with components $Y_{i,j}$ is obtained by appropriate quantization using one of the 52 available quantizer levels (a.k.a. quantization step size, $Qstep$) as follows:

$$Y_{i,j} = X_{i,j} round\left(\frac{SF_{ij}}{Qstep}\right), Where\, 0 \leq i, j \leq 3.$$

In the decoder, the received signal Y is scaled with $Qstep$ and SF as the inverse quantization and a part of inverse transform to obtain the inverse transformed block X' with components $X'_{i,j}$:

$$X'_{i,j} = Y_{i,j} Q_{step} SF_{ij}^{-1}, Where\, 0 \leq i, j \leq 3.$$

The 4×4 reconstructed block is: $F' = \widehat{H}_v^T X' \widehat{H}_v$, where the integer inverse transform matrix is given by:

$$\widehat{H}_v = \begin{bmatrix} 1 & 1 & 1 & 1 \\ 1 & \dfrac{1}{2} & \dfrac{-1}{2} & -1 \\ 1 & -1 & -1 & 1 \\ \dfrac{1}{2} & -1 & 1 & -\dfrac{1}{2} \end{bmatrix}$$

In addition, in the hierarchical transform approach for 16×16 intra mode, the 4×4 luma intra DC coefficients are further transformed using a Hadamard transform:

$$\widetilde{H} = \begin{bmatrix} 1 & 1 & 1 & 1 \\ 1 & 1 & -1 & -1 \\ 1 & -1 & -1 & 1 \\ 1 & -1 & 1 & -1 \end{bmatrix}.$$

In 4:2:0 color sampling, for the chroma DC coefficients, the transform matrix is as follows:

$$\widetilde{H} = \begin{bmatrix} 1 & 1 \\ 1 & -1 \end{bmatrix}.$$

In order to increase the compression gain provided by run-length coding, two scanning orders are defined to arrange the quantized coefficients before entropy coding, namely zigzag scan and field scan, as shown in Figure 3-14. While zigzag scanning is suitable for progressively scanned sources, alternate field scanning helps interlaced contents.

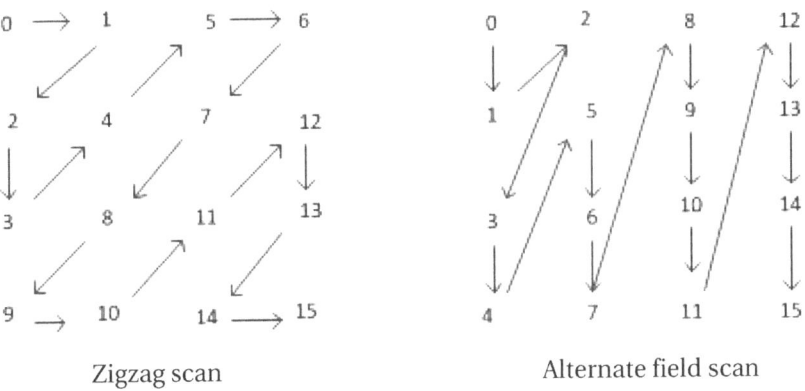

Zigzag scan   Alternate field scan

Figure 3-14. *Zigzag and alternate field scanning orders for 4×4 blocks*

Entropy Coding

Earlier standards provided entropy coding using fixed tables of variable-length codes, where the tables were predefined by the standard based on the probability distributions of a set of generic videos. It was not possible to optimize those Huffman tables for specific video sources. In contrast, AVC uses different VLCs to find a more appropriate code for each source symbol based on the context characteristics. Syntax elements other than the residual data are encoded using the Exponential-Golomb codes. The residual data is rearranged through zigzag or alternate field scanning, and then coded using *context-adaptive variable length codes* (CAVLC) or, optionally for Main and High profiles, using *context-adaptive binary arithmetic codes* (CABAC). Compared to CAVLC, CABAC provides higher coding efficiency at the expense of greater complexity.

CABAC uses an adaptive binary arithmetic coder, which updates the probability estimation after coding each symbol, and thus adapts to the context. The CABAC entropy coding has three main steps:

- **Binarization**: Before arithmetic coding, a non-binary symbol such as a transform coefficient or motion vector is uniquely mapped to a binary sequence. This mapping is similar to converting a data symbol into a variable-length code, but in this case the binary code is further encoded by the arithmetic coder prior to transmission.

- **Context modeling**: A probability model for the binarized symbol, called the context model, is selected based on previously encoded syntax element.

- **Binary arithmetic coding**: In this step, an arithmetic coder encodes each element according to the selected context model, and subsequently updates the model.

Flexible Interlaced Coding

In order to provide enhanced interlaced coding capabilities, AVC supports *macroblock-adaptive frame-field* (MBAFF) coding and *picture-adaptive frame-field* (PAFF) coding techniques. In MBAFF, a macroblock pair structure is used for pictures coded as frames, allowing 16×16 macroblocks in field mode. This is in contrast to MPEG-2, where field mode processing in a frame-coded picture could only support 16×8 half-macroblocks. In case of PAFF, it is allowed to mix pictures coded as complete frames with combined fields with those coded as individual single fields.

In-Loop Deblocking

Visible and annoying blocking artifacts are produced owing to block-based transform in intra and inter prediction coding, and the quantization of the transform coefficients, especially for higher quantization scales. In an effort to mitigate such artifacts at the block boundaries, AVC provides deblocking filters, which also prevents propagation of accumulated coded noise.

A deblocking filter is not new; it was introduced in H.261 as an optional tool, and had some success in reducing temporal propagation of coded noise, as integer-pel accuracy in motion compensation alone was insufficient in reducing such noise. However, in MPEG-1 and MPEG-2, a deblocking filter was not used owing to its high complexity. Instead, the half-pel accurate motion compensation, where the half-pels were obtained by bilinear filtering of integer-pel samples, played the role of smoothing out the coded noise.

However, despite the complexity, AVC uses a deblocking filter to obtain higher coding efficiency. As it is part of the prediction loop, with the removal of the blocking artifacts from the predicted pictures, a much closer prediction is obtained, leading to a reduced-energy error signal. The deblocking filter is applied to horizontal or vertical edges of 4×4 blocks. The luma filtering is performed on four 16-sample edges, and the chroma filtering is performed on two 8-sample edges. Figure 3-15 shows the deblocking boundaries.

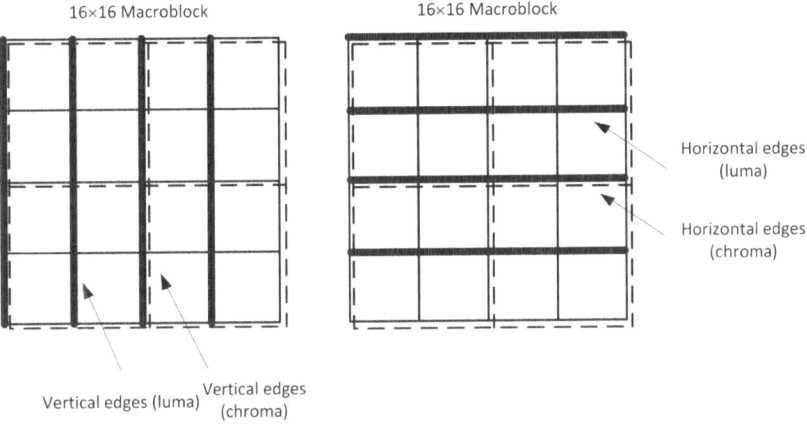

Figure 3-15. *Deblocking along vertical and horizontal boundaries in macroblock*

Error Resilience

AVC provides features for enhanced resilience to channel errors, which include NAL units, redundant slices, data partitioning, flexible macroblock ordering, and so on. Some of these features are as follows:

- **Network Abstraction Layer** (NAL): By defining NAL units, AVC allows the same video syntax to be used in many network environments. In previous standards, header information was part of a syntax element, thereby exposing the entire syntax element to be rendered useless in case of erroneous reception of a single packet containing the header. In contrast, in AVC, self-contained packets are generated by decoupling information relevant to more than one slice from the media stream. The high-level crucial parameters, namely the *Sequence Parameter Set* (SPS) and *Picture Parameter Set* (PPS), are kept in NAL units with a higher level of error protection. An active SPS remains unchanged throughout a coded video sequence, and an active PPS remains unchanged within a coded picture.

- **Flexible macroblock ordering** (FMO): FMO is also known as *slice groups*. Along with *arbitrary slice ordering* (ASO), this technique re-orders the macroblocks in pictures, so that losing a packet does not affect the entire picture. Missing macroblocks can be regenerated by interpolating from neighboring reconstructed macroblocks.

- **Data partitioning** (DP): This is a feature providing the ability to separate syntax elements according to their importance into different packets of data. It enables the application to have *unequal error protection* (UEP).

- **Redundant slices** (RS): This is an error-resilience feature in AVC that allows an encoder to send an extra representation of slice data, typically at lower fidelity. In case the primary slice is lost or corrupted by channel error, this representation can be used instead.

HEVC

The High Efficiency Video Coding (HEVC), or the H.265 standard (ISO/IEC 23008-2), is the most recent joint video coding standard ratified in 2013 by the ITU-T Video Coding Experts Group (VCEG) and ISO/IEC Moving Picture Experts Group (MPEG) standardization organizations. It follows the earlier standard known as AVC or H.264, also defined by the same MPEG and VCEG Joint Collaborative Team on Video Coding (JCT-VC), with a goal of addressing the growing popularity of ever higher resolution videos, high-definition (HD, 1920 × 1080), ultra-high definition (UHD, 4k × 2k), and beyond. In particular, HEVC addresses two key issues: increased video resolution and increased use of parallel processing architectures. As such, HEVC algorithm has a design target of achieving twice the compression efficiency achievable by AVC.

Picture Parititioning and Structure

In earlier standards, macroblocks were the basic coding building block, which contains a 16×16 luma block, and typically two 8×8 chroma blocks for 4:2:0 color sampling. In HEVC, the analogous structure is the *coding tree unit* (CTU), also known as the *largest coding unit* (LCU), containing a luma *coding tree block* (CTB), corresponding chroma CTBs, and syntax elements. In a CTU, the luma block size can be 16×16, 32×32, or 64×64, specified in the bitstream sequence parameter set. CTUs can be further partitioned into smaller square blocks using a tree structure and quad-tree signaling.

The quad-tree specifies the *coding units* (CU), which forms the basis for both prediction and transform. The coding units in a coding tree block are traversed and encoded in Z-order. Figure 3-16 shows an example of ordering in a 64×64 CTB.

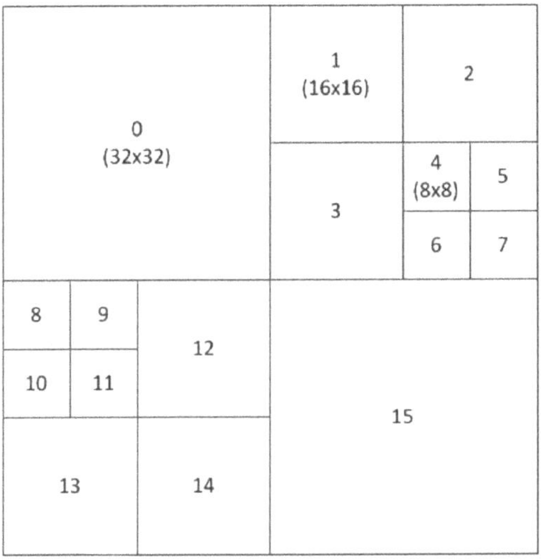

Figure 3-16. *An example of ordering of coding units in a 64×64 coding tree block*

A coding unit has one luma and two chroma *coding blocks* (CB), which can be further split in size and can be predicted from corresponding *prediction blocks* (PB), depending on the prediction type. HEVC supports variable PB sizes, ranging from 64×64 down to 4×4 samples. The prediction residual is coded using the *transform unit* (TU) tree structure. The luma or chroma coding block residual may be identical to the corresponding *transform block* (TB) or may be further split into smaller transform blocks. Transform blocks can only have square sizes 4×4, 8×8, 16×16, and 32×32. For the 4×4 transform of intra-picture prediction residuals, in addition to the regular DCT-based integer transform, an integer transform based on a form of *discrete sine transform* (DST) is also specified as an alternative. This quad-tree structure is generally considered the biggest contributor for the coding efficiency gain of HEVC over AVC.

HEVC simplies coding and does not support any interlaced tool, as interlaced scanning is no longer used in displays and as interlaced video is becoming substantially less common for distribution. However, interlaced video can still be coded as a sequence of field pictures. Metadata syntax is available in HEVC to allow encoders to indicate that interlace-scanned video has been sent by coding one of the following:

- Each *field* (i.e., the even or odd numbered lines of each video frame) of interlaced video as a separate picture

- Each interlaced frame as an HEVC coded picture

This provides an efficient method of coding interlaced video without inconveniencing the decoders with a need to support a special decoding process for it.

Profiles and Levels

There are three profiles defined by HEVC: the *Main* profile, the *Main 10* profile, and the *Still picture* profile, of which currently *Main* is the most commonly used. It requires 4:2:0 color format and imposes a few restrictions; for instance, bit depth should be 8, groups of LCUs forming rectangular tiles must be at least 256×64, and so on (tiles are elaborated later in this chapter in regard to parallel processing tools). Many levels are specified, ranging from 1 to 6.2. A Level-6.2 bitstream could support as large a resolution as 8192×4320 at 120 fps.[3]

Intra Prediction

In addition to the planar and the DC prediction modes, intra prediction supports 33 directional modes, compared to eight directional modes in H.264/AVC. Figure 3-17 shows the directional intra prediction modes.

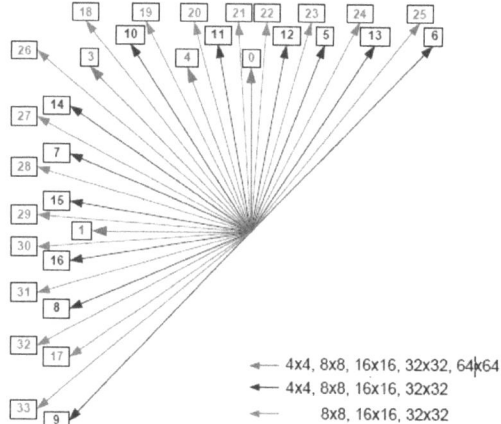

Figure 3-17. *Directional intra prediction modes in HEVC*

Intra prediction in a coding unit exactly follows the TU tree such that when an intra coding unit is coded using an *N*×*N* partition mode, the TU tree is forcibly split at least once, ensuring a match between the intra coding unit and the TU tree. This means that the intra operation is always performed for sizes 32×32, 16×16, 8×8, or 4×4. Similar to AVC, intra prediction requires two one-dimensional arrays that contain the upper and left neighboring samples, as well as an upper-left sample. The arrays are twice as long as the intra block size, extending below and to the right of the block. Figure 3-18 shows an example array for an 8×8 block.

---

[3]*ITU-T Rec. H.265: High Efficiency Video Coding* (Geneva, Switzerland: International Telecommunications Union, 2013).

140		132	131	139	133	150	152	167	168	167	171	189	190	167	178	177	178

139		136	132	138	142	149	137	170	149
138		133	129	151	155	156	171	167	159
135		129	130	153	159	150	157	169	172
142		141	150	160	172	155	151	182	151
147		149	137	170	149	146	132	140	142
132		133	151	181	180	153	135	148	160
135		150	157	189	172	149	145	153	163
120		155	151	182	151	151	155	160	172
120
120
120
120
120
120
120
120

Figure 3-18. Luma intra samples and prediction structure in HEVC

Inter Prediction

For inter-picture prediction, HEVC provides two reference lists, L0 and L1, each with a capacity to hold 16 reference frames, of which a maximum of eight pictures can be unique. This implies that some reference pictures will be repeated. This would facilitate predictions from the same picture with different weights.

Motion Vector Prediction

Motion vector prediction in HEVC is quite complex, as it builds a list of candidate motion vectors and selects one of the candidates from the list using an index of list that is coded in the bitstream. There are two modes for motion vector prediction: *merge* and *advanced motion vector prediction* (AMVP). For each prediction unit (PU), the encoder decides which mode to use and indicates it in the bitstream with a flag. The AMVP process uses a delta motion vector coding and can produce any desired value of motion vector. HEVC subsamples the temporal motion vectors on a 16×16 grid. This means that a decoder only needs to allocate space for two motion vectors (L0 and L1) in the temporal motion vector buffer for a 16×16 pixel area.

Motion Compensation

HEVC specifies motion vectors in a quarter-pel granularity, but uses an eight-tap filter for luma, and a four-tap eighth-pel filter for chroma. This is an improvement over the six-tap luma and bilinear (two-tap) chroma filters defined in AVC. Owing to the longer length of the eight-tap filter, three or four extra pixels on all sides are needed to be read for each block. For example, for an 8×4 block, a 15×11 pixel area needs to be read into the memory, and the impact would be more for smaller blocks. Therefore, HEVC limits the smallest prediction unit to be uni-directional and larger than 4×4. HEVC supports weighted prediction for both uni- and bi-directional PUs. However, the weights are always explicitly transmitted in the slice header; unlike AVC, there is no implicit weighted prediction.

Entropy Coding

In HEVC, entropy coding is performed using *Context-Adaptive Binary Arithmetic Codes* (CABAC) at the CTU level. The CABAC algorithm in HEVC improves upon that of AVC with a few minor enhancements. There are about half as many context-state variables as in AVC, and the initialization process is much simpler. The bitstream syntax is designed such that bypass-coded bins are grouped together as much as possible. CABAC decoding is inherently a sequential operation; therefore, parallelization or fast hardware implementation is difficult. However, it is possible to decode more than one bypass-coded bin at a time. This, together with the bypass-bin grouping, greatly facilitates parallel implementation in hardware decoders.

In-loop Deblocking and SAO

In HEVC, two filters could be applied on reconstructed pixel values: the *in-loop deblocking* (ILD) filter and the *sample adaptive offset* (SAO) filter. Either or both of the filters can be optionally applied across the tile- and slice-boundaries. The in-loop deblocking filter in HEVC is similar to that of H.264/AVC, while the SAO is a new filter and is applied following the in-loop deblock filter.

Unlike AVC, which deblocks at every 4×4 grid edge, in HEVC, deblocking is performed on the 8×8 grid only. All vertical edges in the picture are deblocked first, followed by all horizontal edges. The filter itself is similar to the one in AVC, but in the case of HEVC, only the boundary strengths 2, 1, and 0 are supported. With an 8-pixel separation between the edges, there is no dependency between them, enabling a highly parallelized implementation. For example, the vertical edge can be filtered with one thread per 8-pixel column in the picture. Chroma is only deblocked when one of the PUs on either side of a particular edge is intra-coded.

As a secondary filter after deblocking, the SAO performs a non-linear amplitude mapping by using a lookup table at the CTB level. It operates once on each pixel of the CTB, a total of 6,144 times for each 64×64 CTB. (64×64 + 32×32 + 32×32 = 6144). For each CTB, a filter type and four offset values, ranging from -7 to 7 for 8-bit video for example, are coded in the bitstream. The encoder chooses these parameters with a view toward better matching the reconstructed and the source pictures.

Parallel Processing Syntax and Tools

There are three new features in the HEVC standard to support enhanced parallel processing capability or to modify the structure of slice data for purposes of packetization.

Tiles

There is an option to partition a picture into rectangular regions called *tiles*. Tiles are independently decodable regions of a picture that are encoded with some shared header information. Tiles are specified mainly to increase the parallel processing capabilities, although some error resilience may also be attributed to them. Tiles provide coarse-grain parallelism at picture and sub-picture level, and no sophisticated synchronization of threads is necessary for their use. The use of tiles would require reduced-size line buffers, which is advantageous for high-resolution video decoding on cache-constrained hardware and cheaper CPUs.

Wavefront Parallel Processing

When *wavefront parallel processing* (WPP) is enabled, a slice is divided into rows of CTUs. The first row is processed in a regular manner; but processing of the second row can be started only after a few decisions have been made in the first row. Similarly, processing of the third row can begin as soon as a few decisions have been made in the second row, and so on. The context models of the entropy coder in each row are inferred from those in the preceding row, with a small fixed processing lag. WPP provides fine-grained parallel processing within a slice. Often, WPP provides better compression efficiency compared to tiles, while avoiding potential visual artifacts resulting from the use of tiles.

Slice Segments and Dependent Slices

A sequence of coding tree blocks is called a *slice*. A *picture* constituting a video frame can be split into any number of slices, or the whole picture can be just one slice. In turn, each slice is split up into one or more *slice segments*, each in its own NAL unit. Only the first slice segment of a slice contains the full slice header, and the rest of the segments are referred to as dependent slice segments. As such, a decoder must have access to the first slice segment for successful decoding. Such division of slices allows low-delay transmission of pictures without paying any coding efficiency penalty that would have otherwise incurred owing to many slice headers. For example, a camera can send out a slice segment belonging to the first CTB row so that a playback device on the other side of the network can start decoding before the camera sends out the next CTB row. This is useful in low-latency applications such as video conferencing.

International Standards for Video Quality

Several standards for video quality have been specified by the ITU-T and ITU-R visual quality experts groups. Although they are not coding standards, they are worth mentioning as they relate to the subject of this book. Further, as the IEEE standard 1180 relates to the accuracy of computation of the common IDCT technology used in all the aforementioned standards, it is also briefly described here.

VQEG Standards

In 1997, a small group of video quality experts from the ITU-T and ITU-R study groups formed the Visual Quality Experts Group (VQEG), with a view toward advancing the field of video quality assessment. This group investigated new and advanced subjective assessment methods and objective quality metrics and measurement techniques.

VQEG took a systematic approach to validation testing that typically includes several video databases for which objective models are needed to predict the subjective visual quality, and they defined the test plans and procedures for performing objective model validation. The initial standard was published in 2000 by the ITU-T Study Group 9 as Recommendation J.144, but none of the various methods studied outperformed the well-known *peak signal to noise ratio* (PSNR). In 2003, an updated version of J.144 was published, where four objective models were recommended for cable television services. A mirror standard by the ITU-R Study Group 6 was published as ITU-R Recommendation BT.1683 for baseband television services.

The most recent study aimed at digital video and multimedia applications, known as *Multimedia Phase I* (MM-I), was completed in 2008. The MM-I set of tests was used to validate *full-reference* (FR), *reduced reference* (RR), and *no reference* (NR) objective models. (These models are discussed in Chapter 4 in some detail.) Subjective video quality was assessed using a single-stimulus presentation method and the *absolute category rating* (ACR) scale, where the video sequences including the source video are presented for subjective evaluation one at a time without identifying the videos, and are rated independently on the ITU five-grade quality scale. A *mean opinion score* (MOS) and a *difference mean opinion score* (DMOS) were computed, where the DMOS was the average of the arithmetic difference of the scores of processed videos compared to the scores of the source videos, in order to remove any hidden reference. The software used to control and run the VQEG multimedia tests is known as AcrVQWin.[4] (Details of subjective quality assessment methods and techniques are captured in ITU-T Recommendations P.910, P.912, and so on, and are discussed in Chapter 4.)

IEEE Standard 1180-1990

Primarily intended for use in visual telephony and similar applications where the 8x8 *inverse discrete cosine transform* (IDCT) results are used in a reconstruction loop, the IEEE Standard 1180-1990[5] specifies the numerical characteristics of the 8x8 IDCT. The specifications ensure the compatibility between different implementations of the IDCT. A mismatch error may arise from the different IDCT implementations in the encoders and decoders from different manufacturers; owing to the reconstruction loop in the system,

---

[4]K. Brunnstrom, D. Hands, F. Speranza, and A. Webster, "VQEG Validation and ITU Standardization of Objective Perceptual Video Quality Metrics," *IEEE Signal Processing* (May 2009):96-101; software available at www.acreo.se/acrvqwin.
[5]The IEEE Standard Board, *IEEE Standard Specifications for the Implementations of 8x8 Inverse Discrete Cosine Transform* (New York, USA: Institute of Electrical and Electronics Engineers, 1991).

the mismatch error may propagate over the duration of the video. Instead of restricting all manufacturers to a single strict implementation, the IEEE standard allows a small mismatch for a specific period of time while the video is refreshed periodically using the intra-coded frame—for example, for ITU-T visual telephony applications, the duration is 132 frames.

The standards specify that the mismatch errors shall meet the following requirements:

- For any pixel location, the peak error *(ppe)* shall not exceed 1 in magnitude.

- For any pixel location, the mean square error *(pmse)* shall not exceed 0.06.

- Overall, the mean square error *(omse)* shall not exceed 0.02.

- For any pixel location, the mean error *(pme)* shall not exceed 0.015 in magnitude.

- Overall, the mean error *(ome)* shall not exceed 0.0015 in magnitude.

- For all-zero input, the proposed IDCT shall generate all-zero output.

Overview of Other Industry Standards and Formats

In addition to the international standards defined by the ISO, the ITU, or the *Institute of Electrical and Electronics Engineers* (IEEE), there are standards well known in the industry. A few of these standards are described below.

VC-1

Initially developed as a proprietary video format by Microsoft, the well-known VC-1 format was formally released as the SMPTE 421M video codec standard in 2006, defined by the Society of Motion Pictures and Television Engineers (SMPTE). It is supported by Blu-ray, currently obsolete HD-DVD, Microsoft Windows Media, Microsoft Silverlight framework, Microsoft X-Box 360, and Sony PlayStation 3 video game consoles, as well as various Windows-based video applications. Hardware decoding of VC-1 format is available in Intel integrated processor graphics since second-generation Intel (R) Core (TM) processor (2011) and in Raspberry Pi (2012).

VC-1 uses the conventional DCT-based design similar to the international standards, and supports both progressive and interlaced video. The specification defines three profiles: Simple, Main, and Advanced, and up to five levels. Major tools supported by each profile are shown in Table 3-3.

Table 3-3. *VC-1 Profiles and Levels*

Profiles	Levels	Maximum Bit Rate (kbps)	Resolution @ Frame rate	Tools
Simple	Low	96	176 × 144 @ 15 (QCIF)	Baseline intra frame compression; variable sized, 16-bit and overlapped transform; four motion vectors per macroblock; quarter pixel luma motion compensation
	Medium	384	240 × 176 @ 30, 352 × 288 @ 15 (CIF)	
Main	Low	2,000	320 × 240 @ 24 (QVGA)	In addition to Simple profile: Start codes; quarter-pixel chroma motion compensation; extended motion vectors; loop filter; adaptive quantization; *B*-frames; dynamic resolution change; intensity compensation; range adjustment
	Medium	10,000	720 × 480 @ 30 (480p30), 720 × 576 @ 25 (576p25)	
	High	20,000	1920 × 1080 @ 30 (1080p30)	
Advanced	L0	2,000	352 × 288 @ 30 (CIF)	In addition to Main profile: GOP layer; field and frame coding modes; display metadata
	L1	10,000	720 × 480 @ 30 (NTSC-SD), 720 × 576 @ 25 (PAL-SD)	
	L2	20,000	720 × 480 @ 60 (480p60), 1280 × 720 @ 30 (720p30)	
	L3	45,000	1920 × 1080 @ 24 (1080p24), 1920 × 1080 @ 30 (1080i30), 1280 × 720 @ 60 (720p60)	
	L4	135,000	1920 × 1080 @ 60 (1080p60), 2048 × 1536 @ 24	

VP8

With the acquisition of On2 Technologies, Google became the owner of the VP8 video compression format. In November 2011, the VP8 data format and decoding guide was published as RFC 6386[6] by the Internet Engineering Task Force (IETF). The VP8 codec library software, *libvpx*, is also released by Google under a BSD license. VP8 is currently supported by Opera, FireFox, and Chrome browsers, and various hardware and software-based video codecs, including the Intel integrated processor graphics hardware.

Like many modern video compression schemes, VP8 is based on decomposition of frames into square subblocks of pixels, prediction of such subblocks using previously constructed blocks, and adjustment of such predictions using a discrete cosine transform (DCT), or in one special case, a Walsh-Hadamard transform (WHT). The system aims to reduce data rate through exploiting the temporal coherence of video signals by specifying the location of a visually similar portion of a prior frame, and the spatial coherence by taking advantage of the frequency segragtion provided by DCT and WHT and the tolerance of the human visual system to moderate losses of fidelity in the reconstituted signal. Further, VP8 augments these basic concepts with, among other things, sophisticated usage of contextual probabilities, resulting in a significant reduction in data rate at a given quality.

The VP8 algorithm exclusively specifies fixed-precision integer operations, preventing the reconstructed signal from any *drift* that might have been caused by truncation of fractions. This helps verify the correctness of the decoder implementation and helps avoid inconsistencies between decoder implementations. VP8 works with 8-bit YUV 4:2:0 image formats, internally divisible into 16×16 macroblocks and 4×4 subblocks, with a provision to support a secondary YUV color format. There is also support of an optional upscaling of internal reconstruction buffer prior to output so that a reduced-resolution encoding can be done, while the decoding is performed at full resolution. *Intra* or key frames are defined to provide random access while *inter* frames are predicted from any prior frame up to and including the most recent key frame; no bi-directional prediction is used. In general, the VP8 codec uses three different reference frames for inter-frame prediction: the previous frame, the golden frame, and the *altref* frame to provide temporal scalability.

VP8 codecs apply data partitioning to the encoded data. Each encoded VP8 frame is divided into two or more partitions, comprising an uncompressed section followed by compressed header information and per-macroblock information specifying how each macroblock is predicted. The first partition contains prediction mode parameters and motion vectors for all macroblocks. The remaining partitions all contain the quantized DCT or WHT coefficients for the residuals. There can be one, two, four, or eight DCT/WHT partitions per frame, depending on encoder settings. Details of the algorithm can be found in the RFC 6386.

---

[6]J. Bankoski, J. Koleszar, L. Quillio, J. Salonen, P. Wilkins, and Y. Xu, "VP8 Data Format and Coding Guide, RFC 6386," November 2011, retrieved from http://datatracker.ietf.org/doc/rfc6386/.

An RTP payload specification[7] applicable to the transmission of video streams encoded using the VP8 video codec has been proposed by Google. The RTP payload format can be used both in low-bit-rate peer-to-peer and high-bit-rate video conferencing applications. The RTP payload format takes the frame partition boundaries into consideration to improve robustness against packet loss and to facilitate error concealment. It also uses advanced reference frame structure to enable efficient error recovery and temporal scalability. Besides, marking of the non-reference frames is done to enable servers or media-aware networks to discard appropriate data as needed.

The IETF Internet Draft standard for browser *application programming interface* (API), called the *Web Real Time Communication* (WebRTC),[8] specifies that if VP8 is supported, then the *bilinear* and the *none* reconstruction filters, a frame rate of at least 10 frames per second, and resolutions ranging from 320×240 to 1280×720 must be supported. Google Chrome, Mozilla, FireFox, and Opera browsers support the WebRTC APIs, intended for browser-based applications including video chatting. Google Chrome operating system also supports WebRTC.

VP9

The video compression standard VP9 is a successor to the VP8 and is also an open standard developed by Google. The latest specification was released in February 2013 and is currently available as an Internet-Draft[9] from the IETF; the final specification is not ratified yet. The VP9 video codec is developed specifically to meet the demand for video consumption over the Internet, including professional and amateur-produced video-on-demand and conversational video content. The WebM media container format provides royalty-free, open video compression for HTML5 video, by primarily using the VP9 codec, which replaces the initially supported codec VP8.

The VP9 draft includes a number of enhancements and new coding tools that have been added to the VP8 codec to improve the coding efficiency. The new tools described in the draft include larger prediction block sizes up to 64×64, various forms of compound inter prediction, more intra prediction modes, one-eighth-pixel motion vectors, 8-tap switchable sub-pixel interpolation filters, improved motion reference generation and motion vector coding, improved entropy coding including frame-level entropy adaptation for various symbols, improved loop filtering, the incorporation of the *Asymmetric Discrete Sine Transform* (ADST), larger 16×16 and 32×32 DCTs, and improved frame-level segmentation. However, VP9 is currently under development and the final version of the VP9 specification may differ considerably from the draft specification, of which some features are described here.

---

[7]P. Westin, H. Lundin, M. Glover, J. Uberti, and F. Galligan, "RTP Payload Format for VP8 Video," Internet draft, February 2014, retrieved from http://tools.ietf.org/html/draft-ietf-payload-vp8-11.

[8]C. Bran, C. Jennings, J. M.Valin, "WebRTC Codec and Media Processing Requirement," Internet draft, March 2012, retrieved from http://tools.ietf.org/html/draft-cbran-rtcweb-codec-02.

[9]A. Grange and H. Alvestrand, "A VP9 Bitstream Overview," Internet draft, February 2013, retrieved from http://tools.ietf.org/html/draft-grange-vp9-bitstream-00.

Picture Partitioning

VP9 partitions the picture into 64 × 64 *superblocks* (SB), which are processed in raster-scan order, from left to right and top to bottom. Similar to HEVC, superblocks can be subdivided down to a minimum of 4×4 using a recursive quad-tree, although 8×8 block sizes are the most typical unit for mode information. In contrast to HEVC, however, the slice structure is absent in VP9.

It is desirable to be able to carry out encoding or decoding tasks in parallel, or to use multi-threading in order to effectively utilize available resources, especially on resource-constrained personal devices like smartphones. To this end, VP9 offers frame-level parallelism via the frame_parallel_mode flag and two- or four-column based tiling, while allowing loop filtering to be performed across tile boundaries. Tiling in VP9 is done in vertical direction only, while each tile has an integral number of blocks. There is no data dependency across adjacent tiles, and any tile in a frame can be processed in any order. At the start of every tile except the last one, a 4-byte size is transmitted, indicating the size of the next tile. This allows a multi-threaded decoder to start a particular decoding thread by skipping ahead to the appropriate tile. There are four tiles per frame, facilitating data parallelization in hardware and software implementations.

Bitstream Features

The VP9 bitstream is usually available within a container format such as WebM, which is a subset of the Matroska Media Container. The container format is needed for random access capabilities, as VP9 does not provide start codes for this purpose. VP9 bitstreams start with a key frame containing all intra-coded blocks, which is also a decoder reset point. Unlike VP8, there is no data partitioning in VP9; all data types are interleaved in superblock coding order. This change is made to facilitate hardware implementations. However, similar to VP8, VP9 also compresses a bitstream using an 8-bit non-adaptive arithmetic coding (a.k.a. *bool-coding*), for which the probability model is fixed and all the symbol probabilities are known a priori before the frame decoding starts. Each probability has a known default value and is stored as a 1 byte data in the frame context. The decoder maintains four such contexts, and the bitstream signals which one to use for the frame decode. Once a frame is decoded, based upon the occurrence of certain symbols in the decoded frame, a context can be updated with new probability distributions for use with future frames, thus providing limited context adaptability.

Each coded frame has three sections:

- **Uncompressed header**: Few bytes containing picture size, loop-filter strength, etc.

- **Compressed header**: Bool-coded header data containing the probabilities for the frame, expressed in terms of differences from default probability values.

- **Compressed frame data**: Bool-coded frame data needed to reconstruct the frame, including partition information, intra modes, motion vectors, and transform coefficients.

In addition to providing high compression efficiency with reasonable complexity, the VP9 bitstream includes features designed to support a variety of specific use-cases involving delivery and consumption of video over the Internet. For example, for communication of conversational video with low latency over an unreliable network, it is imperative to support a coding mode where decoding can continue without corruption even when arbitrary frames are lost. Specifically, the arithmetic decoder should be able to continue decoding of symbols correctly even though frame buffers have been corrupted, leading to encoder-decoder mismatch.

VP9 supports a frame level `error_resilient_mode` flag to allow coding modes where a manageable drift between the encoder and decoder is possible until a key frame is available or an available reference picture is selected to correct the error. In particular, the following restrictions apply under error resilient mode while a modest performance drop is expected:

- At the beginning of each frame, the entropy coding context probabilities are reset to defaults, preventing propagation of forward or backward updates.

- For the motion vector reference selection, the co-located motion vector from a previously encoded reference frame can no longer be included in the reference candidate list.

- For the motion vector reference selection, sorting of the initial list of motion vector reference candidates based on searching the reference frame buffer is disabled.

The VP9 bitstream does not offer any security functions. Integrity and confidentiality must be ensured by functions outside the bistream, although VP9 is independent of external objects and related security vulnerabilities.

Residual Coding

If a block is not a skipped block (indicated at 8×8 granularity), a residual signal is coded and transmitted for it. Similar to HEVC, VP9 also supports different sizes (32×32, 16×16, 8×8, and 4×4) for an integer transform approximated from the DCT. However, depending on specific characteristics of the intra residues, either or both the vertical and the horizontal transform pass can be ADST instead. The transform size is coded in the bitstream such that a 32×16 block using a 8×8 transform would have luma residual made up of a 4×2 grid of 8×8 transform coefficients, and the two 16×8 chroma residuals, each consisting of a 2×1 grid of 8×8 transform coefficients.

Transform coefficients are scanned starting at the upper left corner, following a "curved zig-zag" pattern toward the higher frequencies, while transform blocks with mixed DCT/DST use a scan pattern skewed accordingly.[10] However, the scan pattern is not straightforward and requires a table lookup. Furthermore, each transform coefficient is coded using bool-coding and has several probabilities associated with it, resulting from various parameters such as position in the block, size of the transform, value of neighboring coefficients, and the like.

---

[10]P. Kapsenberg, "How VP9 Works: Technical Details and Diagrams," Doom9's forum, October 8, 2013, retrieved from http://forum.doom9.org/showthread.php?p=1675961.

Inverse quantization is simply a multiplication by one of the four scaling factors for luma and chroma DC and AC coefficients, which remain the same for a frame; block-level QP adjustment is not allowed. Additionally, VP9 offers a lossless mode at frame level using 4×4 Walsh-Hadamard transform.

Intra Prediction

Intra prediction in VP9 is similar to the intra prediction method in AVC and HEVC, and is performed on partitions the same as are the transform block partitions. For example, a 16×8 block with 8×8 transforms will result in two 8×8 luma prediction operations. There are 10 different prediction modes: the DC, the TM (True Motion), vertical, horizontal, and six angular predictions approximately corresponding to the 27, 45, 63, 117, 135, and 153 degree angles. Like other codecs, intra prediction requires two one-dimensional arrays that contain the reconstructed left and upper pixels of the neighboring blocks. For block sizes above 4×4, the second half of the horizontal array contains the same value as the last pixel of the first half. An example is given in Figure 3-19.

105

100	102	100	96	98	101	93	80	80	80	80	80	80	80	80	80

88	87	86	85	86	87	89	91	93
85	85	86	87	89	91	93	94	95
85	87	89	91	93	94	95	95	94
89	91	93	94	95	95	94	93	91
93	94	95	95	94	93	91	90	90
95	95	94	93	91	90	90	90	90
95	93	91	90	90	90	90	90	90
90	90	90	90	90	90	90	90	90

Figure 3-19. *Luma intra samples, mode D27_PRED*

Inter Prediction

Inter prediction in VP9 uses eighth-pixel motion compensation, offering twice the precision of most other standards. For motion compensation, VP9 primarily uses one motion vector per block, but optionally allows a *compound prediction* with two motion vectors per block resulting in two prediction samples that are averaged together. Compound prediction is only enabled in non-displayable frames, which are used as reference frames.[11] VP9 allows these non-displayable frames to be piggy-backed with a displayable frame, together forming a superframe to be used in the container.

---

[11]It is possible to construct a low-cost displayable frame from such references by using 64×64 blocks with no residuals and (0, 0) motion vectors pointing to this reference frame.

VP9 defines a family of three 8-tap filters, selectable at either the frame or block level in the bitstream:

- 8-tap Regular: An 8-tap Lagrangian interpolation filter

- 8-tap Sharp: A DCT-based interpolation filter, used mostly around sharper edges

- 8-tap Smooth (non-interpolating): A smoothing non-interpolating filter, in the sense that the prediction at integer pixel-aligned locations is a smoothed version of the reference frame pixels

A motion vector, points to one of three possible reference frames, known as the *Last*, the *Golden*, and the *AltRef* frames. The reference frame is applied at 8×8 granularity—for example, two 4×8 blocks, each with their own motion vector, will always point to the same reference frame.

In VP9, motion vectors are predicted from a sorted list of candidate reference motion vectors. The candidates are built using up to eight surrounding blocks that share the same reference picture, followed by a temporal predictor of co-located motion vector from the previous frame. If this search process does not fill the list, the surrounding blocks are searched again but this time the reference doesn't have to match. If this list is still not full, then (0, 0) vectors are inferred.

Associated with a block, one of the four motion vector modes is coded:

- NEW_MV: This mode uses the first entry of the prediction list along with a delta motion vector which is transmitted in the bitstream.

- NEAREST_MV: This mode uses the first entry of the prediction list as is.

- NEAR_MV: This mode uses the second entry of the prediction list as is.

- ZERO_MV: This mode uses (0, 0) as the motion vector value.

A VP9 decoder maintains a list of eight reference pictures at all times, of which three are used by a frame for inter prediction. The predicted frame can optionally insert itself into any of these eight slots, evicting the existing frame. VP9 supports reference frame scaling; a new inter frame can be coded with a different resolution than the previous frame, while the reference data is scaled up or down as needed. The scaling filters are 8-tap filters with 16th-pixel accuracy. This feature is useful in variable bandwidth environments, such as video conferencing over the Internet, as it allows for quick and seamless on-the-fly bit-rate adjustment.

Loop Filter

VP9 introduces a variety of new prediction block and transform sizes that require additional loop filtering options to handle a large number of combinations of boundary types. VP9 also incorporates a flatness detector in the loop filter that detects flat regions and varies the filter strength and size accordingly.

The VP9 loop filter is applied to a decoded picture. The loop filter operates on a superblock, smoothing the vertical edges followed by the horizontal edges. The superblocks are processed in raster-scan order, regardless of any tile structure that may be signaled. This is different from the HEVC loop filter, where all vertical edges of the frame are filtered before any horizontal edges. There are four different filters used in VP9 loop filtering: 16-wide, 8-wide, 4-wide, and 2-wide, where on each side of the edge eight, four, two, and one pixels are processed, respectively. Each of the filters is applied according to a threshold sent in the frame header. A filter is attempted with the conditions that the pixels on either side of the edge should be relatively smooth, and there must be distinct brightness difference on either side of the edge. Upon satisfying these conditions, a filter is used to smooth the edge. If the condition is not met, the next smaller filter is attempted. Block sizes 8×8 or 4×4 start with the 8-wide or smaller filter.

Segmentation

In general, the segmentation mechanism in VP9 provides a flexible set of tools that can be used in a targeted manner to improve perceptual quality of certain areas for a given compression ratio. It is an optional VP9 feature that allows a block to specify a segment ID, 0 to 7, to which it belongs. The frame header can convey any of the following features, applicable to all blocks with the same segment ID:

- **AltQ**: Blocks belonging to a segment with the AltQ feature may use a different inverse quantization scale factor than blocks in other segments. This is useful in many rate-control scenarios, especially for non-uniform bit distribution in foreground and background areas.

- **AltLF**: Blocks belonging to a segment with the AltLF feature may use a different smoothing strength for loop filtering. This is useful in application specific targeted smoothing of particular set of blocks.

- **Mode**: Blocks belonging to a segment with an active mode feature are assumed to have the same coding mode. For example, if skip mode is active in a segment, none of the blocks will have residual information, which is useful for static areas of frames.

- **Ref:** Blocks belonging to a segment that have the Ref feature enabled are assumed to point to a particular reference frame (Last, Golden, or AltRef). It is not necessary to adopt the customary transmission of the reference frame information.

- **EOB:** Blocks belonging to a segment with the coefficient end of block (EOB) marker coding feature may use the same EOB marker coding for all blocks belonging to the segment. This eliminates the need to decode EOB markers separately.

- **Transform size:** The block transform size can also be indicated for all blocks in a segment, which may be the same for a segment, but allows different transform sizes to be used in the same frame.

In order to minimize the signaling overhead, the segmentation map is differentially coded across frames. Segmentation is independent of tiling.

Summary

This chapter presented brief overviews of major video coding standards available in the industry. With a view toward guaranteeing interoperability, ease of implementation, and industry-wide common format, these standards specified or preferred certain techniques over others. Owing to the discussions provided earlier in this chapter, these predilections would be easy to understand. Another goal of the video coding standards was to address all aspects of practical video transmission, storage, or broadcast within a single standard. This was accomplished in standards from H.261 to MPEG-2. MPEG-4 and later standards not only carried forward the legacy of success but improved upon the earlier techniques and algorithms.

Although in this chapter we did not attempt to compare the coding efficiencies provided by various standards' algorithms, such studies are available in the literature; for example, those making an interesting comparison between MPEG-2, AVC, WMV-9, and AVS.[12] Over the years such comparisons—in particular, determination of bit-rate savings of a later-generation standard compared to the previous generation, have become popular, as demonstrated by Grois et al.[13] in their comparison of HEVC, VP9, and AVC standards.

---

[12]S. Kwon, A. Tamhankar, and K. Rao, "Overview of H.264/ MPEG-4 Part 10," *Journal of Visual Communication and Image Representation* 17 (April 2006): 186–216.

[13]D. Grois, D. Marpe, A. Mulayoff, B. Itzhaky, and O. Hadar, "Performance Comparison of H.265/ MPEG-HEVC, VP9, and H.264/MPEG-AVC Encoders," *Proceedings of the 30th Picture Coding Symposium.* (San Jose, CA, USA:Institute of Electronics and Electrical Engineers, 2013).

CHAPTER 4

■ ■ ■

Video Quality Metrics

Quality generally indicates excellence, and the universal societal norm is to strive for the highest achievable quality in most fields. However, in case of digital video, a measured, careful approach is taken to allow some deficit in quality that is not always discernible by typical viewers. Such concessions in perceivable visual quality make room for a valuable accomplishment in terms of compression.

Video quality is a characteristic of a video signal passed through a transmission or processing system, representing a measure of perceived degradation with respect to the original source video. Video processing systems usually introduce some distortions or artifacts in the video signal, but the amount involved may differ depending on the complexity of the content and the parameters chosen to process it. The variable degradation may or may not be perceivable or acceptable to an end user. In general, it is difficult to determine what would be an acceptable quality for all end users. However, it remains an important objective of video quality evaluation studies. So understanding various types of visual degradations or artifacts in terms of their *annoyance factors*, and the evaluation of the quality of a video as apparent to the end user, are very important.

In this chapter, we first focus on the careful and intentional information loss due to compression and the resulting artifacts. Then we discuss the various factors involved in the compression and processing of video that influence the compression and that contribute to visual quality degradation.

With these understandings, we move toward measuring video quality and discuss various subjective and objective quality metrics to measure with particular attention to various ITU-T standards. The discussions include attempts to understand relative strengths and weaknesses of important metrics in terms of capturing perceptible deterioration. We further discuss video coding efficiency evaluation metrics and some example standard-based algorithms.

In the final part of this chapter, we discuss the parameters that primarily impact video quality, and which parameters need to be tuned to achieve good tradeoffs beween video quality and compression speed, the knowledge of which is useful in designing some video applications. Although some parameters are dictated by the available system or network resources, depending on the application, the end user may also be allowed to set or tune some of these parameters.

Compression Loss, Artifacts, and Visual Quality

Compression artifacts are noticeable distortions in compressed video, when it is subsequently decompressed and presented to a viewer. Such distortions can be present in compressed signals other than video as well. These distortions are caused by the lossy compression techniques involved. One of the goals of compression algorithms is to minimize the distortion while maximizing the amount of compression. However, depending on the algorithm and the amount of compression, the output has varying levels of diminishing quality or introduction of artifacts. Some quality-assessment algorithms can distinguish between distortions of little subjective importance and those objectionable to the viewer, and can take steps to optimize the final apparent visual quality.

Compression Loss: Quantization Noise

Compression loss is manifested in many different ways and results in some sort of visual impairment. In this section we discuss the most common form of compression loss and its related artifact, namely the quantization noise.

Quantization is the process of mapping a large set of input values to a smaller set—for example, rounding the input values to some unit of precision. A device or an algorithmic function that performs the quantization is called a quantizer. The round-off error introduced by the process is referred to as *quantization error* or the *quantization noise*. In other words, the difference between the input signal and the quantized signal is the quantization error.

There are two major sources of quantization noise in video applications: first, when an analog signal is converted to digital format; and second, when high-frequency components are discarded during a lossy compression of the digital signal. In the following discussion both of these are elaborated.

Quantization of Samples

The digitization process of an image converts the continuous-valued brightness information of each sample at the sensor to a discrete set of integers representing distinct gray levels—that is, the sampled image is quantized to these levels. The entire process of measuring and quantizing the brightnesses is significantly affected by sensor characteristics such as dynamic range and linearity. Real sensors have a limited dynamic range; they only respond to light intensity between some minimum and maximum values. Real sensors are also non-linear, but there may be some regions over which they are more or less linear, with non-linear regions at either end.

The number of various levels of quantizer output is determined by the bits available for quantization at the analog-to-digital converter. A quantizer with n bits represents $N = 2^n$ levels. Typical quantizers use 8-bits, representing 256 gray levels usually numbered between 0 and 255, where 0 corresponds to black and 255 corresponds to white. However, 10-bit or even 16-bit images are increasingly popular. Using more bits brings the ability to perform quantization with a finer step size, resulting in less noise and a closer approximation of the original signal. Figure 4-1 shows an example of a 2-bit or four-level quantized signal, which is a coarse approximation of the input signal, and a 3-bit or eight-level quantized signal, representing a closer approximation of the input signal.

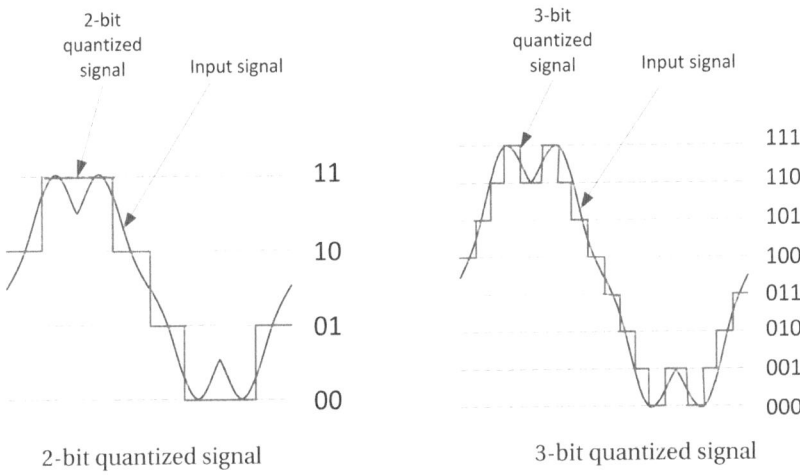

Figure 4-1. *Quantized signals with different bit resolution*

In the case of an image, the difference between the true input brightness of a pixel and the corresponding brightness of the digital level represents the quantization error for that pixel. Quantization errors can take positive or negative values. Note that quantization levels are equally spaced for uniform quantization, but are irregularly spaced for non-uniform (or non-linear) quantization. If the quantization levels are equally spaced with a step size b, the quantization error for a digital image may be approximated as a uniformly distributed signal with zero mean and a variance of $\dfrac{b^2}{12}$.

Such uniform quantizers are typically memoryless—that is, the quantization level for a pixel is computed independently of other pixels.

Frequency Quantization

In frequency quantization, an image or a video frame undergoes a transform, such as the discrete cosine transform, to convert the image into the frequency domain. For an 8×8 pixel block, 64 transform coefficients are produced. However, lossy compression techniques such as those adopted by the standards as described in Chapter 3, perform quantization on these transform coefficients using a same-size quantization matrix, which typically has non-linear scaling factors biased toward attenuating high-frequency components more than low-frequency components. In practice, most high-frequency components become zero after quantization. This helps compression, but the high-frequency components are lost irreversibly. During decompression, the quantized coefficients undergo inverse quantization operation, but the original values cannot be restored. The difference between the original pixel block and the reconstructed pixel block represents the amount of quantization error that was introduced. Figure 4-2 illustrates the concept.

$$\begin{bmatrix} 1260 & -1 & -12 & -5 & 2 & -2 & -3 & 1 \\ -23 & -17 & -6 & -3 & -3 & 0 & 0 & -1 \\ -11 & -9 & -2 & 2 & 0 & -1 & -1 & 0 \\ -7 & -2 & 0 & 1 & 1 & 0 & 0 & 0 \\ -1 & -1 & 1 & 2 & 0 & -1 & 1 & 1 \\ 2 & 0 & 2 & 0 & -1 & 1 & 1 & -1 \\ -1 & 0 & 0 & -1 & 0 & 2 & 1 & -1 \\ -3 & 2 & -4 & -2 & 2 & 1 & -1 & 0 \end{bmatrix}$$

Transform coefficients

$$\begin{bmatrix} 16 & 11 & 10 & 16 & 24 & 40 & 51 & 61 \\ 12 & 12 & 14 & 19 & 26 & 58 & 60 & 55 \\ 14 & 13 & 16 & 24 & 40 & 57 & 69 & 56 \\ 14 & 17 & 22 & 29 & 51 & 87 & 80 & 62 \\ 18 & 22 & 37 & 56 & 68 & 109 & 103 & 77 \\ 24 & 35 & 55 & 64 & 81 & 104 & 113 & 92 \\ 49 & 64 & 78 & 87 & 103 & 121 & 120 & 101 \\ 72 & 92 & 95 & 98 & 112 & 100 & 103 & 99 \end{bmatrix}$$

Quantization matrix

$$\begin{bmatrix} 79 & 0 & -1 & 0 & 0 & 0 & 0 & 0 \\ -2 & -1 & 0 & 0 & 0 & 0 & 0 & 0 \\ -1 & -1 & 0 & 0 & 0 & 0 & 0 & 0 \\ 0 & 0 & 0 & 0 & 0 & 0 & 0 & 0 \\ 0 & 0 & 0 & 0 & 0 & 0 & 0 & 0 \\ 0 & 0 & 0 & 0 & 0 & 0 & 0 & 0 \\ 0 & 0 & 0 & 0 & 0 & 0 & 0 & 0 \\ 0 & 0 & 0 & 0 & 0 & 0 & 0 & 0 \end{bmatrix}$$

Quantized coefficients

$$\begin{bmatrix} 1264 & 0 & -10 & 0 & 0 & 0 & 0 & 0 \\ -24 & -12 & 0 & 0 & 0 & 0 & 0 & 0 \\ -14 & -13 & 0 & 0 & 0 & 0 & 0 & 0 \\ 0 & 0 & 0 & 0 & 0 & 0 & 0 & 0 \\ 0 & 0 & 0 & 0 & 0 & 0 & 0 & 0 \\ 0 & 0 & 0 & 0 & 0 & 0 & 0 & 0 \\ 0 & 0 & 0 & 0 & 0 & 0 & 0 & 0 \\ 0 & 0 & 0 & 0 & 0 & 0 & 0 & 0 \end{bmatrix}$$

Reconstructed coefficients

$$\begin{bmatrix} -4 & -1 & -2 & -5 & 2 & -2 & -3 & 1 \\ 1 & -5 & -6 & -3 & -3 & 0 & 0 & -1 \\ 3 & 4 & -2 & 2 & 0 & -1 & -1 & 0 \\ -7 & -2 & 0 & 1 & 1 & 0 & 0 & 0 \\ -1 & -1 & 1 & 2 & 0 & -1 & 1 & 1 \\ 2 & 0 & 2 & 0 & -1 & 1 & 1 & -1 \\ -1 & 0 & 0 & -1 & 0 & 2 & 1 & -1 \\ -3 & 2 & -4 & -2 & 2 & 1 & -1 & 0 \end{bmatrix}$$

Quantization error

Figure 4-2. *Quantization of a block of transform coefficients*

The quantization matrix is the same size as the block of transform coefficients, which is input to the quantizer. To obtain quantized coefficients, an element-by-element division operation is performed, followed by a rounding to the nearest integer. For example, in Figure 4-2, quantization of the DC coefficient (the upper left element) by doing round(1260/16) gives the quantized coefficient 79. Notice that after quantization, mainly low-frequency coefficients, located toward the upper left-hand corner, are retained, while high-frequency coefficients have become zero and are discarded before transmission. Reconstruction is performed by multiplying the quantized coefficients by the same quantization matrix elements. However, the resultant reconstruction contains the quantization error as shown in Figure 4-2.

Usually quantization of a coefficient in a block depends on how its neighboring coefficients are quantized. In such cases, neighborhood context is usually saved and considered before quantizing the next coefficient. This is an example of a quantizer with memory.

It should be noted that the large number of zeros that appear in the quantized coefficients matrix is not by accident; the quantization matrix is designed in such a way that the high-frequency components–which are not very noticeable to the HVS–are removed from the signal. This allows greater compression of the video signal with little or no perceptual degradation in quality.

Color Quantization

Color quantization is a method to reduce the number of colors in an image. As the HVS is less sensitive to loss in color information, this is an efficient compression technique. Further, color quantization is useful for devices with limited color support. It is common to combine color quantization techniques, such as the nearest color algorithm, with dithering–a technique for randomization of quantization error–to produce an impression of more colors than is actually available and to prevent color banding artifacts where continuous gradation of color tone is replaced by several regions of fewer tones with sudden tone changes.

Common Artifacts

Here are a few common artifacts that are typically found in various image and video compression applications.

Blurring Artifact

Blurring of an image refers to a smoothing of its details and edges, and it results from direct or indirect low-pass filter effects of various processing. Blurring of an object appears as though the object is out of focus. Generally speaking, blurring is an artifact the viewer would like to avoid, as clearer, crisper images are more desirable. But sometimes, blurring is intentionally introduced by using a Gaussian function to reduce image noise or to enhance image structures at different scales. Typically, this is done as a pre-processing step before compression algorithms may be applied, attenuating high-frequency signals and resulting in more efficient compression. This is also useful in edge-detection algorithms, which are sensitive to noisy environments. Figure 4-3 shows an example of blurring.

Image of monochromatic grating with increasing frequency from left to right

Blurred image of monochromatic grating

Figure 4-3. *An example of blurring of a frequency ramp. Low-frequency areas are barely affected by blurring, but the impact is visible in high-freqeuncy regions*

Motion blur appears in the direction of motion corresponding to rapidly moving objects in a still image or a video. It happens when the image being recorded changes position (or the camera moves) during the recording of a single frame, because of either rapid movement of objects or long exposure of slow-moving objects. For example, motion blur is often an artifact in sports content with fast motion. However, in sports contents, motion blur is not always desirable; it can be inconvenient because it may obscure the exact position of a projectile or athlete in slow motion. One way to avoid motion blur is by panning the camera to track the moving objects, so the object remains sharp but the background is blurred instead. Graphics, image, or video editing tools may also generate the motion blur effect for artistic reasons; the most frequent synthetic motion blur is found when computer-generated imagery (CGI) is added to a scene in order to match existing real-world blur or to convey a sense of speed for the objects in motion. Figure 4-4 shows an example of motion blur.

Figure 4-4. *An example of motion blur*

Deinterlacing by the display and telecine processing by studios can soften images, and/or introduce motion-speed irregularities. Also, compression artifacts present in digital video streams can contribute additional blur during fast motion. Motion blur has been a more severe problem for LCD displays, owing to their sample-and-hold nature, where a continuous signal is sampled and the sample values are held for a certain time to eliminate input signal variations. In these displays, the impact of motion blur can be reduced by controlling the backlight.

Block Boundary Artifact

Block-based lossy coding schemes, including all major video and image coding standards, introduce visible artifacts at the boundaries of pixel blocks at low bit rates. In block-based transform coding, a pixel block is transformed to frequency domain using discrete cosine transform or similar transforms, and a quantization process discards the high-frequency coefficients. The lower the bit rate, the more coarsely the block is quantized, producing blurry, low-resolution versions of the block. In the extreme case, only the DC coefficient, representing the average of the data, is left for a block, so that the reconstructed block is only a single color region.

The *block boundary artifact* is the result of independently quantizing the blocks of transform coefficients. Neighboring blocks quantize the coefficients separately, leading to discontinuities in the reconstructed block boundaries. These block-boundary discontinuities are usually visible, especially in the flat color regions such as the sky, faces, and so on, where there are little details to mask the discontinuity. Compression algorithms usually perform deblocking operations to smooth out the reconstructed block boundaries, particularly to use a reference frame that is free from this artifact. Figure 4-5 shows an example of block boundary artifact.

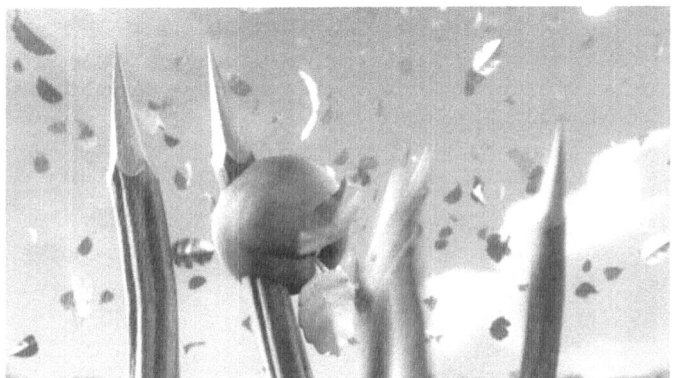

Original video frame

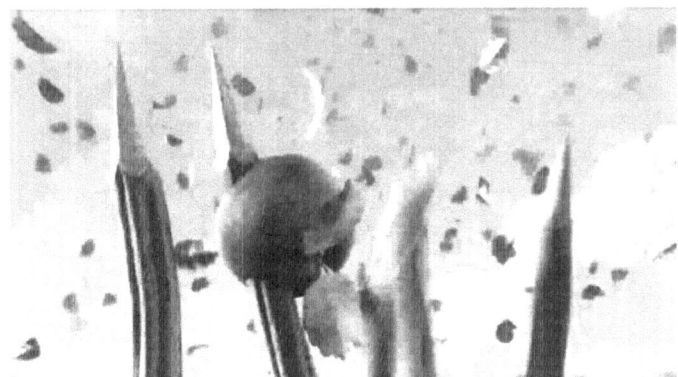

Reconstructed video frame with visible block boundaries

Figure 4-5. *An example of block boundary artifact*

This artifact is so common that many names are popularly used for it. Although the discontinuities may or may not align with the boundaries of macroblocks as defined in the video and image coding standards, *macroblocking* is a common term for this artifact. Other names include *tiling, mosaicing, quilting,* and *checkerboarding.*

Ringing Artifact

Ringing is unwanted oscillation of an output signal in response to a sudden change in the input. Image and video signals in digital data compression and processing are band limited. When they undergo frequency domain techniques such as Fourier or wavelet transforms, or non-monotone filters such as deconvolution, a spurious and visible *ghosting* or *echo* effect is produced near the sharp transitions or object contours. This is due to the well-known Gibb's phenomenon—an oscillating behavior of the filter's impulse response near discontinuities, in which the output takes higher value (*overshoots*) or lower value (*undershoots*) than the corresponding input values, with decreasing magnitude until a steady-state is reached. The output signal oscillates at a fading rate, similar to a bell ringing after being struck, inspiring the name of the *ringing artifact*. Figure 4-6 depicts the oscillating behavior of an example output response showing the Gibb's phenomenon. It also depicts an example of ringing artifact in an image.

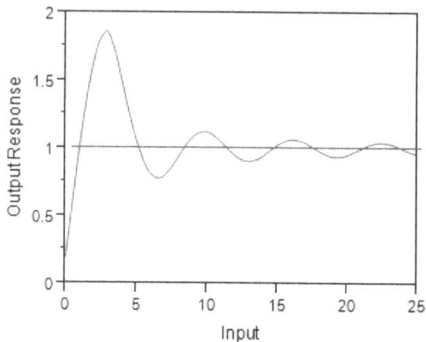

Oscillating output in Gibb's phenomenon

Original image

Image with ringing artifact

Figure 4-6. *An example of ringing artifact*

Aliasing Artifacts

Let us consider a time-varying signal $x(t)$ and its sampled version $x(n) = x(nT)$, with sampling period $T>0$. When $x(n)$ is downsampled by a factor of 2, every other sample is discarded. In the frequency (ω) domain, the Fourier transform of the signal $X(e^{j\omega})$ is stretched by the same factor of 2. In doing so, the transformed signal can in general overlap with its shifted replicas. In case of such overlap, the original signal cannot be unambiguously recovered from its downsampled version, as the overlapped region represents two copies of the transformed signal at the same time. One of these copies is an *alias,* or replica of the other. This overlapping effect is called *aliasing.* Figure 4-7 shows the transform domain effect of downsampling, including aliasing.

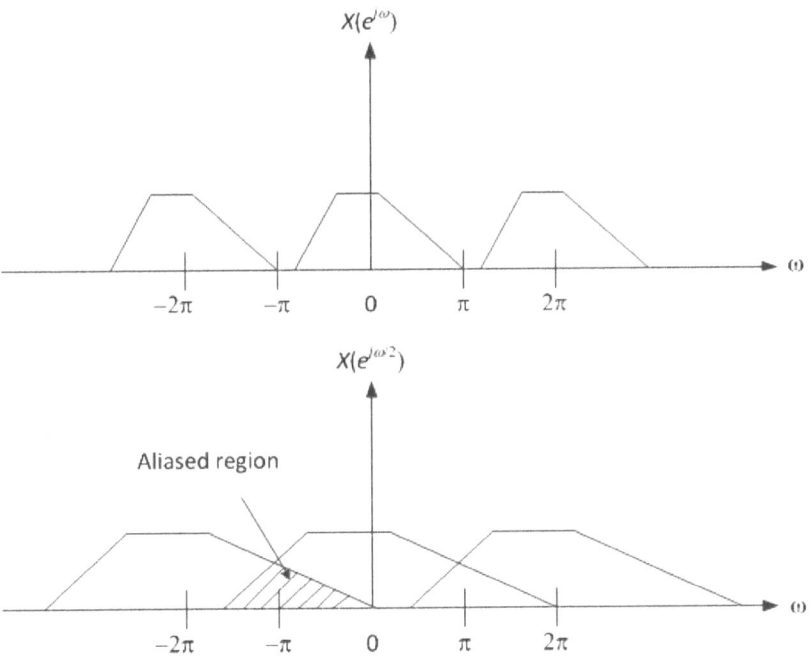

Figure 4-7. *Transform domain effect of downsampling, causing aliasing*

In general, aliasing refers to the artifact or distortion resulting from ambiguous reconstruction of signals from its samples. Aliasing can occur in signals sampled in time—for instance, digital audio—and is referred to as *temporal aliasing.* Aliasing can also occur in spatially sampled signals—for instance, digital images or videos—where it is referred to as *spatial aliasing.*

Aliasing always occurs when actual signals with finite duration are sampled. This is because the frequency content of these functions has no upper bound, causing their Fourier transform representation to always overlap with other transformed functions. On the other hand, functions with bounded frequency content (*bandlimited*) have infinite duration. If sampled at a high rate above the so-called *Nyquist rate*, the original signal can be completely recovered from the samples. From Figure 4-7, it is clear that aliasing can be avoided if the original signal is bandlimited to the region $|\omega| < \dfrac{\pi}{M}$, where M is the downsampling factor. In this case, the original signal can be recovered from the downsampled version using an upsampler, followed by filtering.

Jaggies

Popularly known as *jaggies*, this common form of aliasing artifact produces visible stairlike lines where there should be smooth straight lines or curves in a digital image. These stairs or steps are a consequence of the regular, square layout of a pixel. With increasing image resolution, this artifact becomes less visible. Also, anti-aliasing filters are useful in reducing the visibility of the aliased edges, while sharpening increases such visibility.

Figure 4-8 shows examples of aliasing artifacts such as jaggies and moiré patterns.

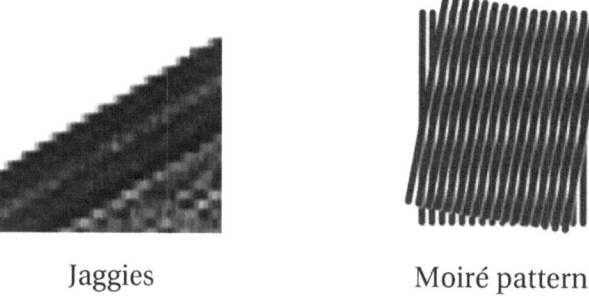

Jaggies   Moiré pattern

Figure 4-8. *Examples of aliasing artifacts*

Moiré Pattern

Due to undersampling of a fine regular pattern, a special case of aliasing occurs in the form of *moiré patterns*. It is an undesired artifact of images produced by various digital imaging and computer graphics techniques—for example, ray tracing a checkered surface. The moiré effect is the visual perception of a distinctly different third pattern, which is caused by inexact superposition of two similar patterns. In Figure 4-8, moiré effect can be seen as an undulating pattern, while the original pattern comprises a closely spaced grid of straight lines.

Flickering Artifacts

Flicker is perceivable interruption in brightness for a sufficiently long time (e.g., around 100 milliseconds) during display of a video. It is a flashing effect that is displeasing to the eye. Flicker occurs on old displays such as cathode ray tubes (CRT) when they are driven at a low refresh rate. Since the shutters used in liquid crystal displays (LCD) for each pixel stay at a steady opacity, they do not flicker, even when the image is refreshed.

Jerkiness

A flicker-like artifact, *jerkiness* (also known as *choppiness*), describes the perception of individual still images in a motion picture. It may be noted that the frequency at which flicker and jerkiness are perceived is dependent upon many conditions, including ambient lighting conditions. Jerkiness is not discernible for normal playback of video at typical frame rates of 24 frames per second or above. However, in visual communication systems, if a video frame is dropped by the decoder owing to its late arrival, or if the decoding is unsuccessful owing to network errors, the previous frame would continue to be displayed. Upon successful decoding of the next error-free frame, the scene on the display would suddenly be updated. This would cause a visible jerkiness artifact.

Telecine Judder

Another flicker-like artifact is the *telecine judder*. In order to convert the 24 fps film content to 30 fps video, a process called telecine, or 2:3 pulldown, is commonly applied. The process converts every four frames of films to five frames of interlaced video. Some DVD or Blu-ray players, line doublers, or video recorders can detect telecine and apply a reverse telecine process to reconstruct the original 24 fps video content. Figure 4-9 shows the telecine process.

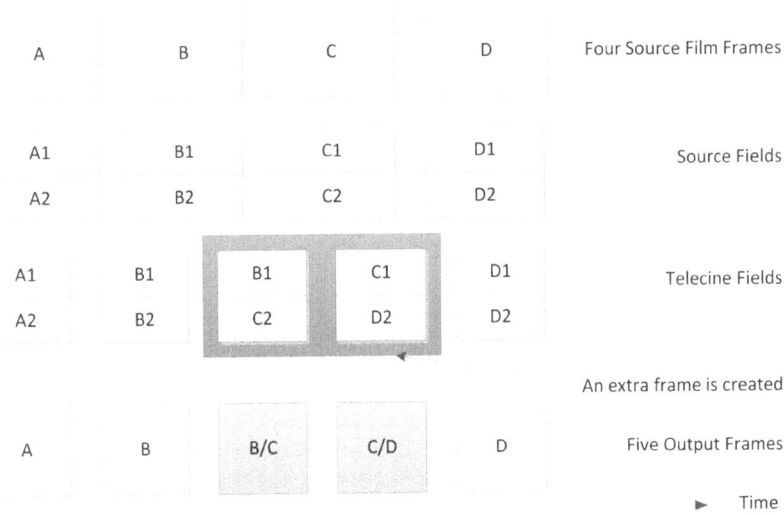

Figure 4-9. *The telecine process*

Notice that by the telecine process two new frames B/C and C/D are created, that were not part of the original set of source frames. Thus, the telecine process creates a slight error in the video signal compared to the original film frames. This used to create the problem for films viewed on NTSC television that they would not appear as smooth as when viewed in a cinema. This problem was particularly visible during slow, steady camera movements that would appear slightly jerky when telecined.

Other Image Artifacts

There are several other artifacts commonly observed in compressed video. Some of these are discussed below. Figure 4-10 shows examples of various image artifacts.

Corruption due to   Sensor noise   Hot pixel noise
transmission error

Figure 4-10. *Examples of various image artifacts*

Corruption due to Transmission Error

Owing to transmission errors in the compressed bitstream ,visible corruption may be observed in the reconstructed signal. Transmission errors can also disrupt the bitstream parsing, leading to partially decoded pictures or decoded pictures with missing blocks. In case of gross errors, decoders may continue to apply updates to the damaged picture for a short time, creating a *ghost image* effect, until the next error-free independently compressed frame is available. Ghosting is a common artifact in open-air television signals.

Image Noise

The camera sensor for each pixel contains one or more light-sensitive photodiodes that convert the incoming light into electrical signals, which is processed into the color value of the pixel in an image. However, this process is not always perfectly repeatable, and there are some statistical variations. Besides, even without incident light, the electrical activity of the sensors may generate some signal. These unwanted signals and variations are the sources of *image noise*. Such noise varies per pixel and over time, and increases with the temperature. Image noise can also originate from film grain.

Noise in digital images is most visible in uniform surfaces, such as in skies and shadows as monochromatic grain, and/or as colored waves (color noise). Another type of noise, commonly called *hot pixel* noise, occurs with long exposures lasting more than a second and appears as colored dots slightly larger than a single pixel. In modern cameras, however, hot pixel noise is increasingly rare.

Factors Affecting Visual Quality

Visual artifacts resulting from loss of information due to processing of digital video signals usually degrade the perceived visual quality. In addition to the visual artifacts described above, the following are important contributing factors that affect visual quality.

- **Sensor noise and pre-filtering:** Sensor noise, as mentioned above, is an undesirable by-product of image capture that affects visual quality. Not only is the noise itself visually disturbing, but its presence also impacts subsequent processing, causing or aggravating further artifacts. For example, pre-filtering is typically done after an image is captured but before encoding. In the pre-filtering stage, aliasing or ringing artifacts can occur; these artifacts would be visible even if lossless encoding is performed.

- **Characteristics of video:** Visual quality is affected by digital video characteristics including bit depth, resolution, frame rate, and frame complexity. Typical video frames use 8 bits for each pixel component, while premium quality videos allocate 10 to 16 bits. Similarly, high-definition video frames are four to six times as large as standard-definition video frames, depending on the format. Ultra-high-definition videos exhibit the highest quality owing to their 24 to 27 times higher pixel resolutions than their standard-definition counterparts.

  Frame rate is another important factor; although the HVS can perceive slow motion at 10 frames per second (fps) and smooth motion at 24 fps, higher frame rates imply smoother motion, especially for fast-moving objects. For example, a moving ball may be blurry at 30 fps, but would be clearer at 120 fps. Very fast motion is more demanding–wing movements of a hummingbird would be blurry at 30 fps, or even at 120 fps; for clear view of such fast motion, 1000 fps may be necessary. Higher frame rates are also used to produce special slow-motion effects. One measure of the complexity of a frame is the amount of details or *spatial business* of the frame. Artifacts in frames with low complexity and low details are generally more noticeable than frames with higher complexity.

  The spatial information (detail) and temporal information (motion) of the video are critical parameters. These play a crucial role in determining the amount of video compression that is possible and, consequently, the level of impairment that is suffered when the scene is transmitted over a fixed-rate digital transmission service channel.

- **Amount of compression:** For compressed digital video the amount of compression matters because compression is usually achieved by trading off visual quality. Highly compressed video has lower visual quality than lightly compressed video. Compression artifacts are noticeable and can be annoying for low-bit-rate video. Also, based on available bits, different amounts of quantization may have been done per block and per frame. The impact of such differences can be visible depending on the frame complexity. Furthermore, although compression techniques such as chroma subsampling take advantage of the HVS characteristics, premium contents with 4:4:4 chroma format have better visual quality compared to 4:2:0 contents.

- **Methods of compression:** Lossless compression retains all the information present in the video signal, so it does not introduce quality degradation. On the other hand, lossy compression aims to control the loss of quality by performing a careful tradeoff between visual quality and amount of compression. In lossy compression, selection of modes also influences the quality. In error-prone environments such as wireless networks, intra modes serve as a recovery point from errors at the expense of using more bits.

- **Multiple passes of processing:** In off-line video applications where real-time processing is not necessary, the video signal may undergo multiple passes. Analyzing the statistics of the first pass, parameters can be tuned for subsequent passes. Such techniques usually produce higher quality in the final resulting signals. However, artifacts due to various processing may still contribute to some quality loss.

- **Multiple generations of compression:** Some video applications may employ multiple generations of compression, where a compressed video signal is decompressed before compressing again with possibly different parameters. This may result in quality degradation owing to the use of different quantization maps for each generation. Typically, after the second generation visual quality deteriorates dramatically. To avoid such quality loss, robust design of quantization parameters is necessary.

- **Post-production:** Post-production effects and scene cuts can cause different portions of the encoded video sequence to have different quality levels.

Video Quality Evaluation Methods and Metrics

Video quality is evaluated for specification of system requirements, comparison of competing service offerings, transmission planning, network maintenance, client-based quality measurement, and so on. Several methods have been proposed in the literature to address the quality evaluation problem for various usages. With many methods and algorithms available, the industry's need for accurate and reliable objective video metrics has generally been addressed by the ITU in several recommendations, each aiming toward particular industries such as standard- and high-definition broadcast TV.

The standardization efforts are being extended with the progress of modern usages like mobile broadcasting, Internet streaming video, IPTV, and the like. Standards address a variety of issues, including definitions and terms of reference, requirements, recommended practices, and test plans. In this section, we focus on the definitions, methods, and metrics for quality-evaluation algorithms. In particular, Quality of Experience (QoE) of video is addressed from the point of view of overall user experience—that is, the viewer's perception—as opposed to the well-known Quality of Service (QoS) measure usually employed in data transmission and network performance evaluation.

There are two approaches to interpreting video quality:

- The first approach is straightforward; the actual visual quality of the image or video content is determined based on *subjective* evaluation done by humans.

- In the second approach is synonymous with the signal fidelity or similarity with respect to a *reference* or *perfect* image in some perceptual space. There are sophisticated models to capture the statistics of the natural video signals; based on these models, *objective* signal fidelity criteria are developed that relate video quality with the amount of information shared between a reference and a distorted video signal.

In the following discussion, both subjective and objective video quality metrics are presented in detail.

Subjective Video Quality Evaluation

Video processing systems perform various tasks, including video signal acquisition, compression, restoration, enhancement, and reproduction. In each of these tasks, aiming for the best video quality under the constraints of the available system resources, the system designers typically make various tradeoffs based on some quality criteria.

An obvious way of measuring quality is to solicit the opinion of human observers or *subjects*. Therefore, the subjective evaluation method of video quality utilizes human subjects to perform the task of assessing visual quality. However, it is impossible to subjectively assess the quality of every video that an application may deal with. Besides, owing to inherent variability in quality judgment among human observers, multiple subjects are usually required for meaningful subjective studies. Furthermore, video quality is affected by viewing conditions such as ambient illumination, display device, and viewing distance. Therefore, subjective studies must be conducted in a carefully controlled environment.

Although the real perceptual video quality can be tracked by using this technique, the process is cumbersome, not automatable, and the results may vary depending on the viewer, as the same visual object is perceived differently by different individuals. Nevertheless, it remains a valuable method in providing ground-truth data that can be used as a reference for the evaluation of automatic or objective video quality-evaluation algorithms.

Objective algorithms estimate a viewer's perception, and the performance of an algorithm is evaluated against subjective test results. Media degradations impact the viewers' perception of the quality. Consequently, it is necessary to design subjective tests that can accurately capture the impact of these degradations on a viewer's perception. These subjective tests require performing comprehensive experiments that produce consistent results. The following aspects of subjective testing are required for accurate evaluation of an objective quality algorithm:

- Viewers should be naïve and non-expert, representing normal users whose perception is estimated by the objective quality models. These viewers vote on the subjective quality as instructed by the test designer. However, for specific applications, such as new codec developments, experienced voters are more suitable.

- The number of voters per sample should meet the subjective testing requirements as described in the appropriate ITU-T Recommendations. Typically a minimum of 24 voters is recommended.

- To maintain consistency and repeatability of experiments, and to align the quality range and distortion types, it is recommended that the experiments contain an anchor pool of samples that best represent the particular application under evaluation. However, it should be noted that even when anchor samples are used, a bias toward different experiments is common, simply because it is not always possible to include all distortion types in the anchor conditions.

Study group 9 (SG9) of ITU-T developed several recommendations, of which the Recommendation BT. 500-13[1] and the P-series recommendations are devoted to subjective and objective quality-assessment methods. These recommendations suggest standard viewing conditions, criteria for the selection of observers and test material, assessment procedures, and data analysis methods. Recommendations P.910[2] through P.913[3] deal with subjective video quality assessment for multimedia applications. Early versions, such as Rec. P.910 and P.911,[4] were designed around the paradigm of a fixed video service for multimedia applications. This paradigm considers video transmission over a reliable link to an immobile *cathode ray tube* (CRT) television located in a quiet and nondistracting environment, such as a living room or office. To accommodate new applications, such as Internet video and distribution quality video, P.913 was introduced.

---

[1] *ITU-R Recommendation BT.500-13: Methodology for the Subjective Assessment of the Quality of Television Pictures* (Geneva, Switzerland: International Telecommunications Union, 2012).

[2] *ITU-T Recommendation P.910: Subjective Video Quality Assessment Methods for Multimedia Applications* (Geneva, Switzerland: International Telecommunications Union, 2008).

[3] *ITU-T Recommendation P.913: Methods for the Subjective Assessment of Video Quality, Audio Quality and Audiovisual Quality of Internet Video and Distribution Quality Television in Any Environment* (Geneva, Switzerland: International Telecommunications Union, 2014).

[4] *ITU-T Recommendation P.911: Subjective Audiovisual Quality Assessment Methods for Multimedia Applications* (Geneva, Switzerland: International Telecommunications Union, 1998).

Ratified in January 2014, Recommendation P.913 describes non-interactive subjective assessment methods for evaluating the one-way overall video quality, audio quality, and/or audio-visual quality. It aims to cover a new paradigm of video—for example, an on-demand video service, transmitted over an unreliable link to a variety of mobile and immobile devices located in a distracting environment, using LCDs and other flat-screen displays. This new paradigm impacts key characteristics of the subjective test, such as the viewing environment, the listening environment, and the questions to be answered. Subjective quality assessment in the new paradigm asks questions that are not considered in the previous recommendations. However, this recommendation does not address the specialized needs of broadcasters and contribution quality television.

The duration, the number and type of test scenes, and the number of subjects are critical for the interpretation of the results of the subjective assessment. P.913 recommends stimuli ranging from 5 seconds to 20 seconds in duration, while 8- to 110-second sequences are highly recommended. Four to six scenes are considered sufficient when the variety of content is respected. P.913 mandates that at least 24 subjects must rate each stimulus in a controlled environment, while at least 35 subjects must be used in a public environment. Fewer subjects may be used for pilot studies to indicate trending.

Subjective Quality Evaluation Methods and Metrics

The ITU-T P-series recommendations define some of the most commonly used methods for subjective quality assessment. Some examples are presented in this section.

Absolute Category Rating

In the *absolute category rating* (ACR) method, the quality judgment is classified into several categories. The test stimuli are presented one at a time and are rated independently on a category scale. ACR is a single-stimulus method, where a viewer watches one stimulus (e.g., video clip) and then rates it. ACR methods are influenced by the subject's opinion of the content—for example, if the subject does not like the production of the content, he may give it a poor rating. The ACR method uses the following five-level rating scale:

5	Excellent
4	Good
3	Fair
2	Poor
1	Bad

A variant of the ACR method is *ACR with hidden reference* (ACR-HR). With ACR-HR, the experiment includes a reference version of each video segment, not as part of a pair but as a freestanding stimulus for rating. During the data analysis the ACR scores are subtracted from the corresponding reference scores to obtain a *differential viewer*

(DV) score. This procedure is known as *hidden reference removal*. The ACR-HR method removes some of the influence of content from the ACR ratings, but to a lesser extent than double-stimulus methods, which are discussed below.

Degradation Category Rating

Also known as the double-stimulus impair scale (DSIS) method, the *degradation category rating* (DCR) presents a pair of stimuli together. The reference stimulus is presented first, followed by a version after it has undergone processing and quality degradation. In this case, the subjects are asked to rate the impairment of the second stimulus with respect to the reference. DCR is minimally influenced by the subject's opinion of the content. Thus, DCR is able to detect color impairments and skipping errors that the ACR method may miss. However, DCR may have a slight bias, as the reference is always shown first. In DCR, the following five-level scale for rating the relative impairment is used:

5	Imperceptible
4	Perceptible but not annoying
3	Slightly annoying
2	Annoying
1	Very annoying

Comparison Category Rating

The *comparison category rating* (CCR) is a double-stimulus method whereby two versions of the same stimulus are presented in a randomized order. For example, half of the time the reference is shown first, and half the time it is shown second, but in random order. CCR is also known as the double-stimulus comparison scale (DSCS) method. It may be used to compare reference video with processed video, or to compare two different impairments. CCR, like DCR, is minimally influenced by the subject's opinion of the content. However, occasionally subjects may inadvertently swap their rating in CCR, which would lead to a type of error that is not present in DCR or ACR. In CCR, the following seven-level scale is used for rating.

-3	Much worse
-2	Worse
-1	Slightly worse
0	The same
1	Slightly better
2	Better
3	Much better

SAMVIQ

Subjective assessment of multimedia video quality (SAMVIQ) is a non-interactive subjective assessment method used for video-only or audio-visual quality evaluation, spanning a large number of resolutions from SQCIF to HDTV. The SAMVIQ methodology uses a continuous quality scale. Each subject moves a slider on a continuous scale graded from zero to 100. This continuous scale is annotated by five quality items linearly arranged: excellent, good, fair, poor, and bad.

MOS

The *mean opinion score* (MOS) is the most common metric used in subjective video quality evaluation. It forms the basis of the subjective quality-evaluation methods, and it serves as a reference for the objective metrics as well. Historically, this metric has been used for decades in telephony networks to obtain the human user's view of the quality of the network. It has also been used as a subjective audio-quality measure. After all the subjects are run through an experiment, the ratings for each clip are averaged to compute either a MOS or a *differential mean opinion score* (DMOS).

The MOS provides a numerical indication of the perceived quality from the user's point of view of the received media after it has undergone compression and/or transmission. The MOS is generally expressed as a single number in the range from 1 to 5, where 1 is the lowest perceived quality, and 5 is the highest perceived quality. MOS is used for single-stimulus methods such as ACR or ACR-HR (using raw ACR scores), where the subject rates a stimulus in isolation. In contrast, the DMOS scores measure a change in quality between two versions of the same stimulus (e.g., the source video and a processed version of the video). ACR-HR (in case of average differential viewer score), DCR, and CCR methods usually produce DMOS scores.

Comparing the MOS values of different experiments requires careful consideration of intra- and inter-experimental variations. Normally, only the MOS values from the same test type can be compared. For instance, the MOS values from a subjective test that use an ACR scale cannot be directly compared to the MOS values from a DCR experiment. Further, even when MOS values from the same test types are compared, the fact that each experiment is slightly different even for the same participants leads to the following limitations:

- A score assigned by subject is rarely always the same, even when an experiment is repeated with the same samples in the same representation order. Usually this is considered as a type of noise on the MOS scores.

- There is a short-term context dependency as subjects are influenced by the short-term history of the samples they have previously scored. For example, following one or two poor samples, subjects tend to score a mediocre sample higher. If a mediocre sample follows very good samples, there is a tendency to score the mediocre sample lower. To average out this dependency, the presentation order should be varied for the individual subjects. However, this strategy does not remove the statistical uncertainty.

- The mid-term contexts associated with the average quality, the quality distribution, and the occurance of distortions significantly contribute to the variations between subjective experiments. For example, if an experiment is composed of primarily low-quality samples, people tend to score them higher, and vice versa. This is because people tend to use the full quality scale offered in an experiment and adapt the scale to the qualities presented in the experiment. Furthermore, individual distortions for less frequent samples are scored lower compared to experiments where samples are presented more often and people become more familiar with them.

- The long-term dependencies reflect the subject's cultural interpretation of the category labels, the cultural attitude to quality, and language dependencies. For example, some people may have more frequent experiences with video contents than others. Also, the expectations regarding quality may change over time. As people become familiar with digital video artifacts, it becomes part of their daily experience.

Although these effects cause differences between individual experiments, they cannot be avoided. However, their impacts can be minimized by providing informative instructions, well-balanced test designs, a sufficient number of participants, and a mixed presentation order.

Objective Video Quality Evaluation Methods and Metrics

Video quality assessment (VQA) studies aim to design algorithms that can automatically evaluate the quality of videos in a manner perceptually consistent with the subjective human evaluation. This approach tracks an *objective* video-quality metric, which is automatable, and the results are verifiable by repeated execution, as they do not require human field trial. However, these algorithms merely attempt to predict human subjective experience and are not perfect; they will fail for certain unpredictable content. Thus, the objective quality evaluation cannot replace subjective quality evaluation; they only aid as a tool in the quality assessment. The ITU-T P.1401[5] presents a framework for the statistical evaluation of objective quality algorithms regardless of the assessed media type.

In P.1401, the recommended statistical metrics for objective quality assessment need to cover three main aspects—accuracy, consistency, and linearity—against subjective data. It is recommended that the prediction error be used for accuracy, the outlier ratio or

---

[5]*ITU-T Recommendation P.1401: Methods, Metrics and Procedures for Statistical Evaluation, Qualification and Composition of Objective Quality Prediction Models* (Geneva, Switzerland: International Telecommunications Union, 2012).

the residual error distribution for consistency, and the Pearson correlation coefficient for linearity. The root mean square of the prediction error is given by:

$$RMSE \text{ of } P_{error} = \sqrt{\frac{1}{N-1}\sum_{i=1}^{N}(MOS(i) - MOS_{predicted}(i))^2} \qquad \text{(Equation 4-1)}$$

where N is the number of samples and N-1 ensures an unbiased estimator for the RMSE.

The distribution of the residual error (MOS–$MOS_{predicted}$) is usually characterized by a binomial distribution. The probability of exhibiting residual error below a pre-established threshold (usually 95% confidence interval) has a mean $P_{th} = (N_{th}/N)$, where N_{th} is number of samples for which residual error remains below the threshold, and a standard deviation

$$\sigma_{th} = \sqrt{\frac{P_{th}(1-P_{th})}{N}} \; .$$

An objective video or image quality metric can play a variety of roles in video applications. Notable among these are the following:

- An objective metric can be used to dynamically monitor and adjust the quality. For example, a network digital video server can appropriately allocate, control, and trade off the streaming resources based on the video quality assessment on the fly.

- It can be used to optimize algorithms and parameter settings of video processing systems. For instance, in a video encoder, a quality metric can facilitate the optimal design of pre-filtering and bit-rate control algorithms. In a video decoder, it can help optimize the reconstruction, error concealment, and post-filtering algorithms.

- It can be used to benchmark video processing systems and algorithms.

- It can be used to compare two video systems solutions.

Classification of Objective Video Quality Metrics

One way to classify the objective video quality evaluation methods is to put them into three categories based on the amount of reference information they require: *full reference* (FR), *reduced reference* (RR), and *no reference* (NR). These methods are discussed below. The FR methods can be further categorized as follows:

- Error sensitivity based approaches

- Structural similarity based approaches

- Information fidelity based aproaches

- Spatio-temporal approaches

- Saliency based approaches

- Network aware approaches

These approaches are discussed in the following subsections. Further, an example metric for each approach is elaborated.

Full Reference

A digital video signal undergoes several processing steps during which video quality may have been traded off in favor of compression, speed, or other criteria, resulting in a distorted signal that is available to the viewer. In objective quality assessment, the fidelity of the distorted signal is typically measured. To determine exactly how much degradation has occurred, such measurements are made with respect to a reference signal that is assumed to have *perfect* quality. However, the reference signal may not be always available.

Full-reference[6] (FR) metrics measure the visual quality degradation in a distorted video with respect to a reference video. They require the entire reference video to be available, usually in unimpaired and uncompressed form, and generally impose precise spatial and temporal alignment, as well as calibration of luminance and color between the two videos. This allows every pixel in every frame of one video to be directly compared with its counterpart in the other video.

Typically, the fidelity is determined by measuring the *distance* between the reference and the distorted signals in a perceptually meaningful way. The FR quality evaluation methods attempt to achieve consistency in quality prediction by modeling the significant physiological and psychovisual features of the HVS and using this model to evaluate signal fidelity. As fidelity increases, perceived quality of the content also increases. Although FR metrics are very effective in analysis of video quality, and are very widely used for analysis and benchmarking, the FR metrics' requirement that the reference be accessible during quality evaluation at the reconstruction end may not be fulfilled in practice. Thus, their usefulness may become limited in such cases.

Reduced Reference

It is possible to design models and evaluation criteria when a reference signal is not fully available. Research efforts in this area generated the various *reduced-reference*[7] (RR) methods that use partial reference information. They extract a number of features from the reference and/or the distorted test video. These features form the basis of the comparison between the two videos, so that the full reference is not necessary. This approach thus avoids the assumptions that must be made in the absence of any reference information, while keeping the amount of reference information manageable.

---

[6]*ITU-T Recommendation J.247: Objective Perceptual Multimedia Video Quality Measurement in the Presence of a Full Reference* (Geneva, Switzerland: International Telecommunications Union, 2009).

[7]*ITU-T Recommendation J.246: Perceptual Visual Quality Measurement Techniques for Multimedia Services over Digital Cable Television Networks in the Presence of a Reduced Bandwidth Reference* (Geneva, Switzerland: International Telecommunications Union, 2008).

No Reference

No-reference (NR) metrics analyze only the distorted test video without depending on an explicit reference video. As a result, NR metrics are not susceptible to alignment issues. The main challenge in NR approaches, however, is to distinguish between the distortions and the actual video signal. Therefore, NR metrics have to make assumptions about the video content and the types of distortions.

Figure 4-11 shows typical block diagrams of the FR and the NR approaches.

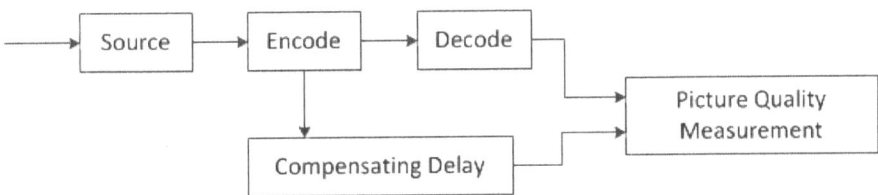

Full reference picture quality measurement

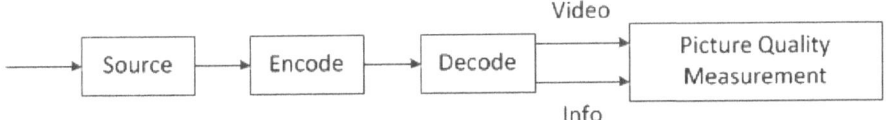

No-reference picture quality measurement

Figure 4-11. *Reference-based classification examples: FR and NR approaches*

An interesting NR quality measurement algorithm is presented in Wang et al.[8] This algorithm considers blurring and blocking as the most significant artifacts generated during the JPEG compression process, and proposes to extract features that can be used to reflect the relative magnitudes of these artifacts. The extracted features are combined to generate a quality prediction model that is trained using subjective experimental results. It is expected that the model would be a good fit for images outside the experimental set as well. While such algorithms are interesting in that an assessment can be made solely on the basis of the available image content without using any reference, it is always better to use an FR approach when the reference is available, so the following discussion will focus on the various FR approaches.

The problem of designing an objective metric that closely agrees with perceived visual quality under all conditions is a hard one. Many available metrics may not account for all types of distortion corrupting an image, or the content of the image, or the strength of the distortion, yet provide the same close agreement with human judgments. As such

[8]Z. Wang, H. R. Sheikh, and A. C. Bovik, "No-Reference Perceptual Quality Assessment of JPEG Compressed Images," *Proceedings of International Conference on Image Processing* 1, (2002): 477-80.

this remains an active research area. Several FR approaches have been taken in the quest for finding a good solution to this problem. Some of these approaches are presented in the following sections.

Error Sensitivity Based Approaches

When an image or video frame goes through lossy processing, a distorted image or video frame is produced. The amount of error or the distortion that is introduced by the lossy process determines the amount of visual quality degradation. Many quality-evaluation metrics are based on the error or intensity difference between the distorted image and the reference image pixels. The simplest and most widely used full-reference quality metric is the *mean squared error* (MSE), along with the related quantity of *peak signal-to-noise ratio* (PSNR). These are appealing because they are simple to calculate, have clear physical meanings, and are mathematically convenient in the context of optimization. However, they are not very well matched to perceived visual quality.

In error sensitivity based image or video quality assessment, it is generally assumed that the loss of perceptual quality is directly related to the visibility of the error signal. The simplest implementation of this concept is the MSE, which objectively quantifies the strength of the error signal. But two distorted images with the same MSE may have very different types of errors, some of which are much more visible than others. In the last four decades, a number of quality-assessment methods have been developed that exploit known characteristics of the human visual system (HVS). The majority of these models have followed a strategy of modifying the MSE measure so that different aspects of the error signal are weighted in accordance with their visibility. These models are based on a general framework, as discussed below.

General Framework

Figure 4-12 depicts a general framework of error sensitivity based approaches of image or video quality assessment. Although the details may differ, most error sensitivity based perceptual quality assessment models can be described with a similar block diagram.

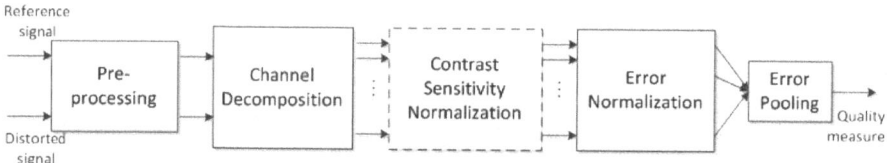

Figure 4-12. *General framework of error sensitivity based approaches*

In Figure 4-12, the general framework includes the following stages:

- **Pre-processing:** In this stage, known malformations are eliminated and the images are prepared so as to perform a fair comparison between the distorted image and the reference image. For example, both images are properly scaled and aligned. If necessary, a color space conversion or gamma correction may be performed that is more appropriate for the HVS. Further, a low-pass filter simulating the point spread function of the eye optics may be applied. Additionally, both images may be modified using a non-linear point operation to simulate light adaptation.

- **Channel Decomposition:** The images are typically separated into subbands or channels that are sensitive to particular spatial and temporal frequency, as well as orientation. Some complex methods try to closely simulate neural responses in primary visual cortex, while others simply use DCT or wavelet transforms for channel decomposition.

- **Contrast Sensitivity Normalization:** The *contrast sensitivity function* (CSF) describes the sensitivity of the HVS to different spatial and temporal frequencies that are present in the visual stimulus. The frequency response of the CSF is typically implemented as a linear filter. Some older methods weigh the signal according to CSF in a stage before channel decomposition, but recent methods use CSF as a base-sensitivity normalization factor for each channel.

- **Error Normalization:** The presence of one image component may decrease or mask the visibility of another, nearby image component, which is in close proximity in terms of spatial or temporal location, spatial frequency, or orientation. Such masking effect is taken into account when the error signal in each channel is calculated and normalized. The normalization process weighs the error signal in a channel by a space-varying visibility threshold. For each channel, the visibility threshold is determined based on the channel's base-sensitivity, as well as the energy of the reference or distorted image coefficients in a spatial neighborhood, either within the same channel or across channels. The normalization process expresses the error in terms of *just noticeable difference* (JND) units. Some methods also consider the effect of saturation of the contrast response.

- **Error Pooling:** In this final stage, the normalized error signals are combined to a single value. To obtain the combined value, typically the Minkowski norm is calculated as follows:

$$E(\{e_{i,j}\}) = \left(\sum_i \sum_j |e_{i,j}|^\beta \right)^{\frac{1}{\beta}}$$
(Equation 4-2)

where $e_{i,j}$ is the normalized error of the jth spatial coefficient in the ith frequency channel, and β is a constant with typical values between 1 and 4.

Limitations

Although error sensitivity based approaches estimate the visibility of the error signal by simulating the functional properties of the HVS, most of these models are based on linear or quasilinear operators that have been characterized using restricted and simplistic stimuli. In practice, however, the HVS is a complex and highly non-linear system. Therefore, error sensitivity based approaches adopt some assumptions and generalizations leading to the following limitations:

- **Quality definition:** As error sensitivity based image or video quality assessment methods only track the image fidelity, lower fidelity does not always mean lower visual quality. The assumption that visibility of error signal translates to quality degradation may not always be valid. Some distortions are visible, but are not so objectionable. Brightening an entire image by globally increasing the luma value is one such example. Therefore, image fidelity only moderately correlates with image quality.

- **Generalization of models:** Many error-sensitivity models are based on experiments that estimate the threshold at which a stimulus is barely visible. These thresholds are used to define error-sensitivity measures such as the contrast sensitivity function. However, in typical image or video processing, perceptual distortion happens at a level much higher than the threshold. Generalization of near-threshold models in suprathreshold psychophysics is thus susceptible to inaccuracy.

- **Signal characteristics:** Most psychophysical experiments are conducted using relatively simple patterns, such as spots, bars, or sinusoidal gratings. For example, the CSF is typically obtained from threshold experiments using global sinusoidal images. However, real-world natural images have much different characteristics from the simple patterns. Therefore, the applicability of the simplistic models may be limited in practice.

- **Dependencies:** It is easy to challenge the assumption used in error pooling that error signals in different channels and spatial locations are independent. For linear channel decomposition methods such as the wavelet transform, a strong dependency exists between intra- and inter-channel wavelet coefficients of natural images. Optimal design of transformation and masking models can reduce both statistical and perceptual dependencies. However, the impact of such design on VQA models is yet to be determined.

- **Cognitive interaction:** It is well known that interactive visual processing such as eye movements influences the perceived quality. Also, cognitive understanding has a significant impact on quality. For example, with different instructions, a human subject may give different scores to the same image. Prior knowledge of or bias toward an image, attention, fixation, and so on may also affect the evaluation of the image quality. However, most error sensitivity based image or video quality assessment methods do not consider the cognitive interactions as they are not well understood and are difficult to quantify.

Peak Signal-to-Noise Ratio

The term *peak signal-to-noise ratio* (PSNR) is an expression for the ratio between the maximum possible power of a signal and the power of distorting noise that affects the quality of its representation after compression, processing, or transmission. Because many signals have a very wide *dynamic range* (ratio between the largest and smallest possible values of a changeable quantity), the PSNR is usually expressed in terms of the logarithmic decibel (dB) scale. The PSNR does not always perfectly correlate with a perceived visual quality, owing to the non-linear behavior of the HVS, but as long as the video content and the codec type are not changed, it is a valid quality measure,[9] as it is a good indicator of the fidelity of a video signal in a lossy environment.

Let us consider a signal f that goes through some processing or transmission and is reconstructed as an approximation $\hat{f}$, where some noise is introduced. Let f_m is the peak or maximum signal value; for n-bit representation of the signal $f_m = 2^n - 1$. For example, in

---

[9]Q. Huynh-Thu, and M. Ghanbari, "Scope of Validity of PSNR in Image/Video Quality Assessment," *Electronic Letters* 44, no. 13 (2008): 800–801.

case of an 8-bit signal $f_m = 255$, while for a 10-bit signal, $f_m = 1023$. PSNR, as a ratio of signal power to the noise power, is defined as follows:

$$PSNR = 10 \log_{10} \frac{(f_m)^2}{MSE} \qquad \text{(Equation 4-3)}$$

where the mean square error (MSE) is given by:

$$MSE = \frac{1}{N} \sum_{i=1}^{N} (f_i - \hat{f}_i)^2 \qquad \text{(Equation 4-4)}$$

where N is the number of samples over which the signal is approximated. Similarly, the MSE for a two-dimensional signal such as image or a video frame with width M and height N is given by:

$$MSE = \frac{1}{M \times N} \sum_{i=1}^{M} \sum_{j=1}^{N} (f(i,j) - \hat{f}(i,j))^2 \qquad \text{(Equation 4-5)}$$

where $f(i, j)$ is the pixel value at location (i, j) of the source image, and $\hat{f}(i,j)$ is the corresponding pixel value in the reconstructed image. PSNR is usually measured for an image plane, such as the luma or chroma plane of a video frame.

Applications

PSNR has traditionally been used in analog audio-visual systems as a consistent quality metric. Digital video technology has exposed some limitations in using the PSNR as a quality metric. Nevertheless, owing to its low complexity and easy measurability, PSNR is still the most widely used video quality metric for evaluating lossy video compression or processing algorithms, particularly as a measure of gain in quality for a specified target bit rate for the compressed video. PSNR is also used in detecting the existence of frame drops or severe frame data corruption and the location of dropped or corrupt frames in a video sequence in automated environments. Such detections are very useful in debugging and optimization of video encoding or processing solutions. Furthermore, PSNR is extensively used as a comparison method between two video coding solutions in terms of video quality.

Advantages

PSNR has the following advantages as a video quality metric.

- PSNR is a simple and easy to calculate picture-based metric. PSNR calculation is also fast and parallelization-friendly—for example, using single instruction multiple data (SIMD) paradigm.

- Since PSNR is based on MSE, it is independent of the direction of the difference signal; either the source or the reconstructed signal can be subtracted from one another yielding the same PSNR output.

- PSNR is easy to incorporate into practical automated quality- measurement systems. This flexibility makes it amenable to a large test suite. Thus, it is very useful in building confidence on the evaluation.

- The PSNR calculation is repeatable; for the same source and reconstructed signals, the same output can always be obtained. Furthermore, PSNR does not depend on the width or height of the video, and works for any resolution.

- Unlike cumbersome subjective tests, PSNR does not require special setup for the environment.

- PSNR is considered to be a reference benchmark for developing various other objective video-quality metrics.

- For the same video source and the same codec, PSNR is a consistent quality indictor, so it can be used for encoder optimization to maximize the subjective video quality and/or the performance of the encoder.

- PSNR can be used separatelyfor luma and chroma channels. Thus, variation in brightness or color between two coding solutions can be easily tracked. In order to determine which solution uses more bits for a given quality level, such information is very useful.

- The popularity of PSNR is not only rooted in its simplicity but also its performance as a metric should not be underestimated. A validation study conducted by the Video Quality Experts Group (VQEG) in 2001 discovered that the nine VQA methods that it tested, including some of the most sophisticated algorithms at that time, were "statistically indistinguishable" from the PSNR.[10]

Limitations

Common criticisms for PSNR include the following.

- Some studies have shown that PSNR poorly correlates with subjective quality.[11]

- PSNR does not consider the visibility differences of two different images, but only considers the numerical differences. It does not take the visual masking phenomenon or the characteristics of the HVS into account–all pixels that are different in two images contribute to the PSNR, regardless of the visibility of the difference.

---

[10]*Final Report from the Video Quality Experts Group on the Validation of Objective Models of Video Quality Assessment,* 2000, available at www.vqeg.org.

[11]B. Girod, "What's Wrong with Mean Squared Error?" in *Visual Factors of Electronic Image Communications,* ed. A. B. Watson, (Cambridge, MA: MIT Press, 1993): 207–20.

- Other objective perceptual quality metrics have been shown to outperform PSNR in predicting subjective video quality in specific cases.[12]

- Computational complexity of the encoder in terms of execution time or machine cycles is not considered in PSNR. Nor does it consider system properties such as data cache size, memory access bandwidth, storage complexity, instruction cache size, parallelism, and pipelining, as all of these parameters contribute to coding complexity of an encoder. Therefore, the comparison of two encoders is quite restricted when PSNR is used as the main criteria.

- PSNR alone does not provide sufficient information regarding coding efficiency of an encoder; a corresponding cost measure is also required, typically in terms of the number of bits used. In other words, saying that a certain video has a certain level of PSNR does not make sense unless the file size or the bit rate for the video is also known.

- PSNR is typically averaged over a frame, and local statistics within the frame are not considered. Also, for a video sequence, the quality may vary considerably from scene to scene, which may not be accurately captured if frame-based PSNR results are aggregated and an averge PSNR is used for the video sequence.

- PSNR does not capture temporal quality issues such as frame delay or frame drops. Additionally, PSNR is only a source coding measure and does not consider channel coding issues such as multi-path propagation or fading. Therefore, it is not a suitable quality measure in lossy network environment.

- PSNR is an FR measure, so reference is required for quality evaluation of a video. However, in practice, an unadulterated reference is not generally available at the reconstruction end. Nevertheless, PSNR remains effective and popular for evaluation, analysis, and benchmarking of video quality.

Improvements on PSNR

Several attempts have been made in literature to improve PSNR. Note that visibility of a given distortion depends on the local content of the source picture. Distortions are usually more objectionable in plain areas and on edges than in *busy* areas. Thus, it is possible to model the visual effect of the distortion itself in a more sophisticated way than

---

[12]*ITU-T Recommendation J.144: Objective Perceptual Video Quality Measurement Techniques for Digital Cable Television in the Presence of a Full Feference* (Geneva, Switzerland: International Telecommunications Union, 2004).

simply measuring its energy, as done in PSNR. For example, a weighting function may be applied in frequency domain, giving more weight to the lower-frequency components of the error than to the higher-frequency components. A new measure, named *just noticeable difference* (JND), has been defined by Sarnoff in 2003, based on a visual discrimination mode.[13]

Moving Picture Quality Metric

PSNR does not take the visual masking phenomenon into consideration. In other words, every single pixel error contributes to the decrease of the PSNR, even if this error is not perceptible. This issue is typically addressed by incorporating some HVS models. In particular, two key aspects of the HVS, namely contrast sensitivity and masking, have been intensively studied in the literature. The first phenomenon accounts for the fact that a signal is detected by the eye only if its contrast is greater than some threshold. The sensitivity of the eye varies as a function of spatial frequency, orientation, and temporal frequency. The second phenomenon is related to the human vision response to a combination of several signals. For example, consider a stimulus consisting of the foreground and the background signals. The detection threshold of the foreground is modified as a function of the contrast from the background.

The *moving picture quality metric* (MPQM)[14] is an error-sensitivity based spatio-temporal objective quality metric for moving pictures that incorporates the two HVS characteristics mentioned above. Following the general framework shown in Figure 4-12, MPQM first decomposes an original video and a distorted version of it into perceptual channels. A channel-based distortion measure is then computed, accounting for contrast sensitivity and masking. After obtaining the distortion data for each channel, the data is combined over all the channels to compute the quality rating. The resulting quality rating is then scaled from 1 to 5 (from bad to excellent). MPQM is known to give good correlation with subjective tests for some videos, but it also yields bad results for others.[15] This is consistent with other error-sensitivity based approaches.

The original MPQM algorithm does not take chroma into consideration, so a variant of the algorithm called the *color MPQM* (CMPQM) has been introduced. In this technique, first the color components are converted to RGB values that are linear with luminance. Then the RGB values are converted to coordinate values corresponding to a luma and two chroma channels. This is followed by the analysis of each component of the original and error sequence by a filter bank. As the HVS is less sensitive to chroma, only nine spatial and one temporal filter is used for these signals. The rest of the steps are similar to those in MPQM.

---

[13]J. Lubin, "A Visual Discrimination Mode for Image System Design and Evaluation," in *Visual Models for Target Detection and Recognition*, ed. E. Peli, (Singapore: World Scientific Publishers, 1995): 207–20.

[14]C. J. Branden-Lambrecht, and O. Verscheure, "Perceptual Quality Measure using a Spatio-Temporal Model of the Human Visual System," in *Proceedings of the SPIE* 2668 (San Jose, CA: SPIE-IS&T, 1996): 450–61.

[15]See http://www.irisa.fr/armor/lesmembres/Mohamed/Thesis.pdf.

Structural Similarity Based Approaches

Natural image signals are highly structured. There are strong dependencies among the pixels of natural images, especially when they are spatially adjacent. These dependencies carry important information about the structure of the objects in a visual scene. The Minkowski error metric used in error sensitivity based approaches does not consider the underlying structure of the signal. Also, decomposition of image signal using linear transforms, as done by most quality measures based on error sensitivity, do not remove the strong dependencies. Structural similarity based quality assessment approaches try to find a more direct way to compare the structures of the reference and the distorted signals. Based on the HVS characteristic that human vision reacts quickly to structural information in the viewing field, these approaches approximate the perceived image distortion using a measure of structural information change. The *Universal Image Quality Index* (UIQI)[16] and the *Structural Similarity Index* (SSIM)[17] are two examples of this category. For a deeper understanding, the SSIM is discussed below in detail.

Structural Similarity Index

Objective methods for assessing perceptual image quality attempt to measure the visible differences between a distorted image and a reference image using a variety of known properties of the HVS. Under the assumption that human visual perception is highly adapted for extracting structural information from a scene, a quality assessment method was introduced based on the degradation of structural information.

The structural information in an image is defined as those attributes that represent the structure of objects in a scene, independent of the luminance and contrast. Since luminance and contrast can vary across a scene, structural similarity index (SSIM) analysis only considers the local luminance and contrast. As these three components are relatively independent, a change in luminance or contrast of an image would not affect the structure of the image.

The system block diagram of the structural similarity index based quality assessment system is shown in Figure 4-13.

---

[16]Z. Wang, and A. C. Bovik, "A Universal Image Quality Index," *IEEE Signal Processing Letters* 9, no. 3 (March 2002): 81–84.
[17]Z. Wang, A. C. Bovik, H. R. Sheikh, and E. P. Simoncelli, "Image Quality Assessment: From Error Visibility to Structural Similarity," *IEEE Transactions on Image Processing* 13, no. 4 (April 2004): 600–12.

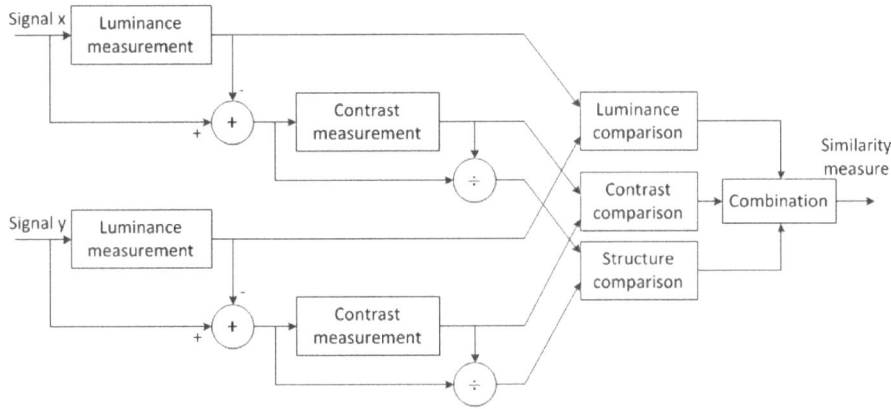

Figure 4-13. *Block diagram of the SSIM measurement system*

As shown in Figure 4-13, the system consists of two nonnegative spatially aligned image signals x and y. If one of the signals has perfect quality, then the similarity measure can serve as a quantitative measurement of the quality of the second signal. The system separates the task of similarity measurement into three comparisons: luminance, contrast, and structure.

Luminance is estimated as the mean intensity (μ) of the image. The luminance comparison function $l(x,y)$ is a function of the mean intensities μ_x and μ_y of images x and y, and can be obtained by comparing their mean intensities. Contrast is estimated as the standard deviation (σ) of an image. The contrast comparison function $c(x,y)$ then reduces to a comparison of σ_x and σ_y. In order to perform the structure comparison, the image is first normalized by dividing the singal by its own standard deviation, so that both images have unit standard deviation. The structure comparison $s(x,y)$ is then done on these normalized signals $(x-\mu_x)/\sigma_x$ and $(y-\mu_y)/\sigma_y$. Combining the results of these three comparisons yields an overall similarity measure::

$$S(x,y) = f(l(x,y), c(x,y), s(x,y)). \qquad \text{(Equation 4-6)}$$

The similarity measure is designed to satisfy the following conditions:

- Symmetry: $S(x,y) = S(y,x)$

- Boundedness: $S(x,y) \leq 1$

- Unity maximum: $S(x,y) = 1$ if and only if $x = y$ (in discrete representations, $x_i = y_i, \forall i = 1, \cdots, N$.)

The luminance comparison function is defined as):

$$l(x,y) = \frac{2\mu_x\mu_y + C_1}{\mu_x^2 + \mu_y^2 + C_1} \qquad \text{(Equation 4-7)}$$

Here, the constant C_1 is introduced to avoid instability when $\mu_x^2 + \mu_y^2$ is close to zero. Specifically, C_1 is chosen as: $C_1 = (K_1 L)^2$, where L is the dynamic range of the pixel values (e.g., 255 for 8-bit grayscale pixels), and the constant $K_1 \ll 1$ is a small constant. Qualitatively, Equation 4-7 is consistent with Weber's law, which is widely used to model light adaptation or luminance masking in the HVS. In simple terms, Weber's law states that the HVS is sensitive to the relative luminance change, and not to the absolute luminance change. If R represents the relative luminance change compared to the background luminance, the distorted signal mean intensity can be substituted by $\mu_y = (1+R)\mu_x$ and Equation 4-7 can be rewritten as ():

$$l(x,y) = \frac{2(1+R)}{1 + (1+R)^2 + \dfrac{C_1}{\mu_x^2}} \qquad \text{(Equation 4-8)}$$

For small values of C_1 with respect to μ_x^2, $l(x,y) = f(R)$, which is consistent with Weber's law.

The contrast comparison function takes a similar form ():

$$c(x,y) = \frac{2\sigma_x \sigma_y + C_2}{\sigma_x^2 + \sigma_y^2 + C_2} \qquad \text{(Equation 4-9)}$$

where $C_2 = (K_2 L)^2$, and $K_2 \ll 1$. Note that with the same amount of contrast change $\Delta\sigma = \sigma_y - \sigma_x$, this measure is less sensitive to a high-base contrast than a low-base contrast. This is consistent with the contrast-masking feature of the HVS.

Structure comparison is conducted after luminance subtraction and variance normalization. Specifically, the two unit vectors $(x - \mu_x)/\sigma_x$ and $(y - \mu_y)/\sigma_y$, are associated with the structure of the two images. The correlation between these two vectors can simply and effectively quantify the structural similarity. Notice that the correlation between $(x - \mu_x)/\sigma_x$ and $(y - \mu_y)/\sigma_y$ is equivalent to the correlation coefficient between x and y. Thus, the structure comparison function is defined as follows:

$$s(x,y) = \frac{\sigma_{xy} + C_3}{\sigma_x \sigma_y + C_3} \qquad \text{(Equation 4-10)}$$

As in the luminance and contrast measures, a small constant C_3 is introduced for stability. In discrete form, σ_{xy} can be estimated as :

$$\sigma_{xy} = \frac{1}{N-1} \sum_{i=1}^{N} (x_i - \mu_x)(y_i - \mu_y) \qquad \text{(Equation 4-11)}$$

The three comparisons of Equations 4-8, 4-9, and 4-10 are combined to yield the resulting similarity measure SSIM between signals x and y:

$$SSIM(x,y) = [l(x,y)]^\alpha [c(x,y)]^\beta [s(x,y)]^\gamma \qquad \text{(Equation 4-12)}$$

where $\alpha > 0$, $\beta > 0$, and $\gamma > 0$ are parameters used to adjust the relative importance of the three components.

The expression is typically used in a simplified form, with $\alpha = \beta = \gamma = 1$ and $C_3 = C_2/2$:

$$SSIM(\boldsymbol{x}, \boldsymbol{y}) = \frac{(2\mu_x \mu_y + C_1)(2\sigma_{xy} + C_2)}{(\mu_x^2 + \mu_y^2 + C_1)(\sigma_x^2 + \sigma_y^2 + C_2)} \qquad \text{(Equation 4-13)}$$

The UIQI[6] is a special case of SSIM with $C_1 = C_2 = 0$. However, it produces unstable results when either $(\mu_x^2 + \mu_y^2)$ or $(\sigma_x^2 + \sigma_y^2)$ is very close to zero.

Information Fidelity Based Approaches

Images and videos generally involve natural scenes, which are characterized using statistical models. Most real-world distortion processes disturb these statistics and make the image or video signals unnatural. This observation led researchers to use *natural scene statistics* (NSS) models in conjunction with a distortion (channel) model to quantify the information shared between a distorted and a reference image, and to show that this shared information is an aspect of signal fidelity that relates well with visual quality. Although in contrast to the HVS error-sensitivity and the structural approaches, the statistical approach, as used in an information-theoretic setting, does not rely on any HVS parameter, or constants requiring optimization, it still yields an FR QA method that is competitive with state-of-the-art QA methods. The visual information fidelity (VIF) is such an information-fidelity based video quality assessment metric.

Visual Information Fidelity

Visual Information Fidelity[18] (VIF) is an information theoretic criterion for image fidelity measurement based on NSS. The VIF measure quantifies the information that could ideally be extracted by the brain from the reference image. Then, the loss of this information to the distortion is quantified using NSS, HVS and an image distortion (channel) model in an information-theoretic framework. It was found that visual quality of images is strongly related to relative image information present in the distorted image, and that this approach outperforms state-of-the-art quality-assessment algorithms. Further, VIF is characterized by only one HVS parameter that is easy to train and optimize for improved performance.

VIF utilizes NSS models for FR quality assessment, and models natural images in the wavelet domain using the well-known Gaussian Scale Mixtures (GSM). Wavelet analysis of images is useful for natural image modeling. The GSM model has been shown to capture key statistical features of natural images, such as linear dependencies in natural images.

Natural images of perfect quality can be modeled as the output of a stochastic source. In the absence of any distortions, this signal passes through the HVS before entering the brain, which extracts cognitive information from it. For distorted images, it is assumed that the reference signal has passed through another *distortion channel* before entering the HVS.

---

[18]H. R. Sheikh and A. C. Bovik, "Image Information and Visual Quality," *IEEE Transactions on Image Processing* 15, no. 2 (2006): 430–44.

The distortion model captures important, and complementary, distortion types: blur, additive noise, and global or local contrast changes. It assumes that in terms of their *perceptual annoyance*, real-world distortions could roughly be approximated locally as a combination of blur and additive noise. A good distortion model is one where the distorted image and the synthesized image look equally perceptually annoying, and the goal of the distortion model is not to model image artifacts but the perceptual annoyance of the artifacts. Thus, even though the distortion model may not be able to capture distortions such as ringing or blocking exactly, it may still be able to capture their perceptual annoyance. However, for distortions other than blur and white noise— for example, for low-bit-rate compression noise—the model fails to adequately reproduce the perceptual annoyance.

The HVS model is also described in the wavelet domain. Since HVS models are duals of NSS models, many aspects of HVS are already captured in the NSS description, including wavelet channel decomposition, response exponent, and masking effect modeling. In VIF, the HVS is considered a distortion channel that limits the amount of information flowing through it. All sources of HVS uncertainty are lumped into one additive white Gaussian stationary noise called the *visual noise*.

The VIF defines mutual informations $I(\vec{C}^N; \vec{E}^N \mid s^N)$ and $I(\vec{C}^N; \vec{F}^N \mid s^N)$ to be the information that could ideally be extracted by the brain from a particular subband in the reference and the distorted images, respectively. Intuitively, visual quality should relate to the amount of image information that the brain could extract from the distorted image relative to the amount of information that the brain could extract from the reference image. For example, if the brain can extract 2.0 bits per pixel of information from the distorted image when it can extract 2.1 bits per pixel from the reference image, then most of the information has been retrieved and the corresponding visual quality should be very good. By contrast, if the brain can extract 5.0 bits per pixel from the reference image, then 3.0 bits per pixel information has been lost and the corresponding visual quality should be very poor.

The VIF is given by:

$$VIF = \frac{\sum_{j \in subbands} I(\vec{C}^{N,j}; \vec{F}^{N,j} \mid s^{N,j})}{\sum_{j \in subbands} I(\vec{C}^{N,j}; \vec{E}^{N,j} \mid s^{N,j})}$$

(Equation 4-14)

where the sum is performed over the subbands of interest, and $\vec{C}^{N,j}$ represent N elements of the random field Cj that describes the coefficients from subband j, and so on.

The VIF has many interesting properties. For example, VIF is bounded below by zero, indicating all information is lost in the distortion channel. If a test image is just a copy of itself, it is not distorted at all, so the VIF is unity. Thus, VIF is always in the range [0,1]. Interestingly, a linear contrast enhancement of the reference image that does not add noise to it will result in a VIF value larger than unity, thereby signifying that the enhanced image has a superior visual quality to the reference image. This is a unique property not exhibited by other VQA metrics.

Spatio-Temporal Approaches

Traditional FR objective quality metrics do not correlate well with temporal distortions such as frame drops or jitter. Spatio-temporal approaches are more suitable for video signals as they consider the motion information between video frames, thereby capturing temporal quality degradation as well. As a result, these algorithms generally correlate well with the HVS. As an example of this approach, the spatio-temporal video SSIM (stVSSIM) is described.

Spatio-Temporal Video SSIM

Spatio-temporal video SSIM[19] (stVSSIM) algorithm is a full-reference VQA algorithm based on the *motion-based video integrity evaluation*[20] (MOVIE) algorithm. MOVIE utilizes a multi-scale spatio-temporal Gabor filter bank to decompose the videos and to compute motion vectors. However, MOVIE has high computational complexity, making practical implementations difficult. So, stVSSIM proposes a new spatio-temporal metric to address the complexity issue. The stVSSIM algorithm was evaluated on VQEG's full-reference data set (Phase I for 525 and 625 line TV signals) and was shown to perform extremely well in terms of correlation with human perception.

For spatial quality assessment, stVSSIM uses the single-scale structural similarity index (SS-SSIM) as it correlates well with human perception of visual quality. For temporal quality assessment, stVSSIM extends the SS-SSIM to the spatio-temporal domain and calls it SSIM-3D. Motion information is incorporated in the stVSSIM using a block-based motion estimation algorithm, as opposed to optical flow, as used in MOVIE. Further, a method to completely avoid block motion estimation is introduced, thereby reducing computational complexity.

For spatial quality assessment, SS-SSIM is computed on a frame-by-frame basis. The spatial-quality measure is applied on each frame and the frame-quality measure is computed using the percentile approach. As humans tend to rate images with low-quality regions with greater severity, using a percentile approach would enhance algorithm performance. So, *Percentile-SSIM* or *P-SSIM* is applied on the scores obtained for each frame. Specifically, the frame-quality measure is:

$$S_{frame} = \frac{1}{|\varphi|} \sum_{i \in \varphi} SSIM(i) \qquad \text{(Equation 4-15)}$$

where the set of the lowest 6 percent of SSIM values from the frame and SSIM(i) the SS-SSIM score is at pixel location i.

The spatial score for the video is computed as the mean of the frame-level scores and is denoted as S_{video}.

---

[19]A. K. Moorthy and A. C. Bovik, "Efficient Motion Weighted Spatio-Temporal Video SSIM Index," in *Proceedings of SPIE-IS&T Electronic Imaging* 7527 (San Jose, CA: SPIE-IS&T, 2010): 1–9.
[20]K. Seshadrinathan and A. C. Bovik, "Motion-based Perceptual Quality Assessment of Video," in *Proceedings of the SPIE* 7240 (San Jose, CA: SPIE-IS&T, 2009): 1–12.

Temporal quality evaluation utilizes three-dimensional structural similarity (SSIM-3D) for a section of the video and performs a weighting on the resulting scores using motion information derived from motion vectors. In this case, a video is viewed as a three-dimensional signal. If x and y are the reference and the distorted video, a volume section is defined around a pixel location (i,j,k) with spatial dimensions (α, β) while the volume temporally encompasses γ frames. Here, (i,j) correspond to the spatial location and k corresponds to the frame number. The SSIM-3D is then expressed as a 3-D extension of the SSIM as follows:

$$SSIM_{3D} = \frac{(2\mu_{x(i,j,k)}\mu_{y(i,j,k)} + C_1)(2\sigma_{x(i,j,k)y(i,j,k)} + C_2)}{(\mu^2_{x(i,j,k)} + \mu^2_{y(i,j,k)} + C_1)(\sigma^2_{x(i,j,k)} + \sigma^2_{y(i,j,k)} + C_2)} \quad \text{(Equation 4-16)}$$

To compute the 3-D mean μ, the variance σ^2, and the co-variance σ_{xy}, the sections x and y are weighted with a weighting factor w for each dimension (i,j,k). The essence of stVSSIM is evaluating spatio-temporal quality along various orientations at a pixel, followed by a weighting scheme that assigns a spatio-temporal quality index to that pixel. The weighting factor depends on the type of filter being used—one out of the four proposed spatio-temporal filters (vertical, horizontal, left, and right).

To incorporate motion information, block motion estimation is used, where motion vectors are computed between neighboring frames using the Adaptive Rood Pattern Search (ARPS) algorithm operating on 8×8 blocks. Once motion vectors for each pixel (i,j,k) are available, spatio-temporal SSIM-3D scores are weighted. To avoid weighting that uses floating point numbers, a greedy weighting is performed. In particular, the spatio-temporal score at pixel (i,j,k) is selected from the scores produced by the four filters based on the type of filter that is closest to the direction of motion at pixel (i,j,k). For example, if the motion vector at a pixel were $(u,v) = (0,2)$, the spatio-temporal score of that pixel would be the SSIM-3D value produced by the vertical filter. If the motion vector is equidistant from two of the filter planes, the spatio-temporal score is the mean of the SSIM-3D scores of the two filters. In case of zero motion, the spatio-temporal score is the mean of all four SSIM-3D values.

The temporal score for the video is computed as the mean of the frame-level scores and is denoted as T_{video}. The final score for the video is given by $S_{video} \times T_{video}$.

Saliency Based Approaches

Quality-assessment methods suitable for single images are also typically used for video. However, these methods do not consider the motion information of the video sequence. As a result, they turn out to be poor evaluation metrics for video quality. In addition, most VQA algorithms ignore the human visual attention mechanism, which is an important HVS characteristic.

Human eyes usually focus on edges with high-contrast or *salient* areas that are different from their neighboring areas. Recognizing this fact, saliency based approaches of video quality evaluation treat the distortion occuring in the salient areas asymmetrically compared to that occuring in other areas. One such approach is SVQA.[21]

---

[21]Q. Ma, L. Zhang, and B. Wang, "New Strategy for Image and Video Quality Assessment," *Journal of Electronic Imaging* 19, no. 1 (2010): 1–14.

Saliency-based Video Quality Assessment

In a saliency based video quality-assessment (SVQA) approach, a spatial saliency map is extracted from reference images or video frames using a fast frequency domain method called the *phase spectrum of quaternion Fourier transform* (PQFT). When the inverse Fourier transform is taken of an image phase spectrum, salient areas are easily recognizable. The saliency map is used as weights to adjust other objective VQA criteria, such as PSNR, MSSIM, VIF, and so on. Similarly, temporal weights are determined from adjacent frames.

Given a reference image and a corresponding distorted image, the saliency map of the reference image can be obtained using the PQFT. Then, an improved quality assessment index, called the saliency-based index (S-index), is determined by weighting the original index by the saliency map. For example, if p_i is the luma value of the ith pixel in the salient area, the pixel saliency weight w_i is given by the following:

$$w_i = \frac{p_i + b}{\frac{1}{M \times N} \sum_{i=1}^{M \times N} (p_i + b)}$$ (Equation 4-17)

where b is a small constant to keep $w_i > 0$, and M and N are the width and height of the image, respectively. Therefore, this weighting takes into account the non-salient areas as well. However, pixels in the salient area have large weights. Using these weights, the *saliency-based PSNR* (SPSNR) is written as:

$$SPSNR(x,y) = 10\log\frac{255^2}{SMSE(x,y)},$$

$$SMSE(x,y) = \frac{1}{M \times N} \sum_{i=1}^{M \times N} (x_i - y_i)^2 w_i.$$ (Equation 4-18)

Thus, the distortion of pixels in the salient area is given more importance than pixels in other areas. *Saliency-based MSSIM* (SMSSIM) and *saliency-based VIF* (SVIF) are also defined in a similar manner.

As SVQA deals with video signals instead of images only, the following considerations are taken into account:

- HVS is sensitive to motion information, but less sensitive to the background. Therefore, distortion of moving objects is very important. SVQA differentiates between a fixed camera and a moving camera while locating a moving object.

- As frames are played out in real time, human eyes can only pay attention to a much smaller area in an image, compared to when looking at a fixed image. This is considered in intraframe weights.

- Due to *motion masking* effect, visual sensitivity is depressed during large-scale scene changes or rapid motion of objects. Therefore, frames should be weighted differently based on motion masking. This is considered in interframe weights.

- Considering spatio-temporal properties of video sequences, saliency weights in both spatial and temporal domains contribute to the final quality index.

In SVQA, the intraframe weight uses PQFT to calculate the saliency map for pixels with non-zero motion. Non-zero motion is represented as the difference image between two adjacent video frames. This establishes the first consideration of moving objects. In a short interval, the saliency map is allowed to have a square area of processing, thus addressing the second consideration. As the interframe weight is based on motion masking, the third consiration for weighting is also addressed. Finally, it is noteworthy that both intraframe weight and interframe weight are considered together in SVQA. Figure 4-14 shows the SVQA framework.

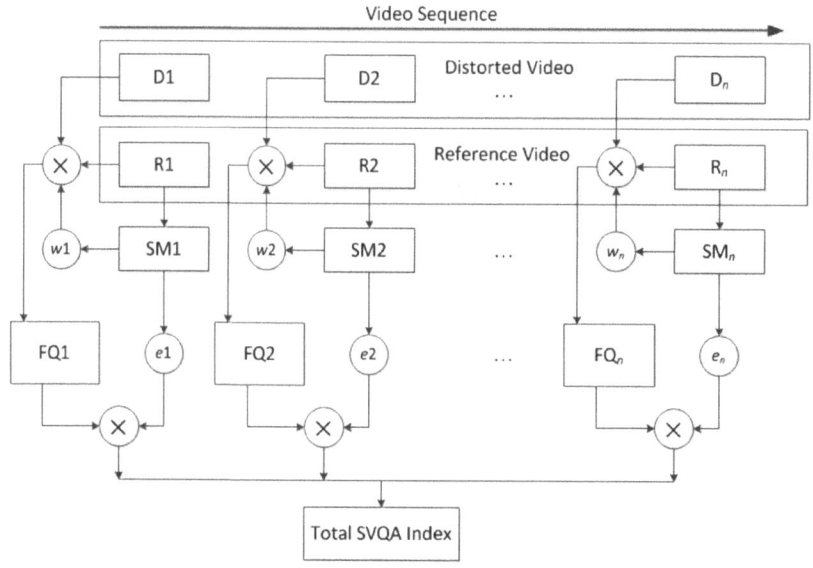

Figure 4-14. *The SVQA framework*

According to Figure 4-14, the flow of SVQA is as follows:

1. The reference and distorted video sequences are divided into frames. Each sequence is composed of n images: D_1 to D_n for distorted frames, R_1 to R_n for reference frames, as shown on the upper part of Figure 4-14.

2. Considering camera shift, where all pixels are moving, a binary function is defined to detect such motion. If the luma difference between pixels co-located in adjacent frames exceeds a threshold, a movement is detected. If the movement is detected for all pixels, a camera movement is understood; otherwise, object movement is considered on a static background. The quaternion for a frame is constructed as a weighted combination of motion information and three color channel information.

3. The *saliency map* (SM) is calculated for each reference video frame using the PQFT with motion, denoted as SM_1 to SM_n.

4. Based on the SM, the intraframe weights ($w_1, ..., w_n$) are calculated for n frames.

5. The frame quality FQ_i is calculated for the ith frame using any of the saliency-based metrics such as SPSNR, SMSSIM, or SVIF.

6. Based on the SM, the interframe weights ($e_1, ..., e_n$) are calculated for n frames.

7. The SVQA index, the measure of quality of the entire video is calculated using the following equation:

$$SVQA = \frac{\sum_{i=1}^{n} FQ_i e_i}{\sum_{i=1}^{n} e_i}$$

(Equation 4-19)

where n is the number of video frames.

Network-Aware Approaches

The objective video quality metrics such as PSNR neither perfectly correlate with perceived visual quality nor take the packet loss into account in lossy network environments such as multihop wireless mesh networks. While PSNR and similar metrics may work well for evaluating video quality in desktop coding applications and streaming over wired networks, remarkable inaccuracy arises when they are used to evaluate video quality over wireless networks.

For instance, in a wireless environment, it could happen that a video stream with a PSNR around 38dB (typically considered medium-high quality in desktop video coding applications) is actually perceived to have the same quality as the original undistorted video. This is because wireless video applications typically use the *User Datagram Protocol* (UDP), which does not guarantee reliable transmissions and may trade packet loss for satisfying delay requirements. Generally, in *wireless local area networks* (WLAN) consisting of unstable wireless channels, the probability of a packet loss is much higher than that in wired networks. In such environments, losing consecutive packets may cause the loss of an entire frame, thereby degrading the perceived video quality further than in desktop video coding applications.

Modified PSNR

Aiming to handle video frame losses, *Modified PSNR* (MPSNR) was proposed.[22] Two objective metrics are derived based on linear regression of PSNR against subjective MOS.

---

[22]A. Chan, K. Zeng, P. Mohapatra, S.-J. Lee, and S. Banerjee, "Metrics for Evaluating Video Streaming Quality in Lossy IEEE 802.11 Wireless Networks," *Proceedings of IEEE INFOCOM*, (San Diego, CA: March 2010): 1–9.

The first metric, called *PSNR-based Objective MOS* (POMOS), predicts the MOS from the mean PSNR, while achieving a correlation of 0.87 with the MOS. The second metric, called *Rate-based Objective MOS* (ROMOS), adds streaming network parameters such as the frame loss rate, and achieves a higher correlation of 0.94 with the MOS.

Frame losses are prevalent in wireless networks, but are not accounted for in the traditional PSNR calculations. Due to packet losses during streaming, a frame can be missing, which is typically unrecognizable by a human viewer. However, a missing frame causes the wrong frame to be compared against the original frame during PSNR calculation. Such off-position comparisons result in low PSNR values. A straightforward way to fix this is to introduce timing information into the source video. But such modification of source video is undesirable.

To determine if any frame is missing, an alternative approach is to match the frame with the original frame. The algorithm assumes that the sum of PSNRs of all frames is maximized when all frames are matching, and it uses this sum to determine the mismatching frame. In particular, MPSNR matches each frame in a streamed video to a frame in the reference video so that the sum of PSNR of all frame pairs is maximized. A moving window is used to determine the location of the matching frame. If frame j in the streamed video matches frame k belonging to the window in the reference video, it is considered that the frames (k-j) are missing. A frame in the streamed video need only be compared with, at most, g frames in the reference video, where g is the number of frames lost.

In addition to PSNR, the MPSNR measures the following video streaming parameters:

- Distorted frame rate (d): the percentage of mismatched frames in a streaming video.

- Distorted frame PSNR ($dPSNR$): the mean PSNR value of all the mismatched frames.

- Frame loss rate (l): the percentage of lost frames in a streaming video. It is calculated by comparing the total number of frames in the received streamed video with that in the reference video.

Once the corresponding frames in a streamed video and the reference video are matched, and the PSNR of each frame in the streamed video is calculated, all the above parameters are readily available.

In the MPSNR model, this method of matching is applied to a training set of videos, and the average PSNR for a window W is calculated. Experimental results show that the average PSNR exhibits a linear relationship with subjective MOS. Therefore, a linear model of the average PSNR can be used to predict the MOS score. The linear model is given as:

$$POMOS = 0.8311 + 0.0392\,(average\ PSNR) \qquad \text{(Equation 4-20)}$$

Note that average PSNR is used in this model. Since the average PSNR of a perfectly matching frame is infinity (or a very high value), it affects the prediction of MOS. To mitigate this problem, another linear model is proposed that does not use the PSNR values:

$$ROMOS = 4.367 - 0.5040\frac{d}{dPSNR} - 0.0517l.$$ (Equation 4-21)

Noise-Based Quality Metrics

An interesting approach to quality evaluation is to evaluate the noise introduced instead of the signal fidelity.

Noise Quality Measure

In the *noise quality measure*[23] (NQM), a degraded image is modeled as an original image that has been subjected to linear frequency distortion and additive noise injection. These two sources of degradation are considered independent and are decoupled into two quality measures: a distortion measure (DM) resulting from the effect of frequency distortion, and a noise quality measure (NQM) resulting from the effect of additive noise.

The NQM is based on a contrast pyramid and takes into account the following:

- The variation in contrast sensitivity with distance, image dimensions, and spatial frequency

- The variation in the local brightness mean

- The contrast interaction between spatial frequencies

- The contrast masking effects

For additive noise, the non-linear NQM is found to be a better measure of visual quality than the PSNR and linear quality measures.

The DM is computed in three steps. First, the frequency distortion in the degraded image is found. Second, the deviation of this frequency distortion from an all-pass response of unity gain (no distortion) is computed. Finally, the deviation is weighted by a model of the frequency response of the HVS, and the resulting weighted deviation is integrated over the visible frequencies.

Objective Coding Efficiency Metrics

Measuring coding efficiency is another way to look at the tradeoff between visual quality and bit-rate cost in video coding applications. In this section we discuss the popular BD metrics for objective determination of coding efficiency.

---

[23]N. Damera-Venkata, T. D. Kite, W. S. Geisler, B. L. Evans, and A. C. Bovik, "Image Quality Assessment Based on a Degradation Model," *IEEE Transactions on Image Processing* 9, no. 4 (April 2000): 636–50.

BD-PSNR, BD-SSIM, BD-Bitrate

Bjøntegaard delta PSNR[24] (BD-PSNR) is an objective measure of coding efficiency of an encoder with respect to a reference encoder. It was proposed by Gisle Bjøntegaard in April 2001 in the Video Coding Expert Group's (VCEG) meeting. BD-PSNR considers the relative differences between two encoding solutions in terms of number of bits used to achieve a certain quality. In particular, BD-PSNR calculates the average PSNR difference between two rate-distortion (R-D) curves over an interval. This metric is a good indication of visual quality of encoded video, as it considers the cost (i.e., bits used) to achieve a certain visual quality of the decoded video, represented by the popular objective measure PSNR. Improvements to the BD-PSNR model can be performed by using $\log_{10}(bitrate)$ instead of simply the bit rate when plotting R-D data points, resulting in *straighter* R-D curves and more uniformly spaced data points across the axes.

BD-PSNR uses a third order logarithmic polynomial to approximate a given R-D curve. The reconstructed distortion in PSNR is given as:

$$D_{PSNR} = D(r) = a + br + cr^2 + dr^3 \qquad \text{(Equation 4-22)}$$

where $r = \log(R)$, R is the output bit rate, and a, b, c, and d are fitting parameters.

This model is a good fit to R-D curves and there is no problem with singular points, as could have happened for a model with $(r + d)$ in the denominator. The above equation can be solved with four R-D data points obtained from actual encoding, and the fitting parameters a, b, c, and d can be determined. Thus, this equation can be used to interpolate the two R-D curves from the two encoding solutions, and the delta PSNR between the two curves can be obtained as:

$$BD\ PSNR = \frac{1}{(r_H - r_L)} \int_{r_L}^{r_H} (D_2(r) - D_1(r)) dr \qquad \text{(Equation 4-23)}$$

where $r_H = \log(R_H)$, $r_L = \log(R_L)$ are the high and low ends, respectively, of the output bit rate range, and $D_1(r)$ and $D_2(r)$ are the two R-D curves.

Similarly, the interpolation can also be done on the bit rate as a function of SNR:

$$r = a + bD + cD^2 + dD^3 \qquad \text{(Equation 4-24)}$$

where $r = \log(R)$, R is the output bit rate, a, b, c, and d are fitting parameters, and D is the distortion in terms of PSNR. From this the BD-bit rate can be calculated in a similar fashion as is done for PSNR above:

$$BD\ Bit\ rate = \frac{1}{D_H - D_L} \int_{D_L}^{D_H} (r_2 - r_1) dD \qquad \text{(Equation 4-25)}$$

---

[24]B. Bjøntegaard, *Calculation of Average PSNR Differences between RD curves (VCEG-M33)* (Austin, TX: ITU-T VCEG SG16 Q.6, 2001).

Therefore, from BD-PSNR calculations, both of the following can be obtained:

- Average PSNR difference in dB over the whole range of bit rates
- Average bit rate difference in percent over the whole range of PSNR

If the distortion measure is expressed in terms of SSIM instead of PSNR, BD-SSIM can be obtained in the same manner. BD-PSNR/BD-SSIM calculation depends on interpolating polynomials based on a set of rate-distorion data points. Most implementations of BD-PSNR use exactly four rate-distortion data points for polynomial interpolation, resulting in a single number for BD-PSNR.

Advantages

BD metrics have the advantage that they are compact and in some sense more accurate representations of the quality difference compared to R-D curves alone. In case of a large number of tests, BD metrics can readily show the difference between two encoding solutions under various parameters. Further, BD metrics can consolidate results from several tests into a single chart, while showing video quality of one encoding solution with respect to another; these presentations can effectively convey an overall picture of such quality comparisons.

Limitations

The BD metrics are very useful in comparing two encoding solutions. However, for ultra-high-definition (UHD) video sequences, the BD metrics can give unexpected results.[25] The behavior appears owing to polynomial curve-fitting and the high-frequency noise in the video sequences. Standard polynomial interpolation is susceptible to Runge's phenomenon (problematic oscillation of the interpolated polynomial) when using high-degree polynomials. Even with just four data points (third degree polynomial), some interpolated curves see oscillation that can result in inaccurate BD-PSNR evaluations.

Alternative interpolation methods such as splines reduce the error caused by Runge's phenomenon and still provide curves that fit exactly through the measured rate-distortion data points. There are video examples where using piecewise cubic spline interpolation improves the accuracy of BD-PSNR calculation by nearly 1 dB over polynomial interpolation.

When oscillation occurs from polynomial interpolation, the resulting BD-PSNR calculation can be dramatically skewed. Figure 4-15 shows the polynomial interpolation problem in rate-PSNR curves from two sample encoding. The charts show the difference between polynomial interpolation and cubic spline interpolation and the BD-PSNR values using each method.

---

[25] Sharp Corporation, "On the Calculation of PSNR and Bit Rate Differences for the SVT Test Data," ITU SG16, Contibution 404, April 2008, available at http://www.docstoc.com/docs/101609255/On-the-calculation-of-PSNR-and-bit-rate-differences-for-the-SVT-test.

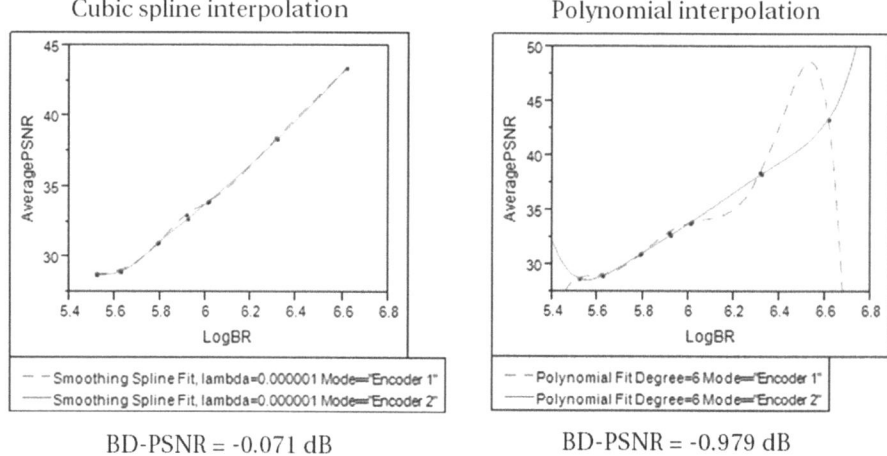

Figure 4-15. *Polynomial interpolation issue in R-D curves*

The average PSNR and bit rates correlate very closely between the two encoders, indicating that the BD-PSNR value achieved using polynomial interpolation would not be an accurate representation of the quality difference between the two encoders.

Additionally, BD-PSNR does not consider the coding complexity, which is a critical problem for practical video applications, especially for those on handheld devices whose computational capability, memory, and power supply are all limited. Such limitations are addressed by considering a generalized BD-PSNR metric that includes complexity in addition to rate and distortion. The generalized metric is presented in the next section.

Generalized BD-PSNR

The *Generalized BD-PSNR*[26] (GBD-PSNR) is a coding efficiency measure developed by generalizing BD-PSNR from R-D curve fitting to rate-complexity-distortion (R-C-D) surface fitting. GBD-PSNR involves measurement of coding complexity, R-C-D surface fitting, and the calculation of differential PSNR between two R-C-D surfaces.

In general, coding complexity is multi-dimensional and requires consideration of several factors, including the computational complexity measured by executing time or machine cycles, data cache size, memory access bandwidth, storage complexity, instruction cache size, parallelism, and pipelining. However, in practice, it is difficult to simultaneously account for all these dimensions. A widely used alternative is the coding time on a given platform. Not only does it indicate the computational complexity, but it also partially reflects the contributions from other complexity dimensions such as memory access in the coding process.

---

[26]X. Li, M. Wien, and J.-R. Ohm, "Rate-Complexity-Distortion Evaluation for Hybrid Video Coding," *Proceedings of IEEE International Conference on Multimedia and Expo*, (July 2010): 685–90.

In order to perform R-C-D surface fitting, the R-C-D function is defined as follows:

Definition 4-1. The rate-complexity-distortion function $D(R, C)$ is the infimum of distortion D such that the rate-complexity-distortion triplet (R, C, D) is in the achievable rate-complexity-distortion region of the source for a given rate-complexity pair (R, C).

Therefore, $D(R, C)$ is non-increasing about R and C, respectively. Similar to convex R-D function, the R-C-D function is convex as well. Based on these properties, $D(R, C)$ can be approximated using an exponential model. To obtain a good tradeoff between accuracy and fitting complexity while keeping backward compatibility with BD-PSNR, $D(R, C)$ is approximated as:

$$D(R,C) = a_0 r^3 + a_1 r^2 + a_2 r + a_3 + a_4 c^2 + a_5 c \qquad \text{(Equation 4-26)}$$

where, $a_0, \ldots, a_5$ are fitting parameters, $r = \log(r)$, $c = \log(C)$, R is the output bit rate, C is the coding complexity, and D is the distortion in terms of PSNR. To fit an R-C-D surface with this equation, at least six (R, C, D) triplets from actual coding are necessary. However, in practice, a higher number of (R, C, D) triplets will lead to a better accuracy. Typically 20 data points are used to fit such a surface.

Similar to BD-PSNR, the average differential PSNR between two R-C-D surfaces can be calculated as:

$$\Delta P_{GBD} = \frac{\int_{r_L}^{r_H} \int_{c_L}^{c_H} (D_2(r,c) - D_1(r,c)) dc dr}{(r_H - r_L)(c_H - c_L)} \qquad \text{(Equation 4-27)}$$

where DP_{GBD} is the GBD-PSNR, $D_1(r, c)$ and $D_2(r, c)$ are the two fitting functions for the two R-C-D surfaces, $r_H, r_L, c_H,$ and c_L are the logarithmic forms of $R_H, R_L, C_H,$ and C_L which bound the overlapped R-C region from the actual coding results.

Due to the complexity nature of some algorithms, the two R-C-D surfaces may have no R-C intersection. In this extreme case, the GBD-PSNR is undefined.

Limitations

The dynamic range of coding complexity covered by GBD-PSNR is sometimes limited. This happens when the coding complexity of the two encoders are so different that there is only a relatively small overlapped region by the two R-C-D surfaces.

Also, the coding complexity is platform and implementation dependent. Although GBD-PSNR shows a good consistency over different platforms, slightly different GBD-PSNR value may still be obtained on different platforms.

Examples of Standards-based Measures

There are a few objective quality measures based on the ITU-T standards.

Video Quality Metric

The video quality metric (VQM)[27] is an objective measurement for perceived video quality developed at the National Telecommunications and Information Administration (NTIA). Owing to its excellent performance in the VQEG Phase 2 validation tests, the VQM methods were adopted by the American National Standards Institute (ANSI) as a national standard, and by ITU as ITU-T Rec. J. 144.[28] The VQM measures the perceptual effects of video impairments, including blurring, jerkiness, global noise, block distortion, and color distortion, and combines them into a single metric. The testing results show that VQM has a high correlation with subjective video quality assessment.

The algorithm takes a source video clip and a processed video clip as inputs and computes the VQM in four steps:

1. *Calibration:* In this step the sampled video is calibrated in preparation for feature extraction. The spatial and temporal shift, the contrast and the brightness offset of the processed video are estimated and corrected with respect to the source video.

2. *Quality Features Extraction*: In this step, using a mathematical function, a set of quality features that characterize perceptual changes in the spatial, temporal, and color properties are extracted from spatio-temporal subregions of video streams.

3. *Quality Parameters Calculation:* In this step, a set of quality parameters that describe perceptual changes in video quality are computed by comparing the features extracted from the processed video with those extracted from the source video.

4. *VQM Calculation:* VQM is computed using a linear combination of parameters calculated from the previous steps.

VQM can be computed using various models based on certain optimization criteria. These models include television model, video conferencing model, general model, developer model, and PSNR model. The general model uses a linear combination of seven parameters. Four of these parameters are based on features extracted from spatial gradients of the luma component, two parameters are based on features extracted from the vector formed by the two chroma components, and the last parameter is based on contrast and absolute temporal information features, both extracted from the luma component. Test results show a high correlation coefficient of 0.95 between subjective tests and the VQM general model (VQMG).[27]

---

[27]M. Pinson, and S. Wolf, "A New Standardized Method for Objectively Measuring Video Quality," *IEEE Transactions on Broadcasting* 50, no. 3 (September 2004): 312–22.
[28]*ITU-T Recommendation J.144: Objective Perceptual Video Quality Measurement Techniques for Digital Cable Television in the Presence of a Full Reference* (Geneva, Switzerland: International Telecommunications Union, 2004).

ITU-T G.1070 and G.1070E

The ITU Recommendation G.1070[29] is a standard computational model for *quality of experience* (QoE) planning. Originally developed for two-way video communication, G.1070 model has been widely used, studied, extended, and enhanced. In G.1070, the visual quality model is based on several factors, including frame rate, bit rate, and packet-loss rate. For a fixed frame rate and a fixed packet-loss rate, a decrease in bit rate would result in a corresponding decrease in the G.1070 visual quality. However, a decrease in bit rate does not necessarily imply a decrease in quality. It is possible that the underlying video content is of low complexity and easy to encode, and thus results in a lower bit rate without corresponding quality loss. G.1070 cannot distinguish between these two cases.

Given assumptions about the coding bit rate, the frame rate, and the packet-loss rate, the G.1070 video quality estimation model can be used to generate an estimate, typically in the form of a quality score, of the perceptual quality of the video that is delivered to the end user. This score is typically higher for higher bit rates of compressed videos, and lower for lower bit rates of compressed videos.

To calculate the G.1070 visual quality estimate, a typical system includes a data collector or estimator that is used to analyze the encoded bitstream, extract useful information, and estimate the bit rate, frame rate, and packet-loss rate. From these three estimates, a G.1070 Video Quality Estimator computes the video quality estimate according to a function defined in Section 11.2 of Rec. G.1070.

Although the G.1070 model is generally suitable for estimating network-related aspects of the perceptual video quality, such as the expected packet-loss rate, information about the content of the video is generally not considered. For example, a video scene with a complex background and a high level of motion, and another scene with relatively less activity or texture, may have dramatically different perceived qualities even if they are encoded at the same bit rate and frame rate. Also, the coding bit rate required to achieve high-quality coding of an easy scene may be relatively low. Since the G.1070 model generally gives low scores for low-bit-rate videos, this model may unjustifiably penalize such easy scenes, notwithstanding the fact that the perceptual quality of that video scene may actually be high. Similarly, the G.1070 score can overestimate the perceptual quality of video scenes. Thus, the G.1070 model may not correlate well with subjective quality scores of the end users.

To address such issues, a modified G.1070 model, called the G.1070E was introduced.[30] This modified model takes frame complexity into consideration, and provides frame complexity estimation methods. Based on the frame complexity, bit-rate normalization is then performed. Finally, the G.1070 Video Quality Estimator uses the normalized bit rate along with the estimated frame rate and packet-loss rate to yield the video quality estimate.

---

[29]*ITU-T Recommendation G.1070: Opinion Model for Video-Telephony Applications* (Geneva, Switzerland: International Telecommunications Union, 2012).

[30]B. Wang, D. Zou, R. Ding, T. Liu, S. Bhagavathi, N. Narvekar, and J. Bloom, "Efficient Frame Complexity Estimation and Application to G.1070 Video Quality Monitoring," *Proceedings of 2011 Third International Workshop on Quality of Multimedia Experience* (2011): 96–101.

The G.1070E is a no-reference compressed domain objective video-quality measurement model. Experimental results show that the G.1070E model yields a higher correlation with subjective MOS scores and can reflect the quality of video experience much better than G.1070.

ITU-T P.1202.2

The ITU-T P.1202 series of documents specifies models for monitoring the video quality of IP-based video services based on packet-header and bitstream information. Recommendation ITU-T P.1202.2[31] specifies the algorithmic model for the higher-resolution application area of ITU-T P.1202. Its applications include the monitoring of performance and quality of experience (QoE) of video services such as IPTV. The Rec. P.1202.2 and has two modes: Mode 1, where the video bitstreams are parsed and not decoded into pixels, and Mode 2, where the video bitstreams are fully decoded into pixels for analyzing.

The Rec. P.1202.2 is a no-reference video-quality metric. An implementation of the algorithm has the following steps:

1. Extraction of basic parameters such as frame resolution, frame level quantization parameter, frame size, and frame number.

2. Aggregation of basic parameters into internal picture level to determine frame complexity.

3. Aggregation of basic parameters into model level to obtain video sequence complexity, and quantization parameter at the video sequence level.

4. Quality estimation model to estimate the MOS as:

$$P.1202.2\ MOS = f(\text{frame QP, frame resolution, frame size, frame number})$$

(Equation 4-28)

Studies have found that the P.1202.2 algorithm's estimated MOS has similar Pearson linear correlation coefficient and Spearman ranked order correlation coefficient to VQEG JEG's (Joint Effort Group) estimated MOS, which uses the following linear relationship:[32]

$$VQEG\ JEG\ MOS = -0.172 \times \text{frame QP} + 9.249$$

(Equation 4-29)

However, both of these results are worse than MS-SSIM. It is also found that P.1202.2 does not capture compression artifacts well.

---

[31]*ITU-T Recommendation P.1202.2: Parametric Non-intrusive Bitstream Assessment of Video Media Streaming Quality – Higher Resolution Application Area* (Geneva, Switzerland: International Telecommunications Union, 2013).

[32]L. K. Choi, Y. Liao, B. O'Mahony, J. R. Foerster, and A. C. Bovik, "Extending the Validity Scope of ITU-T P.1202.2," in *Proceedings of the 8th International Workshop on Video Processing and Quality Metrics for Consumer Electronics* (Chandler, AZ: VPQM, 2014), retrieved from www.vpqm.org.

Therefore, an improved FR MOS estimator is proposed based on MS-SSIM. In particular, an MS-SSIM-based remapping function is developed. The resulting estimated MOS is a function of MS-SSIM and the frame parameters, such as frame level quantization parameter, frame size, frame type, and resolution. The algorithm first performs devices and content analysis, followed by spatial complexity computation.

Then, a non-linear model fitting is performed using logistic function. These results, along with the MS-SSIM values, are provided to the MOS estimator to calculate the estimated MOS. Experimental results show that for a set of tests, the estimated MOS has a Pearson correlation coefficient >0.9 with MOS, which is much better than that given by MS-SSIM (0.7265).

Measurement of Video Quality

We elaborate on important considerations for video quality measurement, for both subjective and objective measurements. Further, for clarity we discuss the objective measurements from typical application point of view.

Subjective Measurements

The metrics used in subjective measurement are MOS and DMOS. However, after obtaining the raw scores, they cannot be directly used. To eliminate bias, the following measurement procedure is generally used.

Let s_{ijk} denote the score assigned by subject i to video j in session k. Usually, two sessions are held. In the processing of the raw scores, difference scores d_{ijk} are computed per session by subtracting the quality assigned by the subject to a video from the quality assigned by the same subject to the corresponding reference video in the same session. Computation of difference scores per session helps account for any variability in the use of the quality scale by the subject between sessions. The difference scores are given as:

$$d_{ijk} = s_{ijk} - s_{ijrefk}.$$ (Equation 4-30)

The difference scores for the reference videos are 0 in both sessions and are removed. The difference scores are then converted to Z-scores per session:

$$\mu_{ik} = \frac{1}{N_{ik}} \sum_{j=1}^{N_{ik}} d_{ijk}$$ (Equation 4-31)

$$\sigma_{ik} = \sqrt{\frac{1}{N_{ik}-1} \sum_{j=1}^{N_{ik}} (d_{ijk} - \mu_{ik})^2}$$ (Equation 4-32)

$$z_{ijk} = d_{ijk} - \mu_{ik}\sigma_{ik}$$ (Equation 4-33)

where N_{ik} is the number of test videos seen by subject i in session k.

Every subject sees each test video in the database exactly once, either in the first session or in the second session. The Z-scores from both sessions are then combined to create a matrix $\{z_{ij}\}$. Scores from unreliable subjects are discarded using the procedure specified in the ITU-R BT.500-13 recommendation.

The distribution of the scores is then investigated. If the scores are normally distributed, the procedure rejects a subject whenever more than 5 percent of scores assigned by that subject fall outside the range of two standard deviations from the mean scores. If the scores are not normally distributed, the subject is rejected whenever more than 5 percent of his scores fall outside the range of 4.47 standard deviations from the mean scores. However, in both situations, subjects who are consistently pessimistic or optimistic in their quality judgments are not eliminated.

The Z-scores are then linearly rescaled to lie in the range [0,100]. Finally, the DMOS of each video is computed as the mean of the rescaled Z-scores from the remaining subjects after subject rejection.

Objective Measurements and Their Applications

Objective measurements are very useful in automated environments—for example, in automated quality comparison of two video encoder solutions. Figure 4-16 shows the block diagram of a typical encoder comparison setup using full-reference objective video-quality metrics.

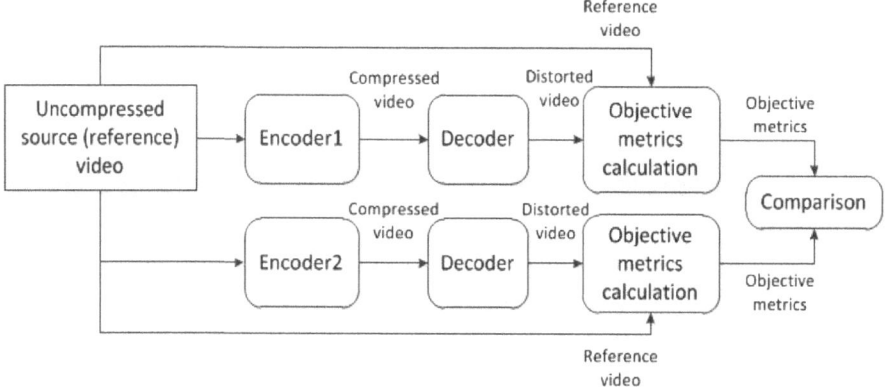

Figure 4-16. *An example of a typical encoder comparison setup using FR objective quality metrics*

Several factors need to be considered for such an application using full reference objective video quality metrics:

- The source and distorted videos need to be aligned in time so that the same video frame is compared for quality.

- The same decoder implementation should be used, eliminating any measurement variability owing to the decoding process.

- To ensure a fair comparison, the encoder parameters must be the same or as close as possible.

- No pre-processing is assumed before the encoding process. Although it is possible to use a pre-processing step before each encoder, in that case the same pre-processor must be used.

Notice that such a setup can take advantage of automation and use an enormous set of video clips for comparison of different encoder implementations, thus exposing the strengths and weaknesses of each encoder under various workload complexities. Such source comparisons without considering network or channel errors are ideal for a fair comparison. However, in practical applications, such as mobile video recording using two different devices, where the recorded videos are stored and decoded before computing objective quality metrics, quality comparison should be done in similar environments as much as possible. For example, in wireless network environment, the packet-loss rate or bit-error rate should be similar.

Objective measures are also extensively used to determine frame drops in video applications. For example, as the (distorted) video is consumed, frame drops can be detected if the PSNR between the source and the distorted video is tracked frame by frame. In low-distortion environments, the consumed video would reasonably match the source; so the PSNR would also be a typical number (e.g., 25–40 dB) depending on the lossy characteristics of the various channels that introduce errors. However, in case of a frame drop, the wrong frame would be compared against the source, and a very low PSNR would be obtained, indicating the frame drop. This effect is exaggerated when the video contains frequent scene changes.

The same concept can be applied to detect sudden low-quality frames with corruption or other artifacts in a video. Such corruption can happen owing to network errors or encoder issues. But a sudden drop in PSNR or other objective measures can indicate the location of the corruption in the video in an automated environment.

Parameters to Tune

In visual communication applications, video codecs are one of the main sources of distortions. Since video decoders must follow certain specifications as defined by various standards, decoders generally do not significantly contribute to video-quality degradation. However, encoders are free to design and implement algorithms to control the amount of compression and thereby the amount of information loss depending on various considerations for system resources, application requirements, and the application environment. Therefore, in the video-encoding applications, there are several parameters that dictate the amount of information loss and thus influence the final video quality. Some of these parameters are adjustable at the algorithm level by the system architects; some are tunable by the implementors, while few parameters are usually available to the users for tuning.

Parameters that Impact Video Quality

It is very important to understand the impact of the following parameters on final visual quality, particularly for benchmarking, optimization, or comparative analysis of the video encoding solutions.

- **Number of lines in the vertical display resolution:** High-definition television (HDTV) resolution is 1,080 or 720 lines. In contrast, standard-definition digital television (DTV) is 480 lines (for NTSC, where 480 out of 525 scanlines are visible) or 576 lines (for PAL/SECAM, where 576 out of 625 scanlines are visible). For example, the so-called *DVD quality* is standard definition, while Blu-ray discs are high definition. An encoder may choose to reduce the resolution of the video as needed, depending on the available number of bits and the target quality level. However, recent encoders typically process the full resolution of video in most applications.

- **Scanning type:** Digital video uses two types of image scanning pattern: progressive scanning or interlaced scanning. Progressive scanning redraws all the lines of a video frame when refreshing the frame, and is usually denoted as 720p or 1080p, for example. Interlaced scanning draws a *field*—that is, every other line of the frame at a time—so the *odd numbered* lines are drawn during the first refresh operation and then the remaining *even numbered* lines are drawn during a second refreshing. Thus, the interlaced refresh rate is double that of the progressive referesh rate. Interlaced scanned video is usually denoted as 480i or 1080i, for example.

  Movement of object makes a difference in perceived quality of interlaced scanned video. On a progressively scanned display, interlaced video yields better quality for still objects in frames owing to the higher refresh rate, but loses up to half of the resolution and suffers *combing* artifacts when objects in a frame is moving. Note that combing artifacts only occur when two fields are woven together to form a single frame and then displayed on a progressive display. Combing artifacts do not occur when interlaced content is shown on an interlaced display and when different deinterlacing algorithms such as *bob* are used for display on progressive monitors.

  In practice, two interlaced fields formulate a single frame because the two fields consisting of the odd and even lines of one frame are temporally shifted. Frame pulldown and segmented frames are special techniques that allow transmitting full frames by means of interlaced video stream. For appropriate reconstruction and presentation at the receiving end of a transmission system, it is necessary to track whether the top or bottom field is transmitted first.

- **Number of frames or fields per second (Hz):** In Europe, 50 Hz television broadcasting system is more common, while in the United States, it is 60 Hz. The well-known 720p60 format is 1280×720 pixels, progressive encoding with 60 frames per second (60 Hz). The 1080i50/1080i60 format is 1920×1080 pixels, interlaced encoding with 50/60 fields, (50/60 Hz) per second. If the frame/field rate is not properly maintained, there may be visible *flickering* artifact. Frame drop and frame jitter are typical annoying video-quality issues resulting from frame-rate mismanagement.

- **Bit rate:** The amount of compression in digital video can be controlled by allocating a certain number of bits for each second's worth of video. The bit rate is the primary defining factor of video quality. Higher bit rate typically implies higher quality video. Efficient bit allocation can be done by taking advantage of skippable macroblocks and is based on the spatio-temporal complexity of macroblocks. The amount of quantization is also determined by the available bit rate, thereby highly impacting the *blocking* artifact at transform block boundaries.

- **Bit-rate control type:** The bit-rate control depends on certain restrictions of the transmission system and the nature of the application. Some transmission systems have fixed channel bandwidth and need video contents to be delivered at a constant bit rate (CBR), while others allow a variable bit rate (VBR), where the amount of data may vary per time segment. CBR means the decoding rate of the video is constant. Usually a decoding buffer is used to keep the decoded bits until a frame's worth of data is consumed instantaneously. CBR is useful in streaming video applications where, in order to meet the requirement of fixed number of bits per second, stuffing bits without useful information may need to be transmitted.

  VBR allows more bits to be allocated for the more complex sections of the video, and fewer bits for the less complex sections. The user specifies a given subjective quality value, and the encoder allocates bits as needed to achieve the given level of quality. Thus a more perceptually consistent viewing experience can be obtained using VBR. However, the resulting compressed video still needs to fit into the available channel bandwidth, necessitating a maximum bit rate limit. Thus, the VBR encoding method typically allows the user to specify a bit-rate range indicating a maximum and/or minimum allowed bit rate. For storage applcations, VBR is typically more appropriate compared to CBR.

  In addition to CBR and VBR, the average bit rate (ABR) encoding may be used to ensure the output video stream achieves a predictable long-term average bit rate.

155

- **Buffer size and latency:** As mentioned above, the decoding buffer temporarily stores the received video as incoming bits that may arrive at a constant or variable bit rate. The buffer is drained at specific time instants, when one frame's worth of bits are taken out of the buffer for display. The number of bits that are removed is variable depending on the frame type (intra or predicted frame). Given that the buffer has a fixed size, the bit arrival rate and the drain rate must be carefully maintained such that the buffer does not overflow or be starved of bits. This is typically done by the rate control mechanism that governs the amount of quantization and manages the resulting frame sizes. If the buffer overflows, the bits will be lost and one or more frames cannot be displayed, depending on the frame dependency. If it underflows, the decoder would not have data to decode, the display would continue to show the previously displayed frame, and decoder must wait until the arrival of a decoder refresh signal before the situation can be corrected. There is an initial delay between the time when the buffer starts to fill and the time when the first frame is taken out of the buffer. This delay translates to the decoding latency. Usually the buffer is allowed to fill at a level between 50 and 90 percent of the buffer size before the draining starts.

- **Group of pictures structure:** The sequence of dependency of the frames is determined by the frame prediction structure. Recall from Chapter 2 that intra frames are independently coded, and are usually allocated more bits as they typically serve as anchor frames for a group of pictures. Predicted and bi-predicted frames are usually more heavily quantized, resulting in higher compression at the expense of comparatively poor individual picture quality. Therefore, the arrangement of the group of picture is very important. In typical broadcast applications, intra frames are transmitted twice per second. In between two intra frames, the predicted and bi-predicted frames are used so that two bi-predicted frames are between the predicted or intra reference frames. Using more bi-predicted frames does not typically improve visual quality, but such usage depends on applications. Note that, in videos with rapidly changing scenes, predictions with long-term references are not very effective. Efficient encoders may perform scene analysis before determining the final group of pictures structure.

- **Prediction block size:** Intra or inter prediction may be performed using various block sizes, typically from 16×16 down to 4×4. For efficient coding, suitable sizes must be chosen based on the pattern of details in a video frame. For example, an area with finer details can benefit from smaller prediction block sizes, while a flat region may use larger prediction block sizes.

- **Motion parameters:** Motion estimation search type, search area, and cost function play important roles in determining visual quality. A full search algorithm inspects every search location to find the best matching block, but at the expense of very high computational complexity. Studies have suggested that over 50 percent of the encoding computations are spent in the block-matching process. The number of computations also grows exponentially as the search area becomes larger to capture large motions or to accommodate high-resolution video. Further, the matching criteria can be selected from techniques such as sum of absolute difference (SAD) and sum of absolute transformed differences (SATD). Using SATD as the matching criteria provides better video quality at the expense of higher computational complexity.

- **Number of reference pictures:** For motion estimation, one or more reference pictures can be used from lists of forward or backward references. Multiple reference pictures increase the probability of finding a better match, so that the difference signal is smaller and can be coded more efficiently. Therefore, the eventual quality would be better for the same overall number of bits for the video. Also, depending on the video content, a frame may have a better match with a frame that is not an immediate or close neighbor. This calls for long-term references.

- **Motion vector precision and rounding:** Motion compensation can be performed at various precision levels: full-pel, half-pel, quarter-pel, and so on. The higher the precision, the better the probability of finding the best match. More accurate matching results in using fewer bits for coding the error signal, or equivalently, using a finer quantization step for the same number of bits. Thus quarter-pel motion compensation provides better visual quality for the same number of bits compared to full-pel motion compensation. The direction and amount of rounding are also important to keep sufficient details of data, leading to achieving a better quality. Rounding parameters usually differ based on intra or inter type of prediction blocks.

- **Interpolation method for motion vectors:** Motion vector interpolation can be done using different types of filters. Typical interpolation methods employ a bilinear, a 4-tap, or a 6-tap filter. These filters produce different quality of the motion vectors, which leads to differences in final visual quality. The 6-tap filters generally produce the best quality, but are more expensive in terms of processing cycles and power consumption.

- **Number of encoding passes:** Single-pass encoding analyzes and encodes the data *on the fly*. It is used when the encoding speed is most important—for example, in real-time encoding applications. Multi-pass encoding is used when the encoding quality is most important. Multi-pass encoding, typically implemented in two passes, takes longer than single-pass, as the input data goes through additional processing in each pass. In multi-pass encoding, one or more initial passes are used to collect the video characteristics data, and a final pass uses that data to achieve uniform quality at a specified target bit rate.

- **Entropy coding type:** Entropy coding type such as CABAC or CAVLC does not generally impact video quality. However, if there is a bit-rate limit, owing to the higher coding efficiency, CABAC may yield better visual quality, especially for low-target bit rates.

Tradeoff Opportunities

Video encoders usually have tunable parameters to achieve the best possible quality or the best possible encoding speed for that encoder. Some parameters allow the encoder to analyze the input video and collect detailed information of the characteristics of the input video. Based on this information, the encoder makes certain decisions regarding the amount of compression to perform or the encoding mode to be used. Often, multiple passes are used for the analysis and subsequent encoding. Thus, the encoder is able to compress the video efficiently and achieve the best possible quality for the given algorithm. However, such analysis would require time and would slow down the encoding process. Further, the analysis work would increase the power consumption of the encoding device. Therefore, sometimes tuning of the certain parameters to adapt to the given video characteristics is not attempted in order to increase performance, or to meet system resource constraints. Rather, these parameters use pre-defined values for this purpose, thereby reducing analysis work and aiming to achieve the best possible speed.

Most of the parameters mentioned in the above section that affect visual quality also affect the encoding speed. To achieve a good tradeoff between quality and speed for a given video encoder, several parameters can be tuned. Although not all parameters listed here are tunable by the end user, depending on the encoder implementation, some parameters may be exposed to the end-user level.

- **Bit rate, frame rate, resolution:** Videos with high bit rate, frame rate, and resolution usually take longer to encode, but they provide better visual quality. These parameters should be carefully set to accommodate the application requirement. For example, real-time requirements for encode and processing may be met on a certain device with only certain parameters.

- **Motion estimation algorithm:** There are a large number of fast-motion estimation algorithms available in the literature, all of which are developed with a common goal: to increase the speed of motion estimation while providing reasonable quality. Since motion estimation is the most time-consuming part of video encoding, it is very important to choose the algorithm carefully.

- **Motion search range:** For best quality, the motion search range should be set to a high value so that large motions can be captured. On the other hand, the larger the search window, the more expensive is the search in terms of amount of computation to be done. So, a large search area directly impacts the encoding speed, memory bandwidth, frame latency, and power consumption. In addition, the large motion vectors would require more bits to encode. If the difference signal between a source block and the predicted block has substantial energy, it may be worthwhile to encode the block in intra-block mode instead of using the large motion vectors. Therefore, a tradeoff needs to be made between the search parameters and coding efficiency in terms of number of bits spent per decibel of quality gain.

- **Adaptive search:** To achieve better quality, often the motion search algorithms can adapt to the motion characteristics of the video and can efficiently curb the search process to gain significant encoding speed. For example, in order to accelerate motion search, an algorithm can avoid searching the stationary regions, use switchable shape search patterns, and take advantage of correlations in motion vectors. Thus, encoding speed can be increased without resorting to suboptimal search and without sacrificing visual quality.

- **Prediction types:** Predicted and bi-predicted frames introduce various levels of computational complexity and generally introduce visual quality loss in order to achieve compression. However, they also provide visually pleasing appearance of smooth motion. Therefore, prediction type of a frame is an important consideration in tradeoffs between quality and encoding speed.

- **Number of reference frames:** Mutiple reference frames can provide better visual quality than single reference frames, but computing motion vectors from multiple references are more time-consuming. In resource constrained environment, such parameters are important factors in tradeoff considerations.

- **Transform mode and partition size:** A block may use 8×8 or 4×4 sizes for the transform and various partition sizes for the prediction. On some platforms, processing four 4×4 blocks may be slower than processing one 8×8 block. However, depending on the amount of details available in the video, such decision may impact the visual quality, as 4×4 partitions have better adaptability to finer details.

- **Skip conditions:** A block can be skipped if it meets certain criteria. Better skip decisions can be made based on analysis of the quantized transform coefficients characteristics compared to simple heuristics, resulting in better quality. But a large amount of computation is necessary to adopt such complex algorithms. It is a clear tradeoff opportunity for resource-constrained devices.

- **Deblocking filter parameters:** Encoding speed is usually sensitive to deblocking filter parameters. Performing strong deblocking slows the encoding, but depending on the content and the amount of blocking artifact, it may provide significantly better visual quality.

Summary

This chapter discussed visual quality issues and factors impacting the perceptual quality of video to a human observer. First, we studied the various compression and processing artifacts that contribute to visual quality degradation, and various factors that affect visual quality in general. Next, we discussed various subjective and objective quality evaluation methods and metrics with particular attention to various ITU-T standards. We discussed several objective quality evaluation approaches in detail. These approaches are based on various factors: error-sensitivity, structural similarity, information fidelity, spatio-temporal, saliency, network awareness, and noise. We also discussed video coding efficiency evaluation metrics and some examples of standard-based algorithms.

In the final part of this chapter, we covered about the encoding parameters that primarily impact video quality. Tuning some parameters offer good tradeoff opportunities beween video quality and compression speed. These include bit rate, frame rate, resolution, motion estimation parameters, Group of Pictures structure, number of reference frames, and deblocking filter parameters. Some of these parameters may be available to the end user for tuning.

CHAPTER 5

Video Coding Performance

During the period between the 1980s to early 2000s, desktop PCs were the main computing platforms, with separate components such as the CPU, chipset, and discrete graphics cards. In this period, integrated graphics was at its infancy starting with the Intel (R) 810 (TM) chipset, mainly targeting the low-cost market segment, and power consumption was not typically a concern. CPU speed was the overarching differentiator between one generation of platforms and the next. Consequently, when the micro-architecture of a CPU was being designed, one of the key questions was how to achieve higher performance. The traditional way to achieve that was to keep increasing the clock speed. However, growth in transistor speed had been approaching its physical limits, and this implied that the processor clock speed could not continue to increase. In the past few years, the maximum CPU speeds for desktops and tablets began to plateau and are now ranging between 3—3.5 and 1.5—2 GHz, respectively. With the advent of platforms with smaller form factors, keeping the processor frequency limited has become the new norm, while focus has shifted toward lowering the system power consumption and toward more efficient utilization of available system resources.

Digital video applications require huge amounts of processing. Additionally, real-time processing and playback requirements mandate certain capabilities and performance levels from the system. Only a couple of decades ago, real-time video encoding was possible only by using high-performance, special-purpose hardware or massively parallel computing on general-purpose processors, primarily in noncommercial academic solutions. Both hardware and software needed careful performance optimization and tuning at the system and application level to achieve reasonable quality in real-time video. However, with the tremendous improvement in processor speed and system resource utilization in recent years, encoding speed at higher orders of magnitude, with even better quality, can be achieved with today's processors.

This chapter starts with a brief discussion of CPU clock speed and considers why indefinite increases in clock speed are impractical. The discourse then turns to motivations for achieving high video coding speed, and the tradeoffs necessary to achieve such performance. Then we discuss the factors affecting encoding speed, performance bottlenecks that can be encountered, and approaches to optimization. Finally, we present various performance-measurement considerations, tools, applications, methods, and metrics.

CPU Speed and its Limits

The following are the major reasons the CPU clock speed cannot continue to increase indefinitely:

- High-frequency circuits consume power at a rate that increases with frequency; dissipating that heat becomes impossible at a certain point. In 2001, Intel CTO Pat Gelsinger predicted, "Ten years from now, microprocessors will run at 10 GHz to 30 GHz." But for their proportional size, "these chips will produce as much heat as a nuclear reactor."[1] Heat dissipation in high-frequency circuits is a fundamental problem with normal cooling technologies, and indefinite increases in frequency is not feasible from either economic or engineering points of view.

- Contemporary power-saving techniques such as clock gating and power gating do not work with high-frequency circuits. In clock gating, a clock-enable is inserted before each state element such that the element is not clocked if the data remains unchanged. This saves significant charge/discharge that would be wasted in writing the same bit, but it introduces an extra delay into the critical clock path, which is not suitable for high-frequency design. In power gating, large transistors act as voltage sources for various functional blocks of the processor; the functional blocks can potentially be turned off when unused. However, owing to the extra voltage drop in power-gating transistors, the switching speed slows down; therefore, this technique is not amenable to high-frequency design, either.

- Transistors themselves have reached a plateau in speed. While transistors are getting smaller, they are not getting much faster. To understand why, let's consider the following fact from electronics: a thinner gate dielectric leads to a stronger electric field across the transistor channel, enabling it to switch faster. A reduction in transistor gate area means that the gate could be made thinner without adversely increasing the load capacitance necessary to charge up the control node to create the electric field. However,

---

[1]M. Kanellos, "Intel CTO: Chip heat becoming critical issue," *CNET News*, February 5, 2001. Available at news.cnet.com/Intel-CTO-Chip-heat-becoming-critical-issue/2100-1040_3-252033.html.

at 45 nm process technology, the gate dielectric was already approximately 0.9 nm thick, which is about the size of a single silicon-dioxide molecule. It is simply impossible to make this any thinner from the same material. With 22 nm, Intel has made use of the innovative tri-gate technology to combat this limitation. Further, changing the gate dielectric and the connection material helped increase the transistor speed but resulted in an expensive solution. Basically, the easy scaling we have had in the 1980s and 1990s, when every shrink in transistor size would also lead to faster transistors, is not available anymore.

- Transistors are no longer the dominant factor in processor speed. The wires connecting these transistors are becoming the most significant delay factor. As transistors become smaller, the connecting wires become thinner, offering higher resistances and allowing lower currents. Given the fact that smaller transistors are able to drive less current, it is easy to see that the circuit path delay is only partially determined by transistor switching speed. To overcome this, attempts are made during chip design to route the clock and the data signal on similar paths, thus obtaining about the same travel time for these two signals. This works effectively for data-heavy, control-light tasks such as a fixed-function video codec engine. However, the design of general-purpose microprocessors is complex, with irregular interactions and data travels to multiple locations that do not always follow the clock. Not only are there feedback paths and loops but there are also control-heavy centralized resources such as scheduling, branch prediction, register files, and so on. Such tasks can be parallelized using multiple cores, but thinner wires are required when processor frequencies are increased.

Motivation for Improvement

In the video world, *performance* is an overloaded term. In some literature, *encoder performance* refers to the compression efficiency in terms of number of bits used to obtain certain visual quality level. The average bit rate savings of the test encoder compared to a reference encoder is reckoned as the objective coding performance criterion. Examples of this approach can be found in Nguen and Merpe[2] and Grois et al.[3] From another view, *encoder performance* means the encoding speed in frames per

---

[2]T. Nguen and D. Marpe, "Performance Analysis of HEVC-Based Intra Coding for Still Image Compression," in *Proceedings of 2012 Picture Coding Symposium* (Krakow, Poland: IEEE, 2012), 233–36.
[3]D. Grois, D. Marpe, A. Mulayoff, B. Itzhaky, and O. Hadar, "Performance Comparison of H.265/MPEG-HEVC, VP9, and H.264/MPEG-AVC Encoders," in *Proceedings of the 30th Picture Coding Symposium* (San Jose, CA: IEEE, 2013), 394–97. Although this paper mentions software run times in addition to bit rate savings, the encoder implementations are not optimized and cannot be objectively compared.

second (FPS). In this book, we adopt this latter meaning. We also note that *FPS* may be used for different purposes. A video clip generally has an associated frame rate in terms of FPS (e.g., 24 or 30 FPS), which means that the clip is supposed to be played back in real time (i.e., at that specified FPS) to offer the perception of smooth motion. However, when the compressed video clip is generated, the processing and compression tasks can be carried out many times faster than real time; this speed, also expressed in FPS, is referred to as the *encoding speed* or *encoder performance*. Note that in some real-time applications such as video conferencing, where the video frames are only consumed in real time, an encoding speed faster than real time is not necessary but is sufficient, as faster processing allows the processor to go to an idle state early, thereby saving power.

However, there are several video applications and usages where faster than real-time processing is desirable. For example:

- Long-duration video can be compressed in a much shorter time. This is useful for video editors, who typically deal with a large amount of video content and work within specified time limits.

- Video archiving applications can call for compressing and storing large amount of video, and can benefit from fast encoding.

- Video recording applications can store the recorded video in a suitable compressed format; the speedy encoding allows concurrently running encoding and/or non-encoding tasks to share processing units.

- Converting videos from one format to another benefits from fast encoding. For example, several DVDs can be simultaneously converted from MPEG-2 to AVC using popular video coding applications such as Handbrake.

- Video transcoding for authoring, editing, uploading to the Internet, burning to discs, or cloud distribution can take advantage of encoding as fast as possible. In particular, by using multiple times faster than real-time encoding, many cloud-based video distribution-on-demand services can serve multiple requests simultaneously while optimizing the network bandwidth by packaging together multiple bitstreams for distribution.

- Video transrating applications can benefit from fast encoding. Cable, telecommunications, and satellite video distribution is often made efficient by transrating a video to a lower bit rate, thereby accommodating more video programs within the same channel bandwidth. Although the overall delay in a transrating and repacketization system is typically constant and only real-time processing is needed, speedup in the transrating and constituent encoding tasks is still desirable from the point of view of scheduling flexibility and resource utilization.

Typical video applications involve a series of tasks, such as video data capture; compression, transmission, or storage; decompression; and display, while trying to maintain a constant overall system delay. The delay introduced by the camera and display devices is typically negligible; quite often, the decoding, encoding, and processing times become the performance focus. Among these, the decoding tasks are usually specified by the video standards and they need a certain number of operations per second. But the magnitude of computation in video encoding and processing tasks exceeds by a large margin the computational need of the decoding tasks. Therefore, depending on the application requirements, the encoding and processing tasks are usually more appropriate candidates for performance optimization, owing to their higher complexities.

In particular, video encoding requires a large number of signal processing operations—on the order of billions of operations per second. Fortunately, video compression can easily be decomposed into pipelined tasks. Within the individual tasks, the video data can be further disintegrated in either spatial or temporal dimensions into a set of independent sections, making it suitable for parallel processing. Taking advantage of this property, it is possible to obtain faster than real-time video encoding performance by using multiple processing units concurrently. These processing units may be a combination of dedicated special-purpose fixed-function and/or programmable hardware units. The advantage of specialized hardware is that it is usually optimized for specific tasks, so that those tasks are accomplished in a performance- and power-optimized manner. However, programmable units provide flexibility and do not become obsolete easily. Performance tuning for programmable units are also less expensive than the dedicated hardware units. Therefore, efficiently combining the specialized and programmable units into a hybrid solution can deliver an order of magnitude greater than real-time performance, as offered by the recent Intel (R) Core (TM) and Intel (R) Atom (TM) CPUs, where the heavy lifting of the encoding tasks is carried out by the integrated graphics processing units (GPU).

Performance Considerations

In video encoding and processing applications, performance optimization aims to appropriately change the design or implementation to improve the encoding or processing speed. Increasing the processor frequency alone does not yield the best-performing encoding solution, and as discussed before, there is a limit to such frequency increase. Therefore, other approaches for performance enhancement need to be explored. Note that some techniques implement the necessary design or implementation changes relatively cheaply, but others may need significant investment. For example, inexpensive approaches to obtaining higher performance include parallelization of encoding tasks, adjusting schedules of the tasks, optimization of resource utilization for individual tasks, and so on. It is interesting to note that higher performance can also be achieved by using more complex dedicated-hardware units, which in turn is more expensive to manufacture. A general consideration for performance optimization is to judiciously choose the techniques that would provide the highest performance with lowest expense and lowest overhead. However, depending on the nature of the application and available resources, it may be necessary to accommodate large dollar expenditures to provide the expected performance. For example, a bigger cache may cost more money, but it will likely help achieve certain performance objectives. Thus, the tradeoffs for any performance optimization must be well thought out.

Usually performance optimization is not considered by itself; it is studied together with visual quality and aspects of power consumption. For instance, a higher CPU or GPU operating frequency will provide faster encoding speed, but will also consume more energy. A tradeoff between energy consumed and faster encoding speed is thus necessary at the system design and architectural level. For today's video applications running on resource-constrained computing platforms, a balanced tradeoff can be obtained by maximizing the utilization of available system resources when they are active and putting them to sleep when they are not needed, thereby achieving simultaneous power optimization.

However, note that higher encoding speeds can also be achieved by manipulating some video encoding parameters such as the bit rate or quantization parameters. By discarding a large percentage of high-frequency details, less information remains to be processed and the encoding becomes faster. However, this approach directly affects the visual quality of the resulting video. Therefore, a balance is also necessary between visual quality and performance achieved using this technique.

There are three major ways encoding performance can be maximized for a given period of time:

- Ensure that available system resources, including the processor and memory, are fully utilized during the active period of the workload. However, depending on the workload, the nature of resource utilization may be different. For example, an encoding application should run at a 100 percent duty cycle of the processor. As mentioned earlier, such performance maximization can also include considerations for power optimization—for example, by running at 100 percent duty cycle for as long as necessary and quickly going to sleep afterwards. However, for a real-time playback application, it is likely that only a fraction of the resources will be utilized—say, at 10 percent duty cycle. In such cases, performance optimization may not be needed and power saving is likely to be emphasized instead.

- Use specialized resources, if available. As these resources are generally designed for balanced performance and power for certain tasks, this approach would provide performance improvement without requiring explicit tradeoffs.

- Depending on the application requirements, tune certain video parameters to enhance encoding speed. However, encoding parameters also affect quality, compression, and power; therefore, their tradeoffs against performance should be carefully considered.

Maximum Resource Utilization

Applications, services, drivers, and the operating system compete for the important system resources, including processor time, physical memory space and virtual address space, disk service time and disk space, network bandwidth, and battery power. To achieve the best performance per dollar, it is important to maximally utilize the available system resources for the shortest period of time possible. Thus, maximum performance is obtained at the cost of minimum power consumption. Toward this end, the following techniques are typically employed:

- **Task parallelization:** Many tasks are independent of each other and can run in parallel, where resources do not need to wait until all other tasks are done. Parallelization of tasks makes full utilization of the processor. Often, pipelines of tasks can also be formed to keep the resources busy during the operational period, thereby achieving maximum resource utilization. (Task parallelization will be discussed in more detail in a later section.)

- **Registers, caches, and memory utilization:** Optimal use of memory hierarchy is an important consideration for performance. Memory devices at a lower level are faster to access, but are smaller in size; they have higher transfer bandwidth with fewer transfer units, but are more costly per byte compared to the higher level memory devices. Register transfer operations are controlled by the processor at processor speed. Caches are typically implemented as static random access memories (SRAMs) and are controlled by the memory management unit (MMU). Careful use of multiple levels of cache at the system-level programs can provide a balance between data access latency and the size of the data. Main memories are typically implemented as dynamic RAMs (DRAMs), are much larger than the cache, but require slower direct memory access (DMA) operations for data access. The main memory typically has multiple modules connected by a system bus or switching network. Memory is accessed randomly or in a block-by-block basis. In parallel memory organizations, both interleaved and pipelined accesses are practiced: interleaving spreads contiguous memory locations into different memory modules, while access memory modules are overlapped in a pipelined fashion. Performance of data transfer between adjacent levels of memory hierarchy is represented in terms of hit (or miss) ratios—that is, the probability that an information item will be found at a certain memory level. The frequency of memory access and the effective access time depend on the program behavior and choices in memory design. Often, extensive analysis of program traces can lead to optimization opportunities.

- **Disk access optimization:** Video encoding consists of processing large amounts of data. Therefore, often disk I/O speed, memory latency, memory bandwidth, and so on become the performance bottlenecks rather than the processing itself. Many optimization techniques are available in the literature addressing disk access. Use of redundant arrays of inexpensive disks (RAID) is a common but costly data-storage virtualization technique that controls data access redundancy and provides balance among reliability, availability, performance, and capacity.

- **Instruction pipelining:** Depending on the underlying processor architecture, such as complex instruction set computing (CISC) processor, reduced instruction set computing (RISC) processor, very long instruction word (VLIW) processor, vector supercomputer, and the like, the cycles per instruction are different with respect to their corresponding processor clock rates. However, to achieve the minimum number of no operations (NOPs) and pipeline stalls, and thereby optimize the utilization of resources, there needs to be careful instruction pipelining and pipeline synchronization.

Resource Specialization

In addition to maximizing the utilization of resources, performance is enhanced by using specialized resources. Particular improvements in this area include the following:

- **Special media instruction sets:** Modern processors have enhanced instruction sets that include special media instructions possessing inherent parallelism. For example, to calculate the sum of absolute difference (SAD) for a eight 16-bit pixel vector, a 128-bit single instruction multiple data (SIMD) instruction can be used, expending one load and one parallel operation, as opposed to the traditional sequential approach where sixteen 16-bit loads, eight subtractions, eight absolute-value operations, and eight accumulation operations would have been needed. For encoding tasks such as motion estimation, such media instructions play the most important role in speeding up the compute-intensive task.

- **GPU acceleration:** Traditionally, video encoding tasks have been carried out on multi-core CPUs. Operation-intensive tasks such as video encoding often run with high CPU utilization for all cores. For higher resolution videos, the CPU can be pushed beyond its capability so that the task would not be complete in real time. There are several research efforts to employ parallelization techniques on various shared-memory and distributed-memory platforms to deal with this issue, some of which are discussed in the next section. However, it is easy to see that to obtain a desirable and scalable encoding solution, CPU-only solutions are often not sufficient.

  Recent processors such as Intel Core and Atom processors offer hardware acceleration for video encoding and processing tasks by using the integrated processor graphics hardware. While special-purpose hardware units are generally optimized for certain tasks, general-purpose computing units are more flexible in that they can be programmed for a variety of tasks. The Intel processor graphics hardware is a combination of fixed-function and programmable units, providing a balance among speed, flexibility, and scalability. Substantial attention is also paid to optimizing the systems running these graphics hardware for low power consumption, thus providing high performance with reduced power cost. Thus, using hardware acceleration for video encoding and processing tasks is performance and power friendly as long as the real-time supply of input video data is ensured.

  Figure 5-1 shows CPU utilization of a typical encoding session with and without processor graphics hardware—that is, GPU acceleration. From this figure, it is obvious that employing GPU acceleration not only makes the CPU available for other tasks but also increases the performance of the encoding itself. In this example, the encoding speed went up from less than 1 FPS to over 86 FPS.

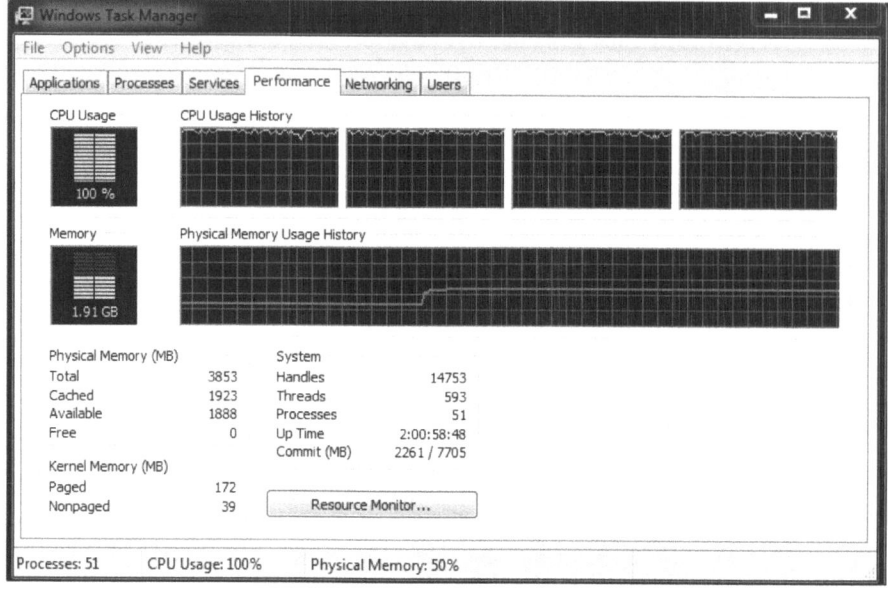

CPU-only encoding

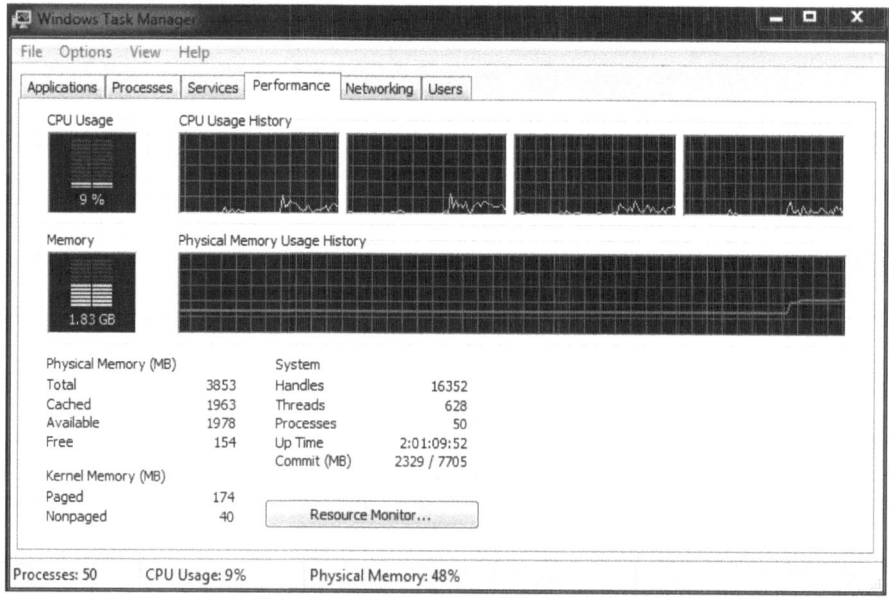

GPU-accelerated encoding

Figure 5-1. *CPU utilization of typical encoding with and without GPU acceleration*

Video Parameters Tuning

To tune the video parameters for optimum performance, it is important to understand the main factors that contribute to performance, and to identify and address the typical performance bottlenecks.

Factors Determining Encoding Speed

Many factors affect the video encoding speed, including system hardware, network configurations, storage device types, nature of the encoding tasks, available parallelization opportunities, video complexity and formats, and hardware acceleration possibilities. Interactions among these factors can make performance tuning complex.

System Configurations

There are several configurable system parameters that affect, to varying degrees, the performance of workloads such as the video encoding speed. Some of these parameters are the following:

- **Number of cores:** The number of processing CPU and GPU cores directly contributes to workload performance. Distributing the workload into various cores can increase the speed of processing. In general, all the processing cores should be in the same performance states for optimum resource utilization. The performance states are discussed in Chapter 6 in detail.

- **CPU and GPU frequencies:** The CPU and GPU core and package clock frequencies are the principal determining factors for the execution speed of encoding tasks. Given that such tasks can take advantage of full hardware acceleration, or can be shared between the CPU and the GPU, utilization of these resources, their capabilities in terms of clock frequencies, the dependences and scheduling among these tasks, and the respective data access latencies are crucial factors for performance optimization.

- **Memory size and memory speed:** Larger memory size is usually better for video encoding and processing tasks, as this helps accommodate the increasingly higher video resolutions without excessive memory paging costs. Higher memory speed, obviously, also significantly contributes to speeding up these tasks.

- **Cache configurations:** Cache memory is a fast memory built into the CPU or other hardware units, or located next to it on a separate chip. Frequently repeated instructions and data are stored in the cache memory, allowing the CPU to avoid loading and storing data from the slower system bus, and thereby improving overall system speed. Cache built into the CPU itself is referred to as Level 1 (L1) cache, while cache residing on a separate chip next to the CPU is called Level 2 (L2) cache. Some CPUs have both L1 and L2 caches built in and designate the cache chip as Level 3 (L3) cache. Use of L3 caches significantly improves the performance of video encoding and processing tasks. Similarly, integrated GPUs have several layers of cache. Further, recent processors with embedded dynamic random access memories (eDRAMs) generally yield 10 to 12 percent higher performance for video encoding tasks.

- **Data access speed:** Apart from scheduling delays, data availability for processing depends on the non-volatile storage speed and storage type. For example, solid-state disk drives (SSDs) provide much faster data access compared to traditional spinning magnetic hard disk drives, without sacrificing reliability. Disk caching in hard disks uses the same principle as memory caching in CPUs. Frequently accessed hard-disk data is stored in a separate segment of RAM, avoiding frequent retrieval from the hard disk. Disk caching yields significantly better performance in video encoding applications where repeated data access is quite common.

- **Chipset and I/O throughput:** Given that uncompressed video is input to the video encoding tasks, nd some processing tasks also output the video in uncompressed formats, often I/O operations become the bottleneck in these tasks, especially for higher resolution videos. In I/O-bound tasks, an appropriately optimized chipset can remove this bottleneck, improving overall performance. Other well-known techniques to improve the efficiency of I/O operations and to reduce the I/O latency include intelligent video data placement on parallel disk arrays, disk seek optimization, disk scheduling, and adaptive disk prefetching.

- **System clock resolution:** The default timer resolution in Windows is 15.625 msec, corresponding to 64 timer interrupts per second. For tasks such as video encoding, where all operations related to a video frame must be done within the specified time frame (e.g., 33 msec for 30 fps video), the default timer resolution is not sufficient. This is because a task may need to wait until the next available timer tick to get scheduled for execution. Since there are often dependences among the encoding tasks, such as DCT transform and variable length coding, scheduling these tasks must carefully consider timer resolution along with the power consumption for optimum performance. In many applications, a timer resolution of 1 msec is typically a better choice.

- **BIOS:** Several performance-related parameters can be adjusted from the BIOS; among them are peripheral component interconnect express (PCIe) latency and clock gating, advanced configuration and power interface (ACPI) settings (e.g., disabling hibernation), CPU configuration (e.g., enabling adjacent cache line prefetch), CPU and graphics power management control (e.g., allowing support for more than two frequency ranges, allowing turbo mode, allowing CPU to go to C-states when it is not fully utilized [details of C-states are discussed in Chapter 6], configuring C-state latency, setting interrupt response time limits, enabling graphics render standby), enabling overclocking features (e.g., setting graphics overclocking frequency), and so on.

- **Graphics driver:** Graphics drivers incorporate various performance optimizations, particularly for hardware-accelerated video encoding and processing tasks. Appropriate and updated graphics drivers would make a difference in attaining the best performance.

- **Operating system:** Operating systems typically perform many optimizations, improving the performance of the run-time environments. They also control priorities of processes and threads. For example, Dalvik and ART (Android RunTime) are the old and new run times, respectively, that execute the application instructions inside Android. While Dalvik is a just-in-time (JIT) run time that executes code only when it is needed, ART—which was introduced in Android 4.4 KitKat and is already available to users—is an ahead-of-time (AOT) run time that executes code before it is actually needed. Comparisons between Dalvik and ART on Android 4.4 have shown that the latter brings enhanced performance and battery efficiency, and will be available as the default run time for devices running Android version 4.5 (Lollipop).

- **Power settings:** In addition to thermal design power (TDP), Intel has introduced a new specification, called the *scenario design power* (SDP) since the third-generation Core and Pentium Y-processors. While TDP specifies power dissipation under worst-case real-world workloads and conditions, SDP specifies power dissipation under a specific usage scenario. SDP can be used for benchmarking and evaluation of power characteristics against specific target design requirements and system cooling capabilities. Generally, processors with higher TDP (or SDP) give higher performance. Therefore, depending on the need, a user can choose to obtain a system with higher TDP. However, on a certain platform, the operating system usually offers different power setting modes, such as high performance, balanced, or power saver. These modes control how aggressively the system will go to various levels of idle states. These modes have a noticeable impact on performance, especially for video encoding and processing applications.

The Nature of Workloads

The nature of a workload can influence the performance and can help pinpoint possible bottlenecks. For example, for video coding applications, the following common influential factors should be considered:

- **Compute-bound tasks:** A task is "compute bound" if it would complete earlier on a faster processor. It is also considered compute bound if the task is parallelizable and can have an earlier finish time with an increased number of processors. This means the task spends the majority of its time using the processor for computation rather than on I/O or memory operations. Depending on the parameters used, many video coding tasks, such as motion estimation and prediction, mode decision, transform and quantization, in-loop deblocking, and so on, may be compute bound. Integrated processor graphics, where certain compute-intensive tasks are performed using fixed-function hardware, greatly helps improve the performance of compute-bound tasks.

- **I/O-bound tasks:** A task is "I/O bound" if it would complete earlier with an increase in speed of the I/O subsystem or the I/O throughput. Usually, disk speed limits the performance of I/O-bound tasks. Reading raw video data from files for input to a video encoder, especially reading higher resolution uncompressed video data, is often I/O bound.

- **Memory-bound tasks:** A task is "memory bound" if its rate of progress is limited by the amount of memory available and the speed of that memory access. For example, storing multiple reference frames in memory for video encoding is likely to be memory bound. The same task may be transformed from compute bound to memory bound on higher frequency processors, owing to the ability of faster processing.

- **Inter-process communication:** Owing to dependences, tasks running on different processes in parallel often need to communicate with each other. This is quite common in parallel video encoding tasks. Depending on the configuration of the parallel platform, interprocess communication may materialize using message passing, using shared memory, or other techniques. Excessive interprocess communication adversely affects the performance and increasingly dominates the balance between the computation and the communication as the number of processes grows. In practice, to achieve improved scalability, parallel video encoder designers need to minimize the communication cost, even at the expense of increased computation or memory operations.

- **Task scheduling:** The scheduling of tasks running in parallel has a huge impact on overall performance, particularly on heterogeneous computing platforms. Heterogeneous multi-core processors with the same instruction set architecture (ISA) are typically composed of small (e.g., in-order) power-efficient cores and big (e.g., out-of-order) high-performance cores. In general, small cores can achieve good performance if the workload inherently has high levels of instruction level parallelism (ILP). On the other hand, big cores provide good performance if the workload exhibits high levels of memory-level parallelism (MLP) or requires the ILP to be extracted dynamically. Therefore, scheduling decisions on such platforms can be significantly improved by taking into account how well a small or big core can exploit the ILP and MLP characteristics of a workload. On the other hand, making wrong scheduling decisions can lead to suboptimal performance and excess energy or power consumption. Techniques are available in the literature to understand which workload-to-core mapping is likely to provide the best performance.[4]

- **Latency:** Latency usually results from communication delay of a remote memory access and involves network delays, cache miss penalty, and delays caused by contentions in split transactions. Latency hiding can be accomplished through four complementary approaches[5]: (i) using *prefetching techniques* which brings instructions or data close to the processor before it is actually needed, (ii) using *coherent caches* supported by hardware to reduce cache misses, (iii) using *relaxed memory consistency models* that allow buffering and pipelining of memory references, and (iv) using *multiple-context* support that allows a processor to switch from one context to another when a long latency operation is encountered. Responsiveness of a system depends on latency. For real-time video communication applications such as video conferencing, latency is an important performance factor, as it significantly impacts the user experience.

---

[4]K. V. Craeynest, A. Jaleel, L. Eeckhout, P. Narvaez, and J. Emer, "Scheduling Heterogeneous Multi-Cores through Performance Impact Estimation," in *Proceedings of 39th Annual International Symposium on Computer Architecture* (Portland, OR: IEEE, June 2012), 213–24.
[5]K. Hwang, *Advanced Computer Architecture: Parallelism, Scalability, Programmability* (Singapore: McGraw-Hill, 1993).

- **Throughput:** Throughput is a measure of how many tasks a system can execute per unit of time. This is also known as the *system throughput*. The number of tasks the CPU can handle per unit time is the *CPU throughput*. As system throughput is derived from the CPU (and other resource) throughput, when multiple tasks are interleaved for CPU execution, CPU throughput is higher than the system throughput. This is due to the system overheads caused by the I/O, compiler, and the operating system, because of which the CPU is kept idle for a fraction of the time. In real-time video communication applications, the smoothness of the video depends on the system throughput. Thus, it is important to optimize all stages in the system, so that inefficiency in one stage does not hinder overall performance.

Encoding Tools and Parameters

It should be noted that not only do the various algorithmic tasks affect the performance, but some video encoding tools and parameters are also important factors. Most of these tools emerged as quality-improvement tools or as tools to provide robustness against transmission errors. Fortunately, however, they usually offer opportunities for performance optimization through parallelization. The tools that are not parallelization friendly can take advantage of algorithmic and code optimization techniques, as described in the following sections. Here are a few important tools and parameters.

Independent data units

To facilitate parallelization and performance gain, implementations of video coding algorithms usually exploit frame-level or group of frame-level independence or divide video frames into independent data units such as slices, slice groups, tiles, or wavefronts.

At the frame level, usually there is little parallelism owing to motion compensation dependences. Even if parallelized, because of the varying frame complexities, the encoding and decoding times generally fluctuate a lot, thus creating an imbalance in resource utilization. Also, owing to dependency structure, the overall latency may increase with frame-level parallelization.

A video frame consists of one or more slices. A slice is a group of macroblocks usually processed in raster-scan order. Figure 5-2 shows a typical video frame partitioned into several slices or groups of slices.

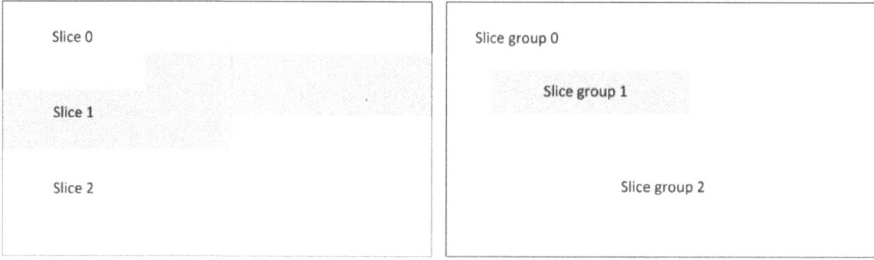

Figure 5-2. Partitioning of a video frame into slices and slice groups

Slices were introduced mainly to prevent loss of quality in the case of transmission errors. As slices are defined as independent data units, loss of a slice is localized and may not impact other slices unless they use the lost slice as a reference. Exploiting the same property of independence, slices can be used in parallel for increased performance. In an experiment using a typical AVC encoder, it was found that four slices per frame can yield a 5 to 15 percent performance gain compared to a single slice per frame, depending on the encoding parameters. However, employing slices for parallelism may incur significant coding efficiency losses. This is because, to keep the data units independent, spatial redundancy reduction opportunities may be wasted. Such loss in coding efficiency may be manifested as a loss in visual quality. For example, in the previous experiment with AVC encoder, four slices per frame resulted in a visual quality loss of ~0.2 to ~0.4 dB compared to a single slice per frame, depending on the encoding parameters. Further, a decoder relying on performance gains from parallel processing of multiple slices alone may not obtain such gain if it receives a video sequence with a single slice per frame.

The concept of slice groups was also introduced as an error-robustness feature. Macroblocks belonging to a slice group are typically mixed with macroblocks from other slice groups during transmission, so that loss of network packets minimally affects the individual slices in a slice group. However, owing to the independence of slice groups, they are good candidates for parallelization as well.

In standards after H.264, the picture can be divided into rectangular tiles—that is, groups of coding tree blocks separated by vertical and horizontal boundaries. Tile boundaries, similarly to slice boundaries, break parse and prediction dependences so that a tile can be processed independently, but the in-loop filters such as the deblocking filters can still cross tile boundaries. Tiles have better coding efficiency compared to slices. This is because tiles allow picture partition shapes that contain samples with a potential higher correlation than slices, and tiles do not have the slice header overhead. But, similar to slices, the coding efficiency loss increases with the number of tiles, owing to the breaking of dependences along partition boundaries and the resetting of CABAC probabilities at the beginning of each partition.

In the H.265 standard, wavefronts are introduced to process rows of coding tree blocks in parallel, each row starting with the CABAC probabilities available after processing the second block of the row above. This creates a different type of dependency, but still provides an advantage compared to slices and tiles, in that no coding dependences are broken at row boundaries. Figure 5-3 shows an example wavefront.

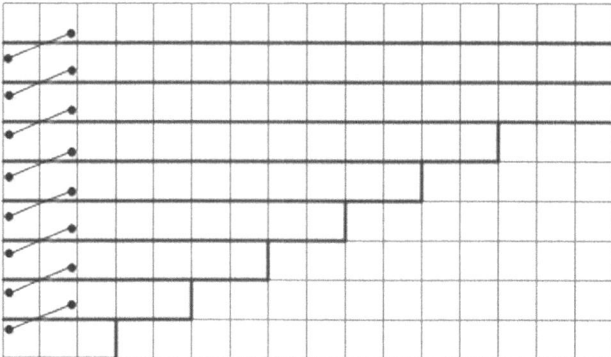

Figure 5-3. *Wavefronts amenable to parallel processing; for the starting macroblock of a row, CABAC probabilities are propagated from the second block of the previous macroblock row*

The CABAC probabilities are propagated from the second block of the previous row without altering the raster-scan order. This reduces the coding efficiency losses and results in only small rate-distortion differences compared to nonparallel bitstreams. However, the wavefront dependencies mean that all the rows cannot start processing at the same time. This introduces parallelization inefficiencies, a situation that is more prominent with more parallel processors.

However, the ramping inefficiencies of wavefront parallel processing can be mitigated by overlapping the execution of consecutive pictures.[6] Experimental results reported by Chi et al. show that on a 12-core system running at 3.33 GHz, for decoding of 3840×2160 video sequences, overlapped wavefronts provide a speedup by a factor of nearly 11, while regular wavefronts and tiles provide reasonable speedup of 9.3 and 8.7, respectively.

GOP structure

The encoding of intra-coded (I) pictures, predicted (P) pictures, and bi-predicted (B) pictures requires different amounts of computation and consequently has different finish times. The pattern of their combination, commonly known as the group of pictures (GOP) structure, is thus an important factor affecting the encoding speed. In standards before the H.264, I-pictures were the fastest and B-pictures were the slowest, owing to added motion estimation and related complexities. However, in the H.264 and later standards, I-pictures may also take a long time because of Intra prediction.

Depending on the video contents, the use of B-pictures in the H.264 standard may decrease the bit rate by up to 10 percent for the same quality, but their impact on performance varies from one video sequence to another, as the memory access frequency varies from -16 to +12 percent.[7] Figure 5-4 shows the results of another experiment

[6]C. C. Chi, M. Alvarez-Mesa, B. Juurlink, G. Clare, F. Henry, et al., "Parallel Scalability and Efficiency of HEVC Parallelization Approaches," *IEEE Transactions on Circuits and Systems for Video Technology* 22, no. 12 (December 2012): 1827–38.

[7]J. Ostermann, J. Bormans, P. List, D. Marpe, M. Narroschke, et al., "Video Coding with H.264/AVC: Tools, Performance and Complexity," *IEEE Circuits and Systems* (First Quarter, 2004): 7–28.

comparing the quality achieved by using no B-picture, one B-picture, and two B-pictures. In this case, using more B-pictures yields better quality. As a rule of thumb, B-pictures may make the coding process slower for a single processing unit, but they can be more effectively parallelized, as a B-picture typically is not dependent on another B-picture unless it is used as a reference—for instance, in a pyramid structure.

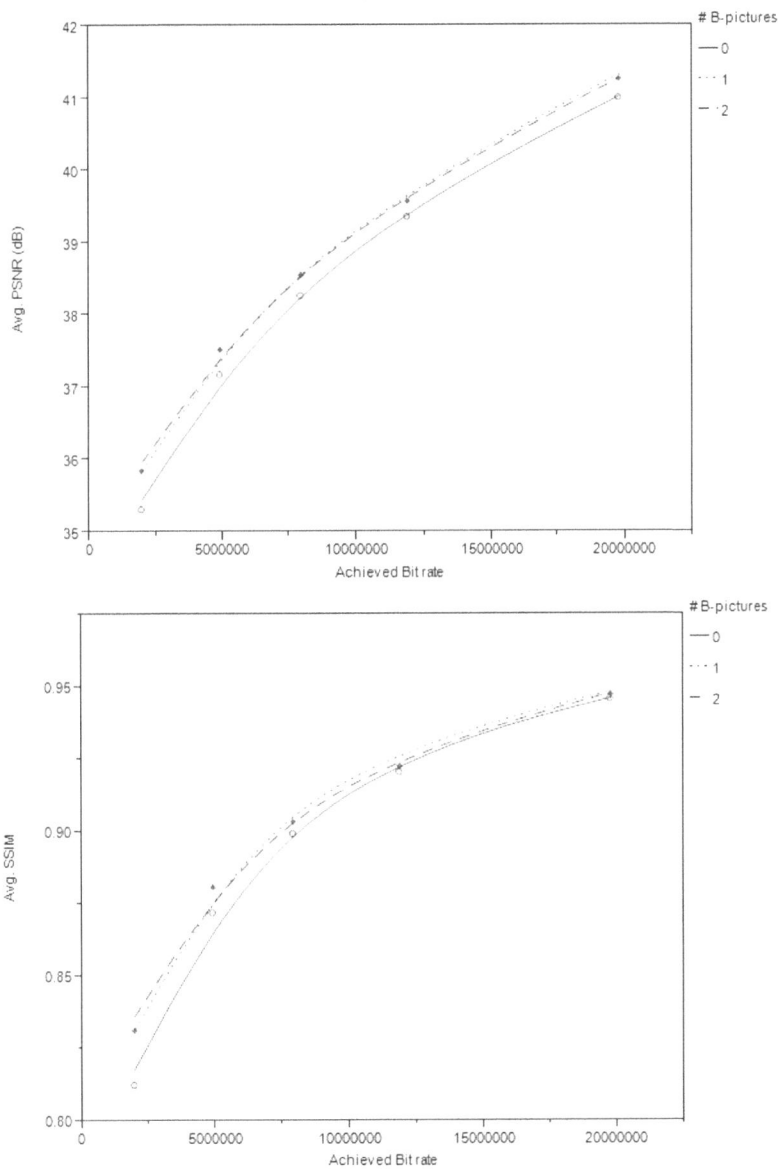

Figure 5-4. *Effect of B-pictures on quality for a 1280×720 H.264 encoded video sequence named park run*

Bit rate control

Using a constant quantization parameter for each picture in a group of pictures is generally faster than trying to control the quantization parameter based on an available bit budget and picture complexity. Extra compute must be done for such control. Additionally, bit rate control mechanisms in video encoders need to determine the impact of choosing certain quantization parameters on the resulting number of bits as they try to maintain the bit rate and try not to overflow or underflow the decoder buffer. This involves a feedback path from the entropy coding unit back to the bit rate control unit, where bit rate control model parameters are recomputed with the updated information of bit usage. Often, this process may go through multiple passes of entropy coding or computing model parameters. Although the process is inherently sequential, algorithmic optimization of bit rate control can be done to improve performance for applications operating within a limited bandwidth of video transmission. For example, in a multi-pass rate control algorithm, trying to reduce the number of passes will improve the performance. An algorithm may also try to collect the statistics and analyze the complexity in the first pass and then perform actual entropy coding in subsequent passes until the bit rate constraints are met.

Multiple reference pictures

It is easy to find situations where one reference picture may yield a better block matching and consequent lower cost of motion prediction than another reference picture. For example, in motion predictions involving occluded areas, a regular pattern of using the immediate previous or the immediate future picture may not yield the best match for certain macroblocks. It may be necessary to search in a different reference picture where that macroblock was visible. Sometimes, more than one reference picture gives a better motion prediction compared to a single reference picture. This is the case, for example, during irregular object motion that does not align with particular grids of the reference pictures. Figure 5-5 shows an example of multiple reference pictures being used.

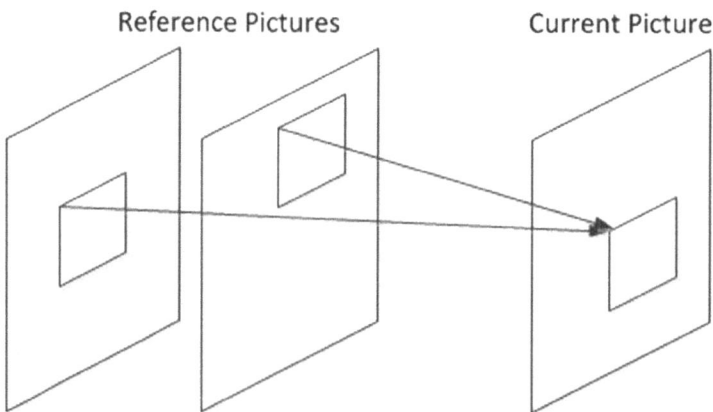

Figure 5-5. *Motion compensated prediction with multiple reference pictures*

To accommodate the need for multiple predictions, in the H.264 and later standards, the multiple reference pictures feature was introduced, resulting in improved visual quality. However, there is a significant performance cost incurred when performing searches in multiple reference pictures. Note that if the searches in various reference pictures can be done in parallel, the performance penalty can be alleviated to some extent while still providing higher visual quality compared to single-reference motion prediction.

R-D Lagrangian optimization

For the encoding of video sequences using the H.264 and later standards, Lagrangian optimization techniques are typically used for choice of the macroblock mode and estimation of motion vectors. The mode of each macroblock is chosen out of all possible modes by minimizing a rate-distortion cost function, where distortion may be represented by the sum of the squared differences between the original and the reconstructed signals of the same macroblock, and the rate is that required to encode the macroblock with the entropy coder. Similarly, motion vectors can be efficiently estimated by minimizing a rate-distortion cost function, where distortion is usually represented by the sum of squared differences between the current macroblock and the motion compensated macroblock, and the rate is that required to transmit the motion information consisting of the motion vector and the corresponding reference frame number. The Lagrangian parameters in both minimization problems are dependent on the quantization parameter, which in turn is dependent on the target bit rate.

Clearly, both of these minimizations require large amounts of computation. While loop parallelization, vectorization, and other techniques can be applied for performance optimization, early exits from the loops can also be made if the algorithm chooses to do so, at the risk of possible non-optimal macroblock mode and motion vectors that may impact the visual quality at particular target bit rates. These parallelization approaches are discussed in the next section.

Frame/field mode for interlaced video

For interlaced video, choice of frame/field mode at the macroblock or picture level significantly affects performance. On the other hand, the interlaced video quality is generally improved by using tools such as macroblock-adaptive or picture-adaptive frame/field coding. It is possible to enhance performance by using only a certain pattern of frame and field coding, but this may compromise the visual quality.

Adaptive deblocking filter

Using in-loop deblocking filters on reconstructed pictures reduces blocky artifacts. Deblocked pictures, therefore, serve as a better-quality reference for intra- and inter-picture predictions, and result in overall better visual quality for the same bit rate. The strength of the deblocking filters may vary and can be adaptive on the three levels: at the slice level, based on individual characteristics of a video sequence; at the block-edge level, based on intra- versus inter-mode decision, motion differences, and the presence of residuals in the two participating neighboring blocks; and at the pixel level, based on an analysis to distinguish between the true edges and the edges created by the blocky artifact. True edges should be left unfiltered, while the edges from quantization should be smoothed out.

In general, deblocking results in bit rate savings of around 6 to 9 percent at medium qualities[8]; equivalently at the same bit rate, the subjective picture quality improvements are more remarkable. Deblocking filters add a massive number of operations per frame and substantially slow down the coding process. Also, it is difficult to parallelize this task because it is not confined to the independent data units, such as slices. This is another example of a tradeoff between visual quality and performance.

Video Complexity and Formats

Video complexity is an important factor that influences the encoding speed. More complex scenes in a video generally take longer to encode, as more information remains to be coded after quantization. Complex scenes include scenes with fine texture details, arbitrary shapes, high motion, random unpredictable motion, occluded areas, and so on. For example, scenes with trees, moving water bodies, fire, smoke, and the like are generally complex, and are often less efficiently compressed, impacting encoding speed as well. On the other hand, easy scenes consisting of single-tone backgrounds and one or two foreground objects, such as head and shoulder-type scenes, are generally prone to better prediction, where matching prediction units can be found early and the encoding can be accelerated. These easy scenes are often generated from applications such as a videophone, video conferencing, news broadcasts, and so on. Frequent scene changes require many frames to be independently encoded, resulting in less frequent use of prediction of the frame data. If the same video quality is attempted, only lower compression can be achieved. With more data to process, performance will be affected.

Video source and target formats are also important considerations. Apart from the professional video contents generated by film and TV studios, typical sources of video include smartphones, point-and-shoot cameras, consumer camcorders, and DVRs/PVRs. For consumption, these video contents are generally converted to target formats appropriate for various devices, such as Apple iPads, Microsoft XBoxes, Sony PSx consoles, and the like, or for uploading to the Internet. Such conversion may or may not use video processing operations such as scaling, denoising, and so on. Thus, depending on the target usage, the complexity of operations will vary, exerting different speed requirements and exhibiting different performance results.

GPU-based Acceleration Opportunities

Applications and system-level software can take advantage of hardware acceleration opportunities, in particular GPU-based accelerations, to speed up the video encoding and processing tasks. Either partial or full hardware acceleration can be used. For example, in a transcoding application, either the decoding or the encoding part or both, along with necessary video processing tasks, can be hardware accelerated for better performance. By employing GPU-based hardware acceleration, typically an order of magnitude faster than real-time performance can be achieved, even for complex videos.

---

[8]Ostermann et al., "Video Coding."

Furthermore, hardware-based security solutions can be used for seamless integration with hardware-accelerated encoding and processing for overall enhancement of the encoding speed of premium video contents. In traditional security solutions, security software would occasionally interrupt and slow down long encoding sessions running on the CPU. However, by employing hardware-based security, improvements can be achieved in both performance and security.

Performance Optimization Approaches

The main video encoding tasks are amenable to performance optimization, usually at the expense of visual quality or power consumption. Some of the techniques may have only trivial impact on power consumption and some may have little quality impact, yet they improve the performance. Other techniques may result in either quality or power impacts while improving performance.

Algorithmic optimizations contribute significantly to speeding up the processing involved in video encoding or decoding. If the algorithm runs on multi-core or multiprocessor environments, quite a few parallelization approaches can be employed. Furthermore, compiler and code optimization generally yield an additional degree of performance improvement. Besides these techniques, finding and removing the performance bottlenecks assists performance optimization in important ways. In the context of video coding, common performance optimization techniques include the following.

Algorithmic Optimization

Video coding algorithms typically focus on improving quality at the expense of performance. Such techniques include the use of B-pictures, multiple-reference pictures, two-pass bit rate control, R-D Langrangian optimization, adaptive deblocking filter, and so on. On the other hand, performance optimization using algorithmic approaches attempt to improve performance in two ways. The first way is by using fast algorithms, typically at the expense of higher complexity, higher power consumption, or lower quality. Joint optimization approaches of performance and complexity are also available in the literature.[9] A second way is to design algorithms that exploit the available parallelization opportunities with little or no quality loss.[10]

---

[9]J. Zhang, Y. He, S. Yang, and Y. Zhong, "Performance and Complexity Joint Optimization for H.264 Video Coding," in *Proceedings of the 2003 International Symposium on Circuits and Systems 2* (Bangkok: IEEE, 2003), 888–91.
[10]S. M. Akramullah, I. Ahmad, and M. L. Liou, "Optimization of H.263 Video Encoding Using a Single Processor Computer: Performance Tradeoffs and Benchmarking," *IEEE Transactions on Circuits and Systems for Video Technology* 11, no. 8 (August 2001): 901–15.

Fast Algorithms

Many fast algorithms for various video coding tasks are available in the literature, especially for the tasks that take longer times to finish. For example, numerous fast-motion estimation algorithms try to achieve an order of magnitude higher speed compared to a full-search algorithm with potential sacrifice in quality. Recent fast-motion estimation algorithms, however, exploit the statistical distribution of motion vectors and only search around the most likely motion vector candidates to achieve not only a fast performance but almost no quality loss as well. Similarly, fast DCT algorithms[11] depend on smart factorization and smart-code optimization techniques. Some algorithms exploit the fact that the overall accuracy of the DCT and inverse DCT is not affected by the rounding off and truncations intrinsic to the quantization process.[12] Fast algorithms for other video coding tasks try to reduce the search space, to exit early from loops, to exploit inherent video properties, to perform activity analysis, and so on, with a view toward achieving better performance. There are several ways to improve the encoding speed using algorithmic optimization.

Fast Transforms

Fast transforms use factorization and other algorithmic maneuvers to reduce the computational complexity in terms of number of arithmetic operations needed to rapidly compute the transform. Fast Fourier Transform (FFT) is a prime example of this, which takes only $O(N \log N)$ arithmetic operations, instead of the $O(N^2)$ operations required in the original N-point Discrete Fourier Transform (DFT) algorithm. For large data sets, the resulting time difference is huge; in fact, the advent of FFT made it practical to calculate Fourier Transform on the fly and enabled many practical applications. Furthermore, instead of floating-point operations, fast transforms tend to use integer operations that can be more efficiently optimized. Typically, fast transforms such as the DCT do not introduce errors so there is no additional impact on the visual quality of the results. However, possible improvements in power consumption because of fewer arithmetic operations are usually not significant, either.

---

[11]E. Feig and S. Winograd, "Fast Algorithms for the Discrete Cosine Transform," *IEEE Transactions on Signal Processing* 40, no. 9 (September 1992): 2174–93.
[12]L. Kasperovich, "Multiplication-free Scaled 8x8 DCT Algorithm with 530 Additions," in *Proceedings of the SPIE 2419, Digital Video Compression: Algorithms and Technologies* (San Jose: SPIE-IS&T, 1995), 499–504.

Fast DCT or its variants are universally used in the video coding standards. In the H.264 and later standards, transform is generally performed together with quantization to avoid loss in arithmetic precision. Nonetheless, as fast transform is performed on a large set of video data, data parallelism approaches can easily be employed to parallelize the transform and improve the performance. A data parallel approach is illustrated in the following example.

Let's consider the butterfly operations in the first stage of DCT (see Figure 2.17), which can be expressed as:

$$\left(u'_0, u'_1, u'_3, u'_2\right) = \left(u_0, u_1, u_3, u_2\right) + \left(u_7, u_6, u_4, u_5\right)$$
$$\left(u'_4, u'_7, u'_5, u'_6\right) = \left(u_3, u_0, u_2, u_1\right) - \left(u_4, u_7, u_5, u_6\right)$$

(Equation 5-1)

Considering each input u_k to be a 16-bit integer, sets of four such inputs can be rearranged into 64-bit wide vectors registers, as shown in Figure 5-6. The rearrangement is necessary to maintain the correspondence of data elements on which operations are performed. This will provide 64-bit wide additions and subtractions in parallel, effectively speeding up this section of operations by a factor of 4. Similarly, wider vector registers can be exploited for further improved performance.

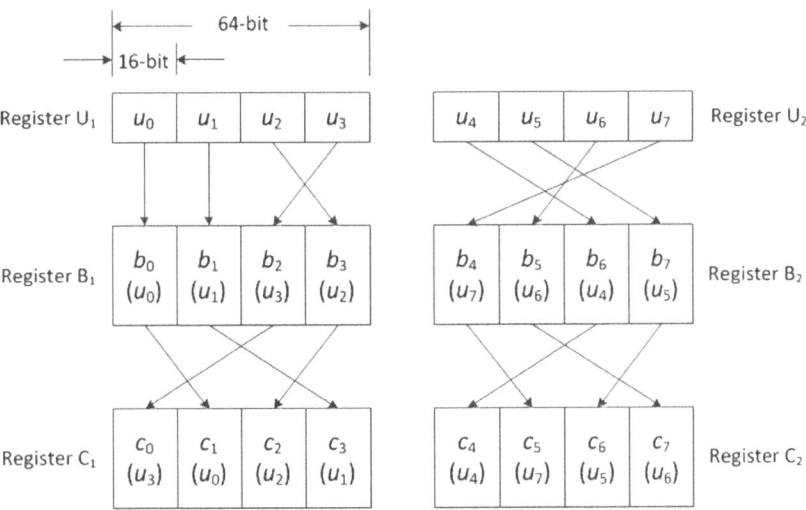

Figure 5-6. *Data rearrangement in 8-point DCT to facilitate data parallelism*

Fast Intra Prediction

In the H.264 and later standards, in addition to the transform, Intra prediction is used in spatial redundancy reduction. However, the Intra frame encoding process has several data-dependent and computationally intensive coding methodologies that limit the overall encoding speed. It causes not only a high degree of computational complexity but also a fairly large delay, especially for the real-time video applications. To resolve these issues, based on the DCT properties and spatial activity analysis, Elarabi and Bayoumi[13] proposed a high throughput, fast and precise Intra mode selection, and a direction-prediction algorithm that significantly reduces the computational complexity and the processing run time required for the Intra frame prediction process. The algorithm achieves ~56 percent better Intra prediction run time compared to the standard AVC implementation (JM 18.2), and ~35 to 39 percent better Intra prediction run time compared to other fast Intra prediction techniques. At the same time, it achieves a PSNR within 1.8 percent (0.72 dB) of the standard implementation JM 18.2, which is also ~18 to 22 percent better than other fast Intra prediction algorithms. In another example, using a zigzag pattern of calculating the 4×4 DC prediction mode, Alam et al.[14] has improved both the PSNR (up to 1.2 dB) and the run time (up to ~25 percent) over the standard implementation.

Fast Motion Estimation

Block matching motion estimation is the most common technique used in inter-picture motion prediction and temporal redundancy reduction. It performs a search to find the best matching block in the reference picture with the current block in the current picture. The estimation process is typically conducted in two parts: estimation with integer pixel-level precision and with fractional pixel-level precision. Often, fractional pixel-level motion search is done with half-pixel and quarter-pixel precision around the best integer pixel position, and the resulting motion vectors are appropriately scaled to maintain the precision.

Motion estimation is the most time-consuming process in the coding framework. It typically takes ~60 to 90 percent of the compute time required by the whole encoding process, depending on the configuration and the algorithm. Thus, a fast implementation of motion estimation is very important for real-time video applications.

[13]T. Elarabi and M. Bayoumi, "Full-search-free Intra Prediction Algorithm for Real-Time H.264/ AVC Decoder," in *Signal Processing, Image Processing and Pattern Recognition* (Jeju, Korea: Springer-Verlag, 2012), 9–16.

[14]T. Alam, J. Ikbal, and T. Alam, "Fast DC Mode Prediction Scheme for Intra 4x4 Block in H.264/AVC Video Coding Standard," *International Journal of Advanced Computer Science and Applications* 3, no. 9 (2012): 90–94.

There are many ways to speed up the motion estimation process. These include:

- Fewer locations can be searched to find the matching block. However, the problem of how to determine which locations to search has been an active area of research for longer than two decades, producing numerous fast-motion estimation algorithms. If the right locations are not involved, it is easy to fall into local minima and miss the global minimum in the search space. This would likely result in nonoptimal motion vectors. Consequently, a higher cost would be incurred in terms of coding efficiency if the block is predicted from a reference block using these motion vectors, compared to when the block is simply coded as Intra. Thus, the block may end up being coded as an Intra block, and fail to take advantage of existing temporal redundancy.

  Recent algorithms typically search around the most likely candidates of motion vectors to find the matching block. Predicted motion vectors are formed based on the motion vectors of the neighboring macroblocks, on the trend of the inter-picture motion of an object, or on the motion statistics. Some search algorithms use different search zones with varying degrees of importance. For example, an algorithm may start the search around the predicted motion vector and, if necessary, continue the search around the co-located macroblock in the reference picture. Experimentally determined thresholds are commonly used to control the flow of the search. The reference software implementation of the H.264 and later standards use fast-search algorithms that depict these characteristics.

- Instead of matching the entire block, partial information from the blocks may be matched for each search location. For example, every other pixel in the current block can be matched with corresponding pixels in the reference block.

- A search can be terminated early based on certain conditions and thresholds that are usually determined experimentally. An example of such early termination can be found in the adaptive motion estimation technique proposed by Zhang et al.,[15] which improves the speed by ~25 percent for the macroblocks in motion, while improves the performance by ~3 percent even for stationary macroblocks by checking only five locations. The average PSNR loss is insignificant at ~0.1 dB.

- Instead of waiting for the reconstructed picture to be available, the source pictures can be used as references, saving the need for reconstruction at the encoder. Although this technique provides significant performance gain, it has the disadvantage that the prediction error is propagated from one frame to the next, resulting in significant loss in visual quality.

[15]D. Zhang, G. Cao, and X. Gu, "Improved Motion Estimation Based on Motion Region Identification," in *2012 International Conference on Systems and Informatics* (Yantai, China: IEEE, 2012), 2034–37.

- Motion estimation is easily parallelizable in a data-parallel manner. As the same block-matching operation such as the SAD is used on all the matching candidates, and the matching candidates are independent of each other, SIMD can easily be employed. Further, motion estimation for each block in the current picture can be done in parallel as long as an appropriate search window for each block is available from the reference picture. Combining both approaches, a single program multiple data (SPMD)-type of parallelization can be used for each picture.

- Using a hierarchy of scaled reference pictures, it is possible to conduct the fractional and integer pixel parts separately in parallel, and then combine the results.

- In bi-directional motion estimation, forward and backward estimations can be done in parallel.

Fast Mode Decision

The H.264 and later standards allow the use of variable block sizes that opens the opportunity to achieve significant gains in coding efficiency. However, it also results in very high computational complexity, as mode decision becomes another important and time-consuming process. To improve the mode decision performance, Wu et al.[16] proposed a fast inter-mode decision algorithm based on spatial homogeneity and the temporal stationarity characteristics of video objects, so that only a few modes are selected as candidate modes. The spatial homogeneity of a macroblock is decided based on its edge intensity, while the temporal stationarity is determined by the difference between the current macroblock and its co-located counterpart in the reference frame. This algorithm reduces 30 percent of the encoding time, on average, with a negligible PSNR loss of 0.03 dB or, equivalently, a bit rate increment of 0.6 percent.

Fast Entropy Coding

Entropy coding such as CABAC is inherently a sequential task and is not amenable to parallelization. It often becomes the performance bottleneck for video encoding. Thus, performance optimization of the CABAC engine can enhance the overall encoding throughput. In one example,[17] as much as ~34 percent of throughput enhancement is achieved by pre-normalization, hybrid path coverage, and bypass bin splitting. Context modeling is also improved by using a state dual-transition scheme to reduce the critical path, allowing real-time ultra-HDTV video encoding on an example 65 nm video encoder chip running at 330 MHz.

---

[16]D. Wu, F. Pan, K. P. Lim, S. Wu, Z. G. Li, et al., "Fast Intermode Decision in H.264/AVC Video Coding," *IEEE Transactions on Circuits and Systems for Video Technology* 15, no. 6 (July 2005): 953–58.

[17]J. Zhou, D. Zhou, W. Fei, and S. Goto, "A High Performance CABAC Encoder Architecture for HEVC and H.264/AVC," in *Proceedings of ICIP* (Melbourne: IEEE, 2013), 1568–72.

Parallelization Approaches

Parallelization is critical for enabling multi-threaded encoding or decoding applications adapted to today's multi-core architectures. Independent data units can easily scale with the parallel units, whereas dependences limit the scalability and parallelization efficiency. Since several independent data units can be found in video data structures, their parallelization is straightforward. However, not all data units and tasks are independent. When there are dependences among some data units or tasks, there are two ways to handle the dependences: by communicating the appropriate data units to the right processors, and by using redundant data structure. It is important to note that the interprocessor communication is an added overhead compared to a sequential (non-parallel, or scalar) processing. Therefore, parallelization approaches are typically watchful of the communication costs, sometimes at the expense of storing redundant data. In general, a careful balance is needed among the computation, communication, storage requirements, and resource utilization for efficient parallelization.

Data Partitioning

The H.264 standard categorizes the syntax elements into up to three different partitions for a priority-based transmission. For example, headers, motion vectors, and other prediction information are usually transmitted with higher priority than the details of the syntax elements representing the video content. Such data partitioning was primarily designed to provide robustness against transmission errors, and was not intended for parallelization. Indeed, parallel processing of the few bytes of headers and many bytes of detailed video data would not be efficient. However, video data can be partitioned in several different ways, making it suitable for parallelization and improved performance. Both uncompressed and compressed video data can be partitioned into independent sections, so both video encoding and decoding operations can benefit from data partitioning.

Data partitioning plays an important role in the parallelization of video encoding. *Temporal* partitioning divides a video sequence into a number of independent subsequences, which are processed concurrently in a pipelined fashion. At least a few subsequences must be available to fill the pipeline stages. This type of partitioning is thus suitable for off-line video encoding.[18] *Spatial* partitioning divides a frame of video into various sections that are encoded simultaneously. Since only one frame is inputted at a time, this type of partitioning is suitable for online and low-delay encoding applications that process video on a frame-by-frame basis. It is clear that parallel encoding of the video subsequences deals with coarser grains of data that can be further partitioned into smaller grains like a section of a single frame, such as slices, slice groups, tiles, or wavefronts.

---

[18]I. Ahmad, S. M. Akramullah, M. L. Liou, and M. Kafil, "A Scalable Off-line MPEG-2 Video Encoding Scheme Using a Multiprocessor System," *Parallel Computing* 27, no. 6 (May 2001): 823–46.

Task Parallelization

The task parallelization approach for video encoding was introduced as early as 1991 for compact disc-interactive applications.[19] This introductory approach took advantage of a multiple instruction multiple data (MIMD) parallel object-oriented computer. The video encoder was divided into tasks and one task was assigned to one or more processors of the 100-node message-passing parallel computer, where a node consisted of a data processor, memory, a communications processor, and I/O interfaces. This approach loosely used task parallelization, where some processors were running tasks with different algorithms, but others were running tasks with the same algorithm at a given time. At a higher level, the tasks were divided into two phases: a motion-estimation phase for prediction and interpolation where motion vectors were searched in each frame, and video compression where it was decided which of these motion vectors (if any) would be used.

The parallelization of the motion estimation phase was not task parallel by itself; it involved assigning each processor its own frame along with the associated reference frames. This process inevitably required copying the reference frames onto several appropriate processors, thus creating a performance overhead. Also, many frames had to have been read before all processors had some tasks to execute. The video compression phase did not have independent frames, so several parts of a frame were processed in parallel. A compression unit made up of a group of processors repeatedly received sets of consecutive blocks to encode. The tasks in the compression unit were mode decision, DCT, quantization, and variable length coding. The resulting bitstream was sent to an output manager running on a separate processor, which combined the pieces from all the compression units and sent the results to the host computer. The compression units reconstructed their own parts of the resulting bitstream to obtain the reference frames.

Note that the quantization parameter depends on the data reduction in all blocks processed previously, and one processor alone cannot compute it. Therefore, a special processor must be dedicated to computation of the quantization parameter, sending the parameter to appropriate compression units and collecting the size of the compressed data from each of the compression units for further calculation. An additional complication arises from the fact that motion vectors are usually differentially coded based on the previous motion vector. But the compression units working independently do not have access to the previous motion vector. To resolve this, compression units must send the last motion vector used in the bitstream to the compression unit that is assigned the next blocks. Figure 5-7 shows the communication structure of the task parallelization approach.

---

[19]F. Sijstermans and J. Meer, "CD-I Full-motion Video Encoding on a Parallel Computer," *Communications of the ACM* 34, no. 4 (April 1991): 81–91.

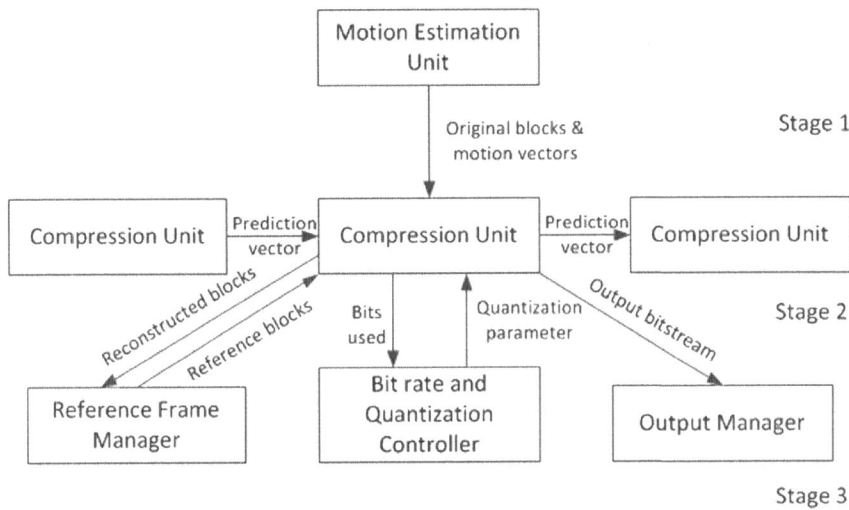

Figure 5-7. *Communication structure in task parallelization*

This idea can be used in video encoding in general, regardless of the video coding standards or the algorithms used. However, the idea can be further improved to reduce the communication overhead. For example, in a system, the processors can identify themselves in the environment and can attach their processor numbers as tags to the data they process. These tags can be subsequently removed by the appropriate destination processors, which can easily rearrange the data as needed. It is important to understand that appropriate task scheduling is necessary in the task parallelization approach, as many tasks are dependent on other tasks, owing to the frame-level dependences.

Pipelining

Pipelines are cascades of processing stages where each stage performs certain fixed functions over a stream of data flowing from one end to the other. Pipelines can be linear or dynamic (nonlinear). *Linear* pipelines are simple cascaded stages with streamlined connections, while in *dynamic* pipelines feedback and/or feed-forward connection paths may exist from one stage to another. Linear pipelines can be further divided into synchronous and asynchronous pipelines. In *asynchronous* pipelines, the data flow between adjacent stages is controlled by a handshaking protocol, where a stage S_i sends a ready signal to the next stage S_{i+1} when it is ready to transmit data. Once the data is received by stage S_{i+1}, it sends an acknowledge signal back to S_i. In *synchronous* pipelines, clocked latches are used to interface between the stages. Upon arrival of a clock pulse, all latches transfer data to the next stage simultaneously. For a k-stage linear pipeline, a multiple of k clock cycles are needed for the data to flow through the pipeline.[20] The number of clock cycles between two initiations of a pipeline is called the *latency* of the pipeline. The *pipeline efficiency* is determined by the percentage of time that each pipeline stage is used, which is called the *stage utilization*.

Video encoding tasks can form a three-stage dynamic pipeline, as shown in Figure 5-7. The first stage consists of the motion-estimation units; the second stage has several compression units in parallel, and the third stage is the output manager. The bit rate and quantization control unit and the reference frame manager can be considered as two delay stages having feedback connections with the second-stage components.

Data Parallelization

If data can be partitioned into independent units, they can be processed in parallel with minimum communication overhead. Video data possess this characteristic. There are a few common data parallelization execution modes, including single instruction multiple data (SIMD), single program multiple data (SPMD), multiple instruction multiple data (MIMD), and so on.

SIMD is a processor-supported technique that allows an operation to be performed on multiple data points simultaneously. It provides data-level parallelism, which is more efficient than scalar processing. For example, some loop operations are independent in successive iterations, so a set of instructions can operate on different sets of data. Before starting execution of the next instruction, typically synchronization is needed among the execution units that are performing the same instruction on the multiple data sets.

SIMD is particularly applicable to image and video applications where typically the same operation is performed on a large number of data points. For example, in brightness adjustment, the same value is added to (or subtracted from) all the pixels in a frame. In practice, these operations are so common that most modern CPU designs include special instruction sets for SIMD to improve the performance for multimedia use. Figure 5-8 shows an example of SIMD technique where two source arrays of eight 16-bit short integers A and B are added simultaneously element by element to produce the result in the destination array C, where the corresponding element-wise sums are written. Using the SIMD technique, a single add instruction operates on 128-bit wide data in one clock cycle.

---

[20]Hwang, *Advanced Computer Architecture.*

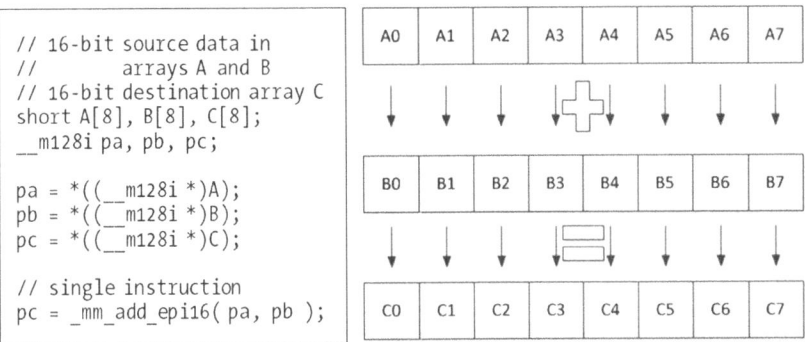

```
// 16-bit source data in
//        arrays A and B
// 16-bit destination array C
short A[8], B[8], C[8];
__m128i pa, pb, pc;

pa = *((__m128i *)A);
pb = *((__m128i *)B);
pc = *((__m128i *)C);

// single instruction
pc = _mm_add_epi16( pa, pb );
```

Figure 5-8. *An example of SIMD technique*

Procedure- or task-level parallelization is generally performed in MIMD execution mode, of which SPMD is a special case. In SPMD, a program is split into smaller independent procedures or tasks, and the tasks are run simultaneously on multiple processors with potentially different input data. Synchronization is typically needed at the task level, as opposed to at the instruction level within a task. Implementations of SPMD execution mode are commonly found on distributed memory computer architectures where synchronization is done using message passing. For a video encoding application, such an SPMD approach is presented by Akramullah et al.[21]

Instruction Parallelization

Compilers translate the high-level implementation of video algorithms into low-level machine instructions. However, there are some instructions that do not depend on the previous instructions to complete; thus, they can be scheduled to be executed concurrently. The potential overlap among the instructions forms the basis of instruction parallelization, since the instructions can be evaluated in parallel. For example, consider the following code:

```
1    R4 = R1 + R2
2    R5 = R1 - R3
3    R6 = R4 + R5
4    R7 = R4 - R5
```

In this example, there is no dependence between instructions 1 and 2, or between 3 and 4, but instructions 3 and 4 depend on the completion of instructions 1 and 2. Thus, instructions 1 and 2 and instructions 3 and 4 can be executed in parallel. Instruction parallelization is usually achieved by compiler-based optimization and by hardware techniques. However, indefinite instruction parallelization is not possible; the parallelization is typically limited by data dependency, procedural dependency, and resource conflicts.

---

[21]S. M. Akramullah, I. Ahmad, and M. L. Liou, "A Data-parallel Approach for Real-time MPEG-2 Video Encoding," *Journal of Parallel and Distributed Computing* 30, no. 2 (November 1995): 129–46.

Instructions in reduced instruction set computer (RISC) processors have four stages that can be overlapped to achieve an average performance close to one instruction per cycle. These stages are instruction fetch, decode, execute, and result write-back. It is common to simultaneously fetch and decode two instructions A and B, but if instruction B has read-after-write dependency on instruction A, the execution stage of B must wait until the write is completed for A. Mainly owing to inter-instruction dependences, more than one instruction per cycle is not achievable in scalar processors that execute one instruction at a time. However, superscalar processors exploit instruction parallelization to execute more than one unrelated instructions at a time; for example, $z=x+y$ and $c=a*b$ can be executed together. In these processors, hardware is used to detect the independent instructions and execute them in parallel.

As an alternative to superscalar processors, very long instruction word (VLIW) processor architecture takes advantage of instruction parallelization and allows programs to explicitly specify the instructions to execute in parallel. These architectures employ an aggressive compiler to schedule multiple operations in one VLIW per cycle. In such platforms, the compiler has the responsibility of finding and scheduling the parallel instructions. In practical VLIW processors such as the Equator BSP-15, the integrated caches are small—the 32 KB data cache and 32 KB instruction cache typically act as bridges between the higher speed processor core and relatively lower speed memory. It is very important to stream in the data uninterrupted so as to avoid the wait times.

To better understand how to take advantage of instruction parallelism in video coding, let's consider an example video encoder implementation on a VLIW platform.[22] Figure 5-9 shows a block diagram of the general structure of the encoding system.

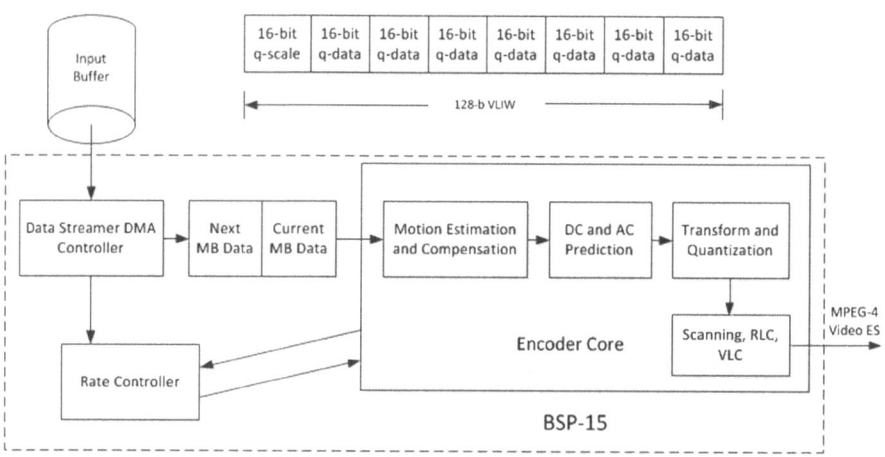

Figure 5-9. *A block diagram of a video encoder on a VLIW platform*

[22]S. M. Akramullah, R. Giduthuri, and G. Rajan, "MPEG-4 Advanced Simple Profile Video Encoding on an Embedded Multimedia System," in *Proceedings of the SPIE 5308, Visual Communications and Image Processing* (San Jose: SPIE-IS&T, 2004), 6–17.

Here, the macroblocks are processed in a pipelined fashion while they go through the different encoding tasks in the various pipeline stages of the encoder core. A *direct memory access* (DMA) controller, commonly known as the *data streamer*, helps prefetch the necessary data. A double buffering technique is used to continually feed the pipeline stages. This technique uses two buffers in an alternating fashion – when the data in one buffer is actively used, the next set of data is loaded onto the second buffer. When processing of the active buffer's data is done, the second buffer becomes the new active buffer and processing of its data starts, while the buffer with used-up data is refilled with new data. Such design is useful in avoiding potential performance bottlenecks.

Fetching appropriate information into the cache is extremely important; care needs to be taken so that both the data and the instruction caches are maximally utilized. To minimize cache misses, instructions for each stage in the pipeline must fit into the instruction cache, while the data must fit into the data cache. It is possible to rearrange the program to coax the compiler to generate instructions that fit into the instruction cache. Similarly, careful consideration of data prefetch would keep the data cache full. For example, the quantized DCT coefficients can be stored in a way so as to help data prefetching in some Intra prediction modes, where only seven coefficients (either from the top row or from the left column) are needed at a given time. The coefficients have a dynamic range (-2048, 2047), requiring 13 bits each, but are usually represented in signed 16-bit entities. Seven such coefficients would fit into two 64-bit registers, where one 16-bit slot will be unoccupied. Note that a 16-bit element relevant for this pipeline stage, such as the quantizer scale or the DC scaler, can be packed together with the quantized coefficients to fill in the unoccupied slot in the register, thereby achieving better cache utilization.

Multithreading

A *thread* is represented by a program *context* comprising a program counter, a register set, and the context status. In a multithreaded parallel computation model, regardless of whether it is run on a SIMD, multiprocessor, or multicomputer, or has distributed or shared memory, a basic unit is composed of multiple threads of computation running simultaneously, each handling a different context on a context-switching basis. The basic structure is as follows:[23] the computation starts with a sequential thread, followed by supervisory scheduling where computation threads begin working in parallel. In case of distributed memory architectures where one or more threads typically run on each processor, interprocessor communication occurs as needed and may overlap among all the processors. Finally, the multiple threads synchronize prior to beginning the next unit of parallel work.

---

[23]G. Bell, "Ultracomputers: A Teraflop before Its Time," *Communications of the ACM* 35, no. 8 (August 1992): 27–47.

Multithreading improves the overall execution performance owing to the facts that a thread, even if stalled, does not prevent other threads from using available resources, and that multiple threads working on the same data can share the cache for better cache usage. However, threads usually work on independent data sets and often interfere with each other when trying to share resources. This typically results in cache misses. In addition, multithreading has increased complexity in terms of synchronization, priorities, and pre-emption handling requirements.

Simultaneously executing instructions from multiple threads is known as *simultaneous multithreading* in general, or Intel Hyper-Threading Technology on Intel processors. To reduce the number of dependent instructions in the pipeline, hyper-threading takes advantage of virtual or logical processor cores. For each physical core, the operating system addresses two logical processors and shares the workload and execution resources when possible.

As performance optimization using specialized media instructions alone is not sufficient for real-time encoding performance, exploiting thread-level parallelism to improve the performance of video encoders has become attractive and popular. Consequently, nowadays multithreading is frequently used for video encoder speed optimization. Asynchronously running threads can dispatch the frame data to multiple execution units in both CPU-based software and GPU-accelerated implementations. It is also possible to distribute various threads of execution between the CPU and the GPU.

Multithreading is often used together with task parallelization, data parallelization, or with their combinations, where each thread operates on different tasks or data sets. An interesting discussion on multithreading as used in video encoding can be found in Gerber et al.,[24] which exploits frame-level and slice-level parallelism using multithreading techniques.

Vectorization

A vector consists of multiple elements of the same scalar data types. The *vector length* refers to the number of elements of the vectors that are processed together, typically 2, 4, 8, or 16 elements.

$$Vector\ length\left(in\ number\ of\ elements\right)= \frac{size\ of\ vector\ registers\left(in\ bits\right)}{Size\ of\ the\ data\ type\left(in\ bits\right)}$$

For example, 128-bit wide vector registers can process eight 16-bit short integers. In this case, vector length is 8. Ideally, vector lengths are chosen by the developer or by the compiler to match the underlying vector register widths.

---

[24]R. Gerber, A. J. C. Bik, K. Smith, and X. Tian, "Optimizing Video Encoding Using Threads and Parallelism," *Embedded*, December 2009. Available at www.embedded.com/design/real-time-and-performance/4027585/Optimizing-Video-Encoding-using-Threads-and-Parallelism-Part-1--Threading-a-video-codec.

Vectorization is a process to convert procedural loops that iterate over multiple pairs of data items and to assign a separate processing unit for each pair. Each processing unit belongs to a *vector lane*. There are the same number of vector lanes as vector lengths, so 2, 4, 8, or 16 data items can be processed simultaneously using as many vector lanes. For example, consider an array A of size 1024 elements is added to an array B, and the result is written to an array C, where B and C are of the same size as A. To implement this addition, a scalar code would use a loop of 1024 iterations. However, if 8 vector lanes are available in the processing units, vectors of 8 elements of the arrays can be processed together, so that only (1024/8) or 128 iterations will be needed. Vectorization is different from thread-level parallelism. It tries to improve performance by using more vector lanes as much as possible. Vector lanes provide additional parallelism on top of each thread running on a single processor core. The objective of vectorization is to maximize the use of available vector registers per core.

Technically, the historic vector-processing architectures are considered separate from SIMD architectures, based on the fact that vector machines used to process the vectors one word at a time through pipelined processors (though still based on a single instruction), whereas modern SIMD machines process all elements of the vector simultaneously. However, today, numerous computational units with SIMD processing capabilities are available at the hardware level, and vector processors are essentially synonymous with SIMD processors. Over the past couple of decades, there has been progressively wider vector registers available for vectorization in each processor core: for example, the 64-bit MMX registers in Pentium to support MMX extensions, 128-bit XMM registers in Pentium IV to support SSE and SSE2 extensions, 256-bit YMM registers in second generation Core processors to support AVX and AVX2 extensions, 512-bit ZMM registers in Xeon Phi co-processors to support MIC extensions. For data-parallelism friendly applications such as video encoding, these wide vector registers are useful.

Conventional programming languages are constrained by their inherent serial nature and don't support the computation capabilities offered by SIMD processors. Therefore, extensions to conventional programming languages are needed to tap these capabilities. Vectorization of the serial codes and vector programming models are developed for this purpose. For example, OpenMP 4.0 supports vector programming models for C/C++ and FORTRAN, and provides language extensions to simplify vector programming, thereby enabling developers to extract more performance from the SIMD processors. The Intel Click Plus is another example that supports similar language extensions.

The auto-vectorization process tries to vectorize a program given its serial constraints, but ends up underutilizing the available computation capabilities. However, as both vector widths and core counts are increasing, explicit methods are developed by Intel to address the trends. With the availability of integrated graphics and co-processors in the modern CPUs, generalized programming models with explicit vector programming capabilities are being added to compilers such as the Intel compiler, GCC, and LLVM, as well as into standards such as OpenMP 4.0. The approach is similar to multithreading, which addresses the availability of multiple cores and parallelizes programs on these cores. Vectorization additionally addresses the availability of increased vector width by explicit vector programming.

Vectorization is useful in video encoding performance optimization, especially for the CPU-based software implementations. Vectors with lengths of 16 elements of pixel data can provide up to 16-fold speed improvement within critical loops—for example, for motion estimation, prediction, transform, and quantization operations. In applications such as video transcoding, some video processing tasks such as noise reduction can take advantage of the regular, easily vectorizable structure of video data and achieve speed improvement.

Compiler and Code Optimization

There are several compiler-generated and manual code optimization techniques that can result in improved performance. Almost all of these techniques offer performance improvement without affecting visual quality. However, depending on the needs of the application, the program's critical path often needs to be optimized. In this section, a few common compiler and code optimization techniques are briefly described. The benefits of these techniques for GPU-accelerated video encoder implementations are usually limited and confined to the application and SDK levels, where the primary encoding tasks are actually done by the hardware units. Nevertheless, some of these techniques have been successfully used in speed optimizations of CPU-based software implementations,[25] resulting in significant performance gains.

Compiler optimization

Most compilers come with optional optimization flags to offer tradeoffs between compiled code size and fast execution speed. For fast speed, compilers typically perform the following:

- **Store variables in registers:** Compilers would store frequently used variables and subexpressions in registers, which are fast resources. They would also automatically allocate registers for these variables.

- **Employ loop optimizations:** Compilers can automatically perform various loop optimizations, including complete or partial loop unrolling, loop segmentation, and so on. Loop optimizations provide significant performance improvements in typical video applications.

- **Omit frame pointer on the call stack:** Often, frame pointers are not strictly necessary on the call stack and can safely be omitted. This usually slightly improves performance.

- **Improve floating-point consistency:** The consistency can be improved, for example, by disabling optimizations that could change floating-point precision. This is a tradeoff between different types of performance optimizations.

---

[25]Akramullah et al., "Optimization."

- **Reduce overhead of function calls:** This can be done, for example, by replacing some function calls with the compiler's intrinsic functions.

- **Trade off register-space saving with memory transaction:** One way to realize such a tradeoff is by reloading pointer variables from memory after each function call. This is another example of choosing between different types of performance optimizations.

Code optimization

Optimizing every part of the software code is not worth the effort. It is more practical to focus on the parts where code optimization will reduce execution time the most. For this reason, profiling and analysis of execution time for various tasks in an application is often necessary.

However, the following techniques often provide significant performance improvement, especially when compilers fail to effectively use the system resources.

- **Reduction of redundant operations:** Careful programming is the key to compact codes. Without loss of functionality, often redundant operations in codes can be reduced or eliminated by carefully reviewing the code.

- **Data type optimization:** Choosing appropriate data types for the program's critical path is important for performance optimization. The data types directly derived from the task definition may not yield optimum performance for various functional units. For example, using scaled floating-point constants and assigning precomputed constants to registers would give better performance than directly using mixed-mode operations of integer and floating-point variables, as defined by most DCT and IDCT algorithms. In some cases such as quantization, or introduction of temporary variables stored in registers, can provide noticeable performance gain.

- **Loop unrolling:** Loop unrolling is the transformation of a loop, resulting in larger loop body size but less iteration. In addition to the automatic compiler optimizer, manual loop unrolling is frequently performed to ensure the right amount of unrolling, as over-unrolling may adversely affect performance. With the CPU registers used more effectively, this process minimizes both the number of load/store instructions and the data hazards arising, albeit infrequently, from inefficient instruction scheduling by the compiler. There are two types of loop unrolling: internal and external. Internal unrolling consists of collapsing some iterations of the innermost loop into larger and more complex statements. These statements require higher numbers of machine instructions, but can be more efficiently scheduled by the compiler optimizer. External loop unrolling consists of moving iterations from outer loops to inner loops through the use of more registers to minimize the number for memory access. In video encoding applications, motion estimation and motion compensated prediction are good candidates to take advantage of loop unrolling.

- **Arithmetic operations optimization:** Divisions and multiplications are usually considered the most cycle-expensive operations. However, in most RISC processors, 32-bit based multiplications take more cycles than 64-bit based multiplications in terms of instruction execution latency and instruction throughput. In addition, floating-point divisions are less cycle-expensive compared to mixed-integer and floating-point divisions. Therefore, it is important to use fewer of these arithmetic operations, especially inside a loop.

Overclocking

Although not recommended, it is possible to operate a processor faster than its rated clock frequency by modifying the system parameters. This process is known as *overclocking*. Although speed can be increased, for stability purposes it may be necessary to operate at a higher voltage as well. Thus, most overclocking techniques result in increasing power consumption and consequently generate more heat, which must be dissipated if the processor is to remain functional. This increases the fan noise and/or the cooling complexity. Contrarily, some manufacturers underclock the processors of battery-powered equipments to improve battery life or implement systems that reduce the frequency when operating under battery. Overclocking may also be applied to a chipset, a discrete graphics card, or memory.

Overclocking allows operating beyond the capabilities of current-generation system components. Because of the increased cooling requirements, the risk of less reliability of operation and potential damage to the component, overclocking is mainly practiced by enthusiasts and hobbyists rather than professional users.

Successful overclocking needs a good understanding of power management. As we will see in Chapter 6, the process of power management is complex in modern processors. The processor hardware and the operating system collaborate to manage the power. In the process, they dynamically adjust the processor core frequencies as appropriate for the current workload. In such circumstances, pushing a certain core to 100 percent frequency may adversely affect the power consumption. In Figure 5-10, the concept is clarified with an example where a typical workload is running on a four-core (eight logical cores) Intel second-generation Core processor.

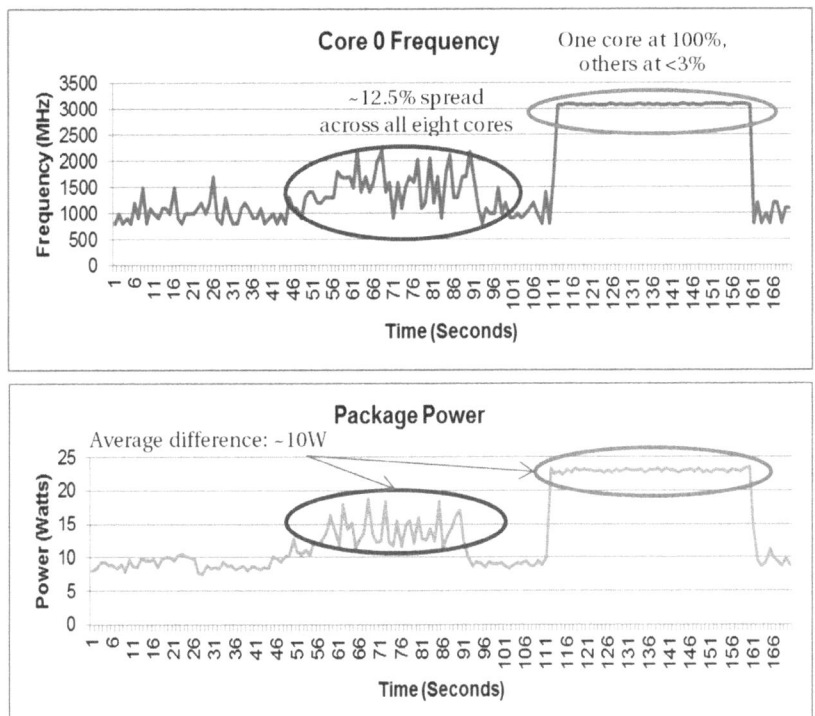

Figure 5-10. *Frequency and power distribution showing the impact of pushing a single core to 100 percent frequency. (Courtesy: J. Feit et al., Intel Corporation, VPG Tech. Summit, 2011)*

In a multi-core processor, if one CPU core is pushed to 100 percent frequency while others are idle, it generally results in higher power consumption. In the example of Figure 5-10, as much as ~10 Watts more power is consumed with a single core running at 100 percent frequency compared to when all eight cores are in use and the average frequency distribution is ~12.5 percent spread across all cores.

Recent Intel processors with integrated graphics allow the hardware-accelerated video encoder to automatically reach the highest frequency state for as long as necessary, and then keep it in idle state when the task is done. Details of this mechanism are discussed in Chapter 6. In a power-constrained environment using modern processors, it is best to leave the frequency adjustment to the hardware and the operating system.

Performance Bottlenecks

Performance bottlenecks occur when system performance is limited by one or more components or stages of the system. Typically, a single stage causes the entire system to slow down. Bottlenecks can be caused either by hardware limitations or inefficient software configurations or both. Although a system may have certain peak performance for a short period of time, for sustainable throughput a system can only achieve performance as fast as its slowest performing component. Ideally, a system should have no performance bottleneck so that the available resources are optimally utilized.

To identify performance bottlenecks, resource utilization needs to be carefully inspected. When one or more resources are underutilized, it is usually an indication of a bottleneck somewhere in the system. Bottleneck identification is an incremental process whereby fixing one bottleneck may lead to discovery of another. Bottlenecks should be identified in a sequential manner, during which only one parameter at a time is identified and varied, and the impact of that single change is captured. Varying more than one parameter at a time could conceal the effect of the change. Once a bottleneck has been eliminated, it is essential to measure the performance again to ensure that a new bottleneck has not been introduced.

Performance-related issues can be found and addressed by carefully examining and analyzing various execution profiles, including:

- Execution history, such as the performance call graphs

- Execution statistics at various levels, including packages, classes, and methods

- Execution flow, such as method invocation statistics

It may be necessary to instrument the code with performance indicators for such profiling. Most contemporary operating systems, however, provide performance profiling tools for run-time and static-performance analysis.

For identification, analysis, and mitigation of performance bottlenecks in an application, the Intel Performance Bottleneck Analyzer[26] framework can be used. It automatically finds and prioritizes architectural bottlenecks for the Intel Core and Atom processors. It combines the latest performance-monitoring techniques with knowledge of static assembly code to identify the bottlenecks. Some difficult and ambiguous cases are prioritized and tagged for further analysis. The tool recreates the most critical paths of instruction execution through a binary. These paths are then analyzed, searching for well-known code-generation issues based on numerous historic performance-monitoring events.

Performance Measurement and Tuning

Performance measurement is needed to verify if the achieved performance meets the design expectations. Furthermore, such measurement allows determination of the actual execution speed of tasks, identification and alleviation of performance bottlenecks, and performance tuning and optimization. It also permits comparison of two tasks—for instance, comparing two video encoding solutions in terms of performance. Thus, it plays an important role in determining the tradeoffs among performance, quality, power use, and amount of compression in various video applications.

Various approaches are available for tuning the system performance of a given application. For instance, compile-time approaches include inserting compiler directives into the code to steer code optimization, using program profilers to modify the object code in multiple passes through the compiler, and so on. Run-time approaches include collecting program traces and event monitoring.

---

[26]E. Niemeyer, "Intel Performance Bottleneck Analyzer," Intel Corporation, August 2011. Retrieved from www.software.intel.com/en-us/articles/intel-performance-bottleneck-analyzer.

Considerations

As configurable system parameters affect the overall performance, it is necessary to fix these parameters to certain values to obtain stable, reliable, and repeatable performance measurements. For example, the BIOS settings, the performance optimization options in the operating system, the options in the Intel graphics common user interface (CUI),[27] and so on must be selected before performance measurements are taken. In the BIOS settings, the following should be considered: the PCIe latency, clock gating, ACPI settings, CPU configuration, CPU and graphics power-management control, C-state latency, interrupt response-time limits, graphics render standby status, overclocking status, and so on.

As we noted in the preceding discussion, workload characteristics can influence the performance. Therefore, another important consideration is the workload parameters. However, it is generally impractical to collect and analyze all possible compile-time and run-time performance metrics. Further, the choice of workloads and relevant parameters for performance measurement is often determined by the particular usage and how an application may use the workload. Therefore, it is important to consider practical usage models so as to select some test cases as key performance indicators. Such selection is useful, for instance, when two video encoding solutions are compared that have performance differences but are otherwise competitive.

Performance Metrics

Several run-time performance metrics are useful in different applications. For example, knowledge of the processor and memory utilization patterns can guide the code optimization. A critical-path analysis of programs can reveal the bottlenecks. Removing the bottlenecks or shortening the critical path can significantly improve overall system performance. In the literature, often system performance is reported in terms of cycles per instruction (CPI), millions of instructions per second (MIPS), or millions of floating-point operations per second (Mflops). Additionally, memory performance is reported in terms of memory cycle or the time needed to complete one memory reference, which is typically a multiple of the processor cycle.

However, in practice, performance tuning of applications such as video coding often requires measuring other metrics, such as the CPU and GPU utilization, processing or encoding speed in frames per second (FPS), and memory bandwidth in megabytes per second. In hardware-accelerated video applications, sustained hardware performance in terms of clocks per macroblock (CPM) can indicate potential performance variability arising from the graphics drivers and the video applications, so that appropriate tuning can be made at the right level for the best performance. Other metrics that are typically useful for debugging purposes include cache hit ratio, page fault rate, load index, synchronization frequency, memory access pattern, memory read and write frequency, operating system and compiler overhead, inter-process communication overhead, and so on.

---

[27]This graphics user interface works on a system with genuine Intel CPUs along with Intel integrated graphics. There are several options available—for example, display scaling, rotation, brightness, contrast, hue and saturation adjustments, color correction, color enhancement, and so on. Some of these options entail extra processing, incurring performance and power costs.

Tools and Applications

The importance of performance measurement can be judged by the large number of available tools. Some performance-analysis tools support sampling and compiler-based instrumentation for application profiling, sometimes with context-sensitive call graph capability. Others support nonintrusive and low-overhead hardware-event-based sampling and profiling. Yet others utilize the hardware-performance counters offered by modern microprocessors. Some tools can diagnose performance problems related to data locality, cache utilization, and thread interactions. In this section, we briefly discuss a couple of popular tools suitable for performance measurement of video applications, particularly the GPU-accelerated applications. Other popular tools, such as Windows Perfmon, Windows Xperf, and Intel Graphics Performance Analyzer, are briefly described in Chapter 6.

VTune Amplifier

The VTune Amplifier XE 2013 is a popular performance profiler developed by Intel.[28] It supports performance profiling for various programming languages, including C, C++, FORTRAN, Assembly, Java, OpenCL, and OpenMP 4.0. It collects a rich set of performance data for hotspots, call trees, threading, locks and waits, DirectX, memory bandwidth, and so on, and provides the data needed to meet a wide variety of performance tuning needs.

Hotspot analysis provides a sorted list of the functions using high CPU time, indicating the locations where performance tuning will yield the biggest benefit. It also supports tuning of multiple threads with *locks and wait* analysis. It enables users to determine the causes of slow performance in parallel programs by quickly finding such common information as when a thread is waiting too long on a lock while the cores are underutilized during the wait. Profiles like hotspot and locks and waits use a software data collector that works on both Intel and compatible processors. The tool also provides advanced hotspot analysis that uses the on-chip Performance Monitoring Unit (PMU) on Intel processors to collect data by hardware event sampling with very low overhead and increased resolution of 1 msec, making it suitable to identify small and quick functions as well. Additionally, the tool supports advanced hardware event profiles like memory bandwidth analysis, memory access, and branch mispredictions to help find tuning opportunities. An optional stack sample collection is supported in the latest version to identify the calling sequence. Furthermore, profiling a remote system and profiling without restarting the application are also supported.

---

[28] The latest version of VTune Amplifier for Systems (2014) is now part of Intel System Studio tool suite at: `https://software.intel.com/en-us/intel-system-studio`.

GPUView

Matthew Fisher and Steve Pronovost originally developed GPUView, which is a tool for determining the performance of the GPU and the CPU. Later, this tool was incorporated into the Windows Performance Toolkit, and can be downloaded as part of the Windows SDK.[29] It looks at performance with regard to direct memory access (DMA) buffer processing and all other video processing on the video hardware. For GPU-accelerated DirectX applications, GPUView is a powerful tool for understanding the relationship between the works done on the CPU and those done on the GPU. It uses an Event Tracing for Windows (ETW) mechanism for measuring and analyzing detailed system and application performance and resource usage. The data-collection process involves enabling trace capture, running the desired test application scenario for which performance analysis is needed and stopping the capture, which saves the data in an event trace log (ETL) file. The ETL file can be analyzed on the same or a different machine using GPUView, which presents the ETL information in a graphic format, as shown in Figure 5-11.

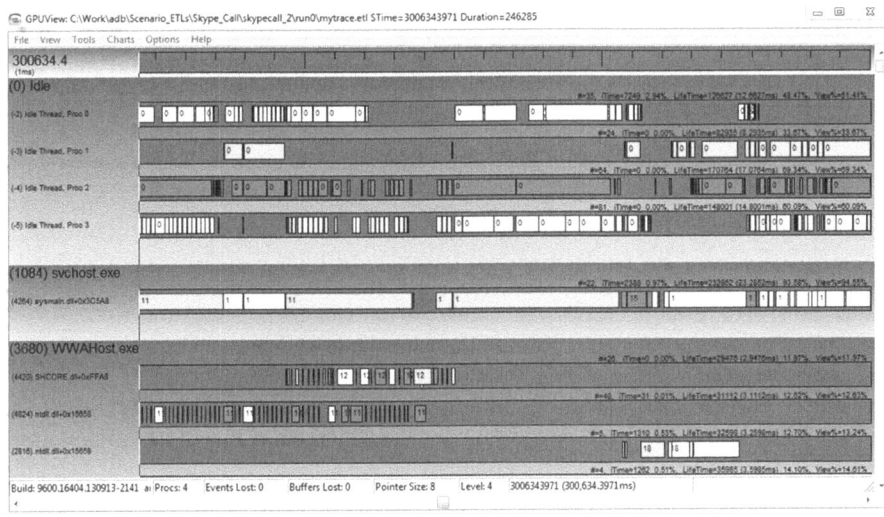

Figure 5-11. *A screenshot from GPUView showing activity in different threads*

GPUView is very useful in analysis and debugging of hardware-accelerated video applications. For example, if a video playback application is observed to drop video frames, the user experience will be negatively affected. In such cases, careful examination of the event traces using GPUView can help identify the issue. Figure 5-12 illustrates an example event trace of a normal video playback, where workload is evenly distributed in regular intervals. The blue vertical lines show the regular *vsync* and red vertical lines show the *present* events.

---

[29]Available from `http://msdn.microsoft.com/en-us/windows/desktop/aa904949.aspx`.

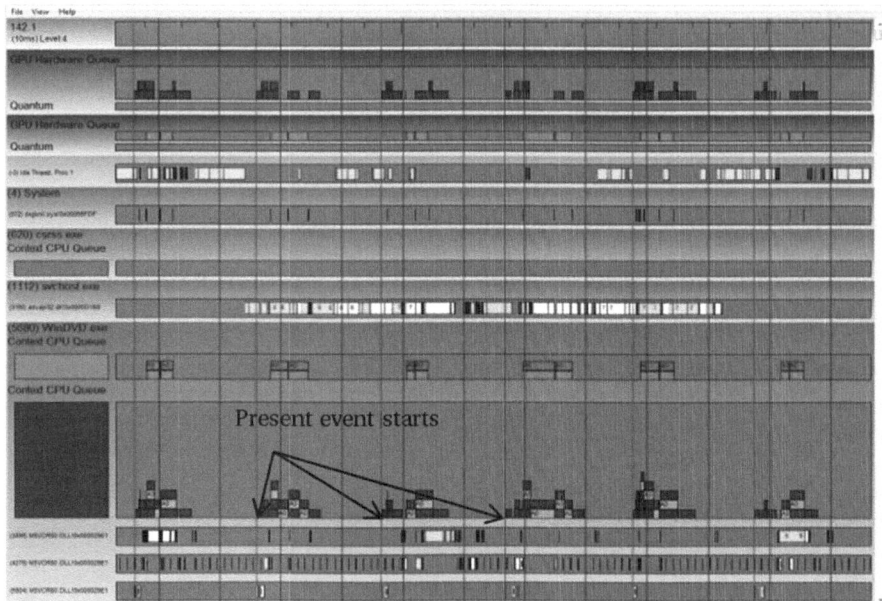

Figure 5-12. *Event trace of a regular video playback*

Figure 5-13 shows event traces of the same video playback application, but when it drops video frames as the frame presentation deadline expires. The profile appears much different compared to the regular pattern seen in Figure 5-12. In the zoomed-in version, the present event lines are visible, from which it is not difficult to realize that there are long delays happening from time to time when the application sends video data packets to the GPU for decoding. Thus it is easy to identify and address the root cause of an issue using GPUView.

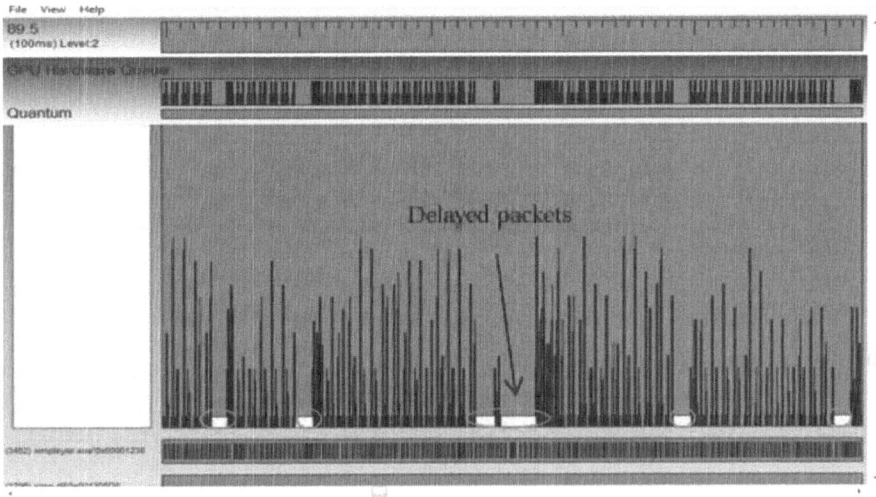

Playback profile zoomed out

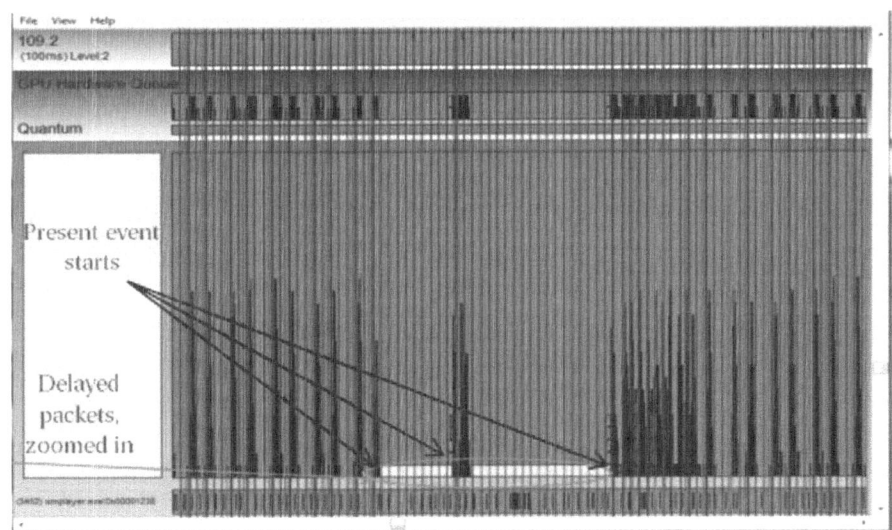

Playback profile zoomed in

Figure 5-13. Event trace of video playback with frame drops

Summary

In this chapter we discussed the CPU clock speed and the extent of the possible increase in clock speed. We noted that the focus in modern processor design has shifted from purely increasing clock speed toward a more useful combination of power and performance. We then highlight the motivation for achieving high performance for video coding applications, and the tradeoffs necessary to achieve such performance.

Then we delved into a discussion of resource utilization and the factors influencing encoding speed. This was followed by a discussion of various performance-optimization approaches, including algorithmic optimization, compiler and code optimization, and several parallelization techniques. Note that some of these parallelization techniques can be combined to obtain even higher performance, particularly in video coding applications. We also discussed overclocking and common performance bottlenecks in the video coding applications. Finally, we presented various performance-measurement considerations, tools, applications, methods, and metrics.

CHAPTER 6

Power Consumption by Video Applications

After discussing video compression, quality, and performance aspects in the previous chapters, in this chapter we turn our attention to another dimension of video application tuning: power consumption. Power consumption needs to be considered together with those other dimensions; tradeoffs are often made in favor of tuning one of these dimensions, based on the needs of the application and with a view toward providing the best user experience. Therefore, we first introduce the concept of power consumption and view its limits on typical modern devices, then we follow with a discussion of common media workloads and usages on consumer platforms. After that, we briefly introduce various criteria for power-aware platform designs.

Within this general setup, the chapter deals with three major topics: power management, power optimization, and power measurement considerations. In regard to power management, we present the standards and management approaches used by the operating system and by the processor. For discussion of power optimization, we present various approaches, including architectural, algorithmic, and system integration optimization. The third topic, power measurement, takes us into the realm of measurement methodologies and considerations.

Besides these three main topics, this chapter also briefly introduces several power measurement tools and applications, along with their advantages and limitations.

Power Consumption and Its Limits

In today's mobile world order, we face ever-increasing desires for compelling user experiences, wearable interfaces, wireless connectivity, all-day computing, and--most critical--higher performance. At the same time, there's high demand for decreasing form factor, lower weight, and quieter battery-powered devices. Such seemingly contradictory requirements present unique challenges: not only do newer mobile devices need extensive battery lives but they also are harder to cool, as they cannot afford bulky fans.

Most important, they need to operate in a limited power envelope. The main concern from the consumer's point of view, however, is having a system with better battery life, which is cost-effective over the device's life. Therefore, power limits are a fundamental consideration in the design of modern computing devices.

Power limits are usually expressed in terms of *thermal design power* (TDP), which is the maximum amount of heat generated for which the cooling requirement is accounted for in the design. For example, TDP is the maximum allowed power dissipation for a platform. The TDP is often broken down into the power consumption of individual components, such as the CPU, the GPU, and so on. Table 6-1 lists typical TDPs for various processor models:

Table 6-1. *Typical Power Envelopes*

Type	TDP (Watts)	Comment
Desktop server/ workstation	47W–120W	High-performance workstations and servers
All-in-one	47W-65W	Common usage
Laptop	35W	
Ultrabook class	17W	Recent areas of focus to decrease TDP
Tablet/Phablet	8W	
Smartphone	<4W	

Although it is important to reduce cost and conserve power for the desktop workstations and servers, these platforms essentially are not limited by the availability of power. Consumer devices and platforms such as portable tablets, phablets, and smartphones, on the other hand, use size-constrained batteries for their power supply.

A major drawback of these devices is that batteries in tablets often drain down before an 8- or 9-hour cross-Atlantic flight is over, and in the case of smartphones, often need a recharge every day. As such, today's market demands over 10 hours of battery life for a tablet to enable users to enjoy a long flight and more than 24 hours for a smartphone for active use. Additionally, the users of these devices and platforms may choose to use them in many different ways, some of which may not be power-efficient. We need, therefore, to understand various power-saving aspects for media applications on these typical consumer platforms.

Power is the amount of energy consumed per unit time, and it is typically expressed in terms of Joules per second, or watts. The switching power dissipated by a chip using static CMOS gates, such as the power consumed by the CPU of a mobile computing device with a capacitance C_{dyn}, running at a frequency f and at a voltage V, is approximately as follows:

$$P = AC_{dyn}V^2f + P_s \qquad \text{(Equation 5.1)}$$

Here, P_s is the static power component introduced mainly due to leakage, and A is an activity constant related to whether or not the processor is active or asleep, or under a gating condition such as clock gating. For a given processor, C_{dyn} is a fixed value; however,

V and f can vary considerably. The formula is not perfect because practical devices as CPUs are not manufactured with 100 percent CMOS and there is special circuitry involved. Also, the static leakage current is not always the same, resulting in variations in the latter part of the equation, which become significant for low-power devices.

Despite the imprecision of the equation, it is still useful for showing how altering the system design will affect power. Running the processor of a device at a higher clock frequency results in better performance; however, as Equation 5-1 implies, at a lower frequency it results in less heat dissipation and consequently lower power consumption. In other words, power consumption not only dictates the performance, it also impacts the battery life.

In today's technology, the power or energy supply for various electronic devices and platforms usually comes from one of three major sources:

- An electrical outlet, commonly known as the "AC power source"

- A so-called *SMPS* unit, commonly known as the "DC power source"

- A rechargeable battery

A switch mode power supply (SMPS) unit rectifies and filters the AC mains input so as to obtain DC voltage, which is then switched on and off at a high frequency—speed in the order of hundreds of KHz to 1 MHz.

The high-frequency switching enables the use of inexpensive and lightweight transformers, inductors, and capacitors circuitry for a subsequent voltage step-down, rectification, and filtering to output a clean and stable DC power supply. Typically, an SMPS is used as a computer power supply.

For mobile usage, rechargeable batteries supply energy to an increasing number of electronic devices, including almost all multimedia devices and platforms. However, because of the change in internal resistance during charging and discharging, rechargeable batteries degrade over time. The lifetime of a rechargeable battery, aka "the battery life," depends on the number of cycles of charge/discharge, until eventually the battery can no longer hold an effective charge.

Batteries are rated in watt-hours (or ampere-hours multiplied by the voltage). Measuring system power consumption in watts gives a good idea of how many hours a battery will work before needing a recharge. This measure of battery life is usually important to consumers of today's electronic devices.

Media Workloads on Consumer Platforms

One of the main goals in designing a typical consumer electronic device or platform is to make it as user-friendly as possible. This implies that an important design consideration is how such devices are to be used. Nowadays, rapid integration of multimedia functionalities in modern mobile devices has become commonplace.

For example, a smartphone is expected to work not only as a wireless phone and a communication device but also should accommodate applications such as calendars, clocks, and calculators, combining to function as a productivity device. It should also serve as a navigation device with a compass, a GPS, and maps; and it should function as an entertainment device, with games and multimedia applications. In addition to these usages, as an educational platform the device is used for digital storytelling or a virtual classroom.

Human interaction with these devices calls for high-resolution cameras, high-speed wireless connection to the Internet, and voice, touch, and gesture input. On the output side, high-fidelity speakers, high-resolution displays, fast processing, and low power consumption are common expectations.

However, supporting multiple high-end functionalities often conflicts with the need to save power. Increases in battery capacity only partially address the problem, as that increase is not sufficient to keep up with the expansion of multimedia integration and enhanced user experience. Analyzing and understanding the nature of these multimedia workloads will help toward achieving the performance and power optimizations within the constraints just mentioned.

In popular consumer electronics press reviews, power data is often measured and analyzed using non-multimedia benchmark workloads, such as Kraken, Sunspider, and Octane Javascript benchmarks. Usually these benchmark applications focus on usage of the device as a computing platform, leaving unexamined the power consumption of the device's other usages. Yet, often these other applications are not optimized for power, or may be optimized to achieve higher performance only on certain platforms and operating systems. This fails to recognize the impact of task migration and resource sharing between the processing units. With the increasing availability of integrated processor graphics platforms, it becomes necessary to include media usages and applications in such analyses.

In the following section, we discuss some of the common media usages and applications.

Media Usages

Multimedia applications are characteristically power-hungry. With the demand for more and more features, requirements for power consumption are increasingly raised to higher levels. Moreover, some usages may need additional instances of an application, or more than one application running at a time.

Mobile devices used as entertainment platforms have typically run two main types of applications: gaming and media. The 2D and 3D video games are the most popular, but many other media applications are also in demand on these devices. Among them, the following are notable:

- Still image capture

- Still image preview/view finder

- Wireless display or Miracast: clone mode or extended mode

- Browser-based video streaming

- Video recording and dual video recording

- Video playback

- Audio playback

- Internet browsing

- Videophone and video chat

- Video conferencing

- Video transcoding

- Video email and multimedia messaging

- Video upload to Internet

- Video editing

- Augmented reality

- Productivity applications

Most of these usages are implemented via special software applications, and some may benefit from hardware acceleration if supported by the platform.

Intel processors are noteworthy for supporting such hardware acceleration through the integrated processor graphics, both serving as a general-purpose computing platform and fulfilling needs for special-purpose applications. The integration of graphics units into the central processor allows mobile devices to eliminate bigger video cards and customized video processors, thereby maintaining a small size suitable for mobile usage.

On many mobile devices, some combinations of multimedia applications are used simultaneously. For example, audio playback may continue when a user is browsing the Internet, or video playback may be complemented with simultaneous video recording. Some of these applications use common hardware blocks for the hardware acceleration of video codec and processing tasks; simultaneous operation of such hardware blocks is interesting from a system resource scheduling and utilization point of view.

One example of such complex system behavior is multi-party video conferencing; another is video delivery over Miracast Wireless Display. Wi-Fi certified Miracast is an industry-standard solution for seamlessly displaying multimedia between devices, without needing cables or a network connection. It enables users to view pictures from a smartphone on a big screen television, or to watch live programs from a home cable box on a tablet. The ability to connect is within the device using Wi-Fi Direct, so separate Wi-Fi access is not necessary.[1]

Figure 6-1 shows a usage model of the Miracast application.

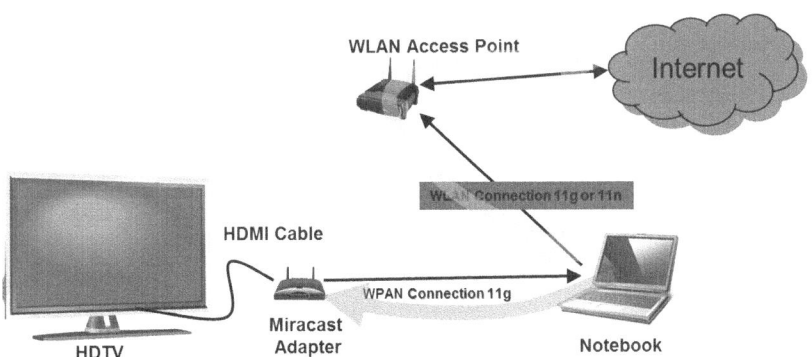

Figure 6-1. *Video delivery over Miracast*

---

[1]For details, see www.wi-fi.org/discover-wi-fi/wi-fi-certified-miracast.

To better understand the power consumption by video applications, let us analyze one of the multimedia usages in detail: video delivery over Miracast Wireless Display.[2]

Wireless Display (WiDi) is an Intel technology originally developed to achieve the same goals as Miracast. Now that Miracast has become the industry standard, it has been supported in Intel (R) Wireless Display (TM) since version 3.5. The goal of this application is to provide a premium-content capable wireless display solution that allows a PC user, or a handheld device user, to remotely display audiovisual content over a wireless link to a remote display. In other words, the notebook or smartphone receives a video from the Internet via the wireless local area network or captures a video using the local camera. The video is played on the device using a local playback application. A special firmware for Miracast Wireless Display then captures the screen of the device and performs hardware-accelerated video encoding so as to send the compressed bit stream via Wi-Fi data exchange technology to a Miracast adapter. The adapter performs a decoding of the bit stream to HDMI format and sends it to the display device through an HDMI cable connection. The end-to-end block diagram is shown in Figure 6-2.

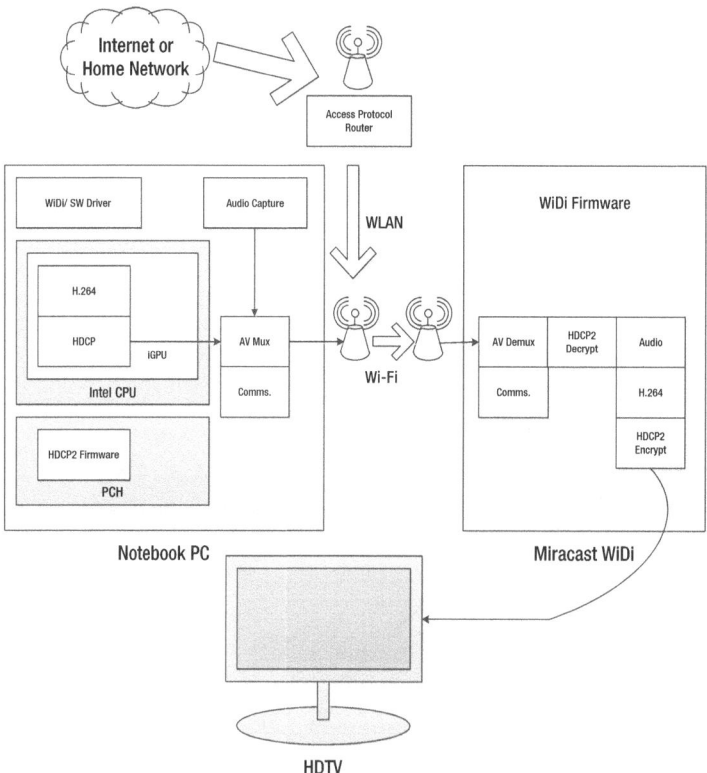

Figure 6-2. *Miracast Wireless Display end to-end block diagram*

---

[2]For details, see www-ssl.intel.com/content/www/us/en/architecture-and-technology/intel-wireless-display.html.

In Figure 6-2, the major power-consuming hardware modules are the CPU, the PCH, the video codec in the integrated GPU, the hardware-accelerated content protection module, the memory, the local display of the notebook, and the remote HDTV display. The Miracast Wireless Display adapter mainly runs the wireless display firmware and consumes a smaller amount of power. The typical distribution of power consumption in this example is shown in Figure 6-3.

%Power consumption

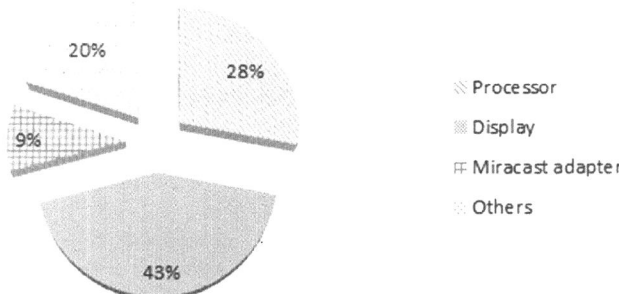

Figure 6-3. *Typical distribution of power consumption by components in a Miracast Wireless Display application*

As can be seen in Figure 6-3, usually the bulk of the power is consumed by the display, which in this example consumes about 1.5 times as much power as the processor. The Miracast adapter itself consumes a moderate amount—in this example, approximately 9 percent of the total power consumed by the application. Due to the complex nature of this application, a careful balance should be maintained between performance needs and power consumption, so that the appropriate optimizations and tradeoffs can be made so as to obtain a satisfactory user experience.

Another common multimedia application is video playback along with associated audio. A detailed analysis of this application is provided in Agrawal et al.[3] Here, we just note that by performing hardware acceleration of the media playback, the overall power consumption is reduced from ~20W to ~5W, while the power consumption profile is also changed significantly. Various tasks within the media playback pipeline get a performance boost from hardware acceleration, as these are offloaded from the CPU to the special-purpose fixed-function hardware with better power-performance characteristics.

Analysis of these applications enables identification of the modules that are prime candidates for power optimization. For example, some power optimization can be achieved by migrating tasks like color-space conversion from the display unit to the GPU. Some Intel platforms are capable of such features, achieving a high level of power optimization. Various power optimization techniques are discussed in detail later in this chapter.

---

[3] A. Agrawal, T. Huff, S. Potluri, W. Cheung, A. Thakur, J. Holland, and V. Degalahal, "Power Efficient Multimedia Playback on Mobile Platform," *Intel Technology Journal* 15, no. 2 (2011): 82–100.

Power-Aware Designs

Power consumption is a function of both hardware and software efficiency. Therefore, performance gains or power savings increasingly depend on improving that efficiency. This is done typically in terms of *performance per watt*, which eventually translates to *performance per dollar*. Performance per watt is the quantity of computation that can be delivered by a computing system for every watt of power consumed. Today's platforms aim to achieve high scores in this measure by incorporating "power awareness" in the design process, as significant consideration is given to the cost of energy in computing environments.

In power-aware designs, typically the power savings are achieved by employing a divide-and-conquer policy: the system is divided into several independent power domains, and only the active domains are supplied with power. Depending on the active state of each domain, intelligent management of power achieves the optimum power solutions. Also, optimization within each domain is done with a view to gaining an edge in power saving and value proposition of a system.

Power awareness is important not only in the hardware design but also in the applications, so as to maximize the performance per dollar. Toward this end, most operating systems provide power-management features. As applications know the utilization pattern of various hardware resources and tasks, better power management can be achieved if the applications can provide appropriate "hints" to the power-management units. Furthermore, power awareness of those applications can yield software-level optimizations, such as context-aware power optimizations, complexity reduction, and memory transfer reduction. These techniques are discussed later in the chapter in regard to power optimization.

Power-Management Considerations

The goal of power management in both computers and computer peripherals, such as monitors and printers, is to turn off the power or switch the system to a low-power state when it is inactive. Power management in computing platforms provides many benefits, including increased battery life, lower heat emission, lower carbon footprint, and prolonged life of devices such as display panels and hard disk drivers.[4]

Power management happens on various constituent hardware devices that may be available in a computer system (aka "the system"); among them are the BIOS, central processing unit (CPU), hard disk drive (HDD), graphics controller, universal serial bus (USB), network, and display. It is also possible to monitor and manage the power use to various parts of memory, such as dynamic random access memory (DRAM) and non-volatile flash memory, but this is more complex and less common. Some examples of power management are listed in Table 6-2; some of these are discussed in subsequent sections of this chapter.

---

[4]M. Vats and I. Verma, *Linux Power Management: IEGD Considerations* (Intel Corporation, 2010). Available at www.intel.com/content/dam/www/public/us/en/documents/white-papers/linux-power-mgmt-paper.pdf.

Table 6-2. *Power Management Features*

Device/ Component	Power Management Features
BIOS	CPU settings (e.g., CPU states enabled, CPU fan throttling), platform settings (e.g., thermal high/low watermarks, chassis fan throttling, etc.)
CPU	HLT (halt instruction in x86 for CPU to halt until next external interrupt is fired), Stop clock, Intel SpeedStep (aka dynamic frequency scaling)
Display	Blanking, dimming, power saver mode, efficient energy use as specified in the Energy Star international standard
Graphics Controller	Power down to intermediate state, power shutoff
Hard drive/ CD-ROM	Spin down
Network/ NIC	Wake on LAN
USB	Power state transition of devices such as mouse, USB drives, etc.; wake on access (e.g., mouse movement)

There may be special power-management hardware or software available. Typically, hardware power management in the processor involves management of various CPU states (aka *C*-states), such as the core *C*-states, the module *C*-states, the package *C*-states, and so on. (Details of the *C*-states are described in the next section.) On the other hand, software power management in the operating system or in the driver involves tasks like CPU core offline, CPU core shielding, CPU load balancing, interrupt load balancing, CPU frequency governing etc. With the introduction of the integrated graphics processing units (iGPU), primarily in Intel platforms, various GPU states are also important considerations for power management.

ACPI and Power Management

The Advanced Configuration and Power Interface (ACPI) specification is an open standard adopted in the industry for system-level configuration and management of I/O devices and resources by the operating system, including power management. Originally proposed by Intel, Microsoft, and Toshiba in 1996, the specification effort was later joined by HP and Phoenix. The latest ACPI specification version 5 was published in 2011.

With wider adoption in the industry, it became necessary to support many operating systems and processor architectures. Toward this end, in 2013, the standards body agreed to merge future developments with the Unified Extensible Firmware Interface (UEFI) forum, which is an alliance of leading technology companies, including the original ACPI participants and major industry players like AMD, Apple, Dell, IBM, and Lenovo. All recent computers and portable computing devices have ACPI support.

ACPI Power States

To the user, a computer system appears as either ON or OFF. However, the system may support multiple power states, as defined in the ACPI specification. ACPI compliance indicates that the system supports the defined power management states, but such compliance does not promise the most power-efficient design. In addition, a system can have power-management features and tuning capabilities without being ACPI compliant.

According to the ACPI specification, the devices of a computer system are exposed to the operating system in a consistent manner. As such, for system-level power management, the ACPI defines power draw states for the individual devices (known as the device states), as well as the overall computer system (known as the global states or the system sleep states). The operating system and the devices can query and set these states. Important system buses such as the peripheral component interconnect (PCI) bus, may take advantage of these states.

Global States

The ACPI specification defines four possible global "Gx" states and six possible sleep "Sx" states for an ACPI-compliant computer-system (Table 6-3). However, some systems or devices may not be capable of all states.

Table 6-3. *ACPI Global States*

State	Gx	Sx	Description
Active	G0	S0	The system is fully usable. CPUs are active. Devices may or may not be active, and can possibly enter a lower power state. There is a subset of S0, called "Away mode," where monitor is off, but background tasks are running.
Sleep	G1	S1	The system appears to be off. Power consumption is reduced. All the processor caches are flushed, and the CPU(s) stops executing instructions. The power to the CPU(s) and RAM is maintained. Nonessential devices may be powered off. This state is rarely used.
		S2	The system appears to be off. CPU is powered off. Dirty cache is flushed to RAM. Similar to S1, the S2 state is also rarely used.
		S3	Commonly known as *standby, sleep,* or *Suspend-to-RAM* (STR). The system appears to be off. System context is maintained on the system DRAM. All power is shut to the noncritical circuits, but RAM power is retained. Transition to S0 takes longer than S2 and S1, respectively.

(contiunued)

Table 6-3. (*contiuned*)

State	Gx	Sx	Description
Hibernation		S4	Known as *hibernation* or *Suspend-to-Disk* (STD). The system appears to be off. Power consumption is reduced to the lowest level. System context is maintained on the disk, preserving the state of the OS, applications, and open documents. Contents of the main memory are saved to non-volatile memory such as a hard drive, and the system is powered down, except for the logic to resume.
Soft Off	G2	S5	The system appears to be off. System context is not maintained. Some components may remain powered, so the computer can wake from input from a keyboard, mouse, LAN, or USB device. The working context can be restored if it is stored on nonvolatile memory. All power is shut, except for the logic required to restart. Full boot is required to restart.
Mechanical Off	G3		The system is completely off and consumes no power. A full reboot is required for the system to return to the active state.

Device States

The power capabilities of all device hardware are not the same. For example, the LAN adapters will have the capability to wake the system; the audio hardware might permit streaming while the system is in standby mode, and so on. Some devices can be subdivided into functional units with independent power control. Furthermore, some devices, such as the keyboard, mice, modems, and LAN adapters, have the capability to wake up the system from a sleep state, while the devices are asleep themselves. Such capability is possible by the fact that the hardware for these devices must draw a small seeping current and be equipped to detect the external wake event.

The ACPI specification defines the device-dependent D-states as shown in Table 6-4.

Table 6-4. *Device States*

Dx	Subset of Dx	Description
D0		Fully ON and operating.
D1-D2		Intermediate power states. Definition varies by device.
D3	Hot	Device is off and unresponsive to bus, but the system is ON. Device is still connected to power. *D3 Hot* has auxiliary power enabling a higher power state. A transition from D0 to D3 implies D3 Hot.
	Cold	No power to the device—both the device and system are OFF. It is possible for the device to consume *trickle* power, but a wake event is needed to move the device and the system back to D0 and/or S0 states.

The ACPI defines the states D0 through D3, and provides a subdivision of D3 into D3 hot and D3 cold. Some devices or operating systems don't distinguish between D3 hot and cold, and treats D3 as having no power to the device. However, Windows 8 explicitly tracks D3 hot and D3 cold. The D1 and D2 states are optional, but if they are used properly, they would provide better cleanliness when the device is idle.

Power Management by the Operating System

Modern operating systems customarily offer many power management features. Linux, Windows, OS X, and Android all support more intelligent power management using the newer ACPI standard, rather than the old BIOS controlled Advanced Power Management (APM). However, all major operating systems have been providing stable support for basic power-management features such as notification of power events to the user space—for example, battery status indication, suspend the CPU when idle, and so on.

In the following sections, we discuss power management by the Linux and the Windows operating systems. In the context of Linux power management, three important components are mentioned: the X Window, the Window Manager, and the Intel Embedded Graphics Driver (IEGD). The Windows power management discourse includes the Windows power requirements, power policy, the Windows driver model, and the Windows driver framework. There's also a brief description of device power management under Windows 8, followed by a discussion on how to deal with power requests.

Linux Power Management

Linux supports both the older APM and the newer ACPI power management implementations. APM focuses on basic system and OS power management, with much of the power-management policy controlled at the BIOS level; whereas an APM driver acts as an interface between the BIOS and the Linux OS, as power-management events pass between the BIOS and OS. Devices are notified of these events so they can respond appropriately.

The ACPI provides greater flexibility in power management and platform configuration, and allows for platform independence and OS control over power-management events. In addition to the power-management policies, ACPI supports policies for responding to thermal events (e.g., fans), physical movement events (e.g., buttons or lids), CPU states, power source (e.g., battery, AC power supply), and the like.

Power-management software manages state transitions along with device drivers and applications. Device drivers are responsible for saving device states before putting them into their low-power states and then restoring the device state when the system becomes active. Generally, applications are not involved in power-management state transitions. A few specialized softwares, such as the IEGD for Linux, deal directly with some devices in order to handle state transitions. Besides the IEGD, there are a few common software technologies in Linux, including the X Window system, the Window managers, and several open-source processes such as /sys/power/state and /proc/acpi/event, which also provide some part of Linux power management.

The X Window

The X Window system is supported by many operating systems, including Linux, Solaris, and HP-UX. It provides graphics capabilities to the OS and supports user-level, system-level, and/or critical standby, suspend, and resume. In the APM implementation, the X-server controls the power-management events. In ACPI implementation, the X Window system handles the graphics messages as a user process, but the system-wide power-management events like suspend/resume are handled by a kernel mode driver.

Window Managers

Window managers on Linux are user-level processes that provide the graphical user interface and also deliver reliable power management of the operating system. Among the many supported windows managers in Linux are two popular window managers, GNOME and KDE. In GNOME, power management uses the hardware abstraction layer (HAL) and involves open-source platform power management built on an open-source messaging interface called DBUS, while KDE3 provides a proprietary power-management solution named KPowersave.

Intel Embedded Graphics Driver

In the Intel Embedded Graphics Driver (IEGD) power management, a kernel mode driver helps the Linux kernel manage the power. It is also responsible for graphics device initialization and resource allocation. In order for you to clearly understand the flow of a power event, here are the main parts of the Suspend to RAM example.[5]

The Suspend to RAM starts when a power-management *event* occurs in the platform, such as when a button is pressed or a window manager option is triggered, and the operating system is notified of the event. Usually the Linux operating system employs a protocol to communicate an event between a software component and the Linux kernel. Using such protocols, the OS (typically via the Window manager) commands the kernel to go to a lower power state, at which point the Linux kernel starts the suspend procedure by notifying the X-Server driver.

The X display must switch to console mode before going into a lower power state. With ACPI implemented in the Linux kernel, this switch happens by the X-Server driver's calling the *Leave virtual terminal* function, when the IEGD process saves the graphics state and registers information. The Linux kernel then freezes all user processes, including the X Window process. Now the kernel is ready to check which devices are ready for the suspend operation, and it calls the suspend function of each device driver (if implemented) in order to put the device clocks to D3 mode--effectively putting all devices into a lower power state. At this point only the Linux kernel code is running, which freezes all other active processors except the one where the code is running.

---

[5]Ibid.

Following execution of the kernel-side suspend code, two ACPI methods--namely, PTS (*Prepare-to-Sleep*) and GTS (*Going-to-Sleep*) are executed, the results of which may not be apparent to the Linux kernel. However, before actually going to sleep, the kernel writes the address of the kernel wakeup code to a location in the Fixed ACPI Description Table (FADT). This enables the kernel to properly wake up upon receiving the restore command.

The restore command usually results from a user event, such as a keystroke, mouse movement, or pressing the power button, which turns the system on. Once on, the system jumps to the BIOS start address, performs housekeeping tasks such as setting up the memory controller, and then scans the ACPI status register to get the indication to RAM that the system was previously suspended. If *video repost* is supported, during resume operation the BIOS also calls this function to re-execute the video BIOS (vBIOS) code, thereby providing a full restart of the vBIOS.

The system then jumps to the address programmed earlier, as indicated by the ACPI register's status and the FADT. The wakeup address leads to the kernel code execution, putting the CPU back into protected mode and restoring the register states. From this point, the rest of the wakeup process traverses the reverse path of the suspend process. The ACPI WAK method is called, all the drivers are resumed, and user space is restarted. If running, the X-server driver calls the *Enter virtual terminal* function, and the IEGD restores the graphics device state and register information. After saving the console mode, the X-server driver re-enters the GUI, thereby completing a successful wakeup.

Windows Power Management

Power management in the Windows operating system, particularly Windows 8, has significant improvements in this area compared to previous Windows versions.

Power Requirements

In versions earlier than Windows 8, the power requirements involved supporting the common ACPI states, such as the S3 and S4 states, mainly on mobile personal computer platforms. However, Windows 8 aimed to standardize on a single power requirement model across all platforms including desktop, server, mobile laptops, tablets, and phones. While one of the goals was to improve the battery life of portable platforms, Windows 8 applies the smartphone power model to all platforms for quick standby-to-ready transitions, and ensures that the hidden applications consume minimal or no resources.

To this end, Windows 8 defines the requirements listed in Table 6-5.[6]

---

[6]J. Lozano, *Windows 8 Power Management.* (StarJourney Training and Seminars, 2013.)

Table 6-5. *Windows 8 Power Requirements*

Requirement Type	Requirements
System Power Requirements	1. Maximum battery life should be achieved with minimum energy consumption.
	2. The delay for startup and shutdown should be minimal.
	3. Power decisions should be intelligently made—for example, a device that is not in a best position to change the system power state should not do so.
	4. Capabilities should be available to adjust fans or driver motors on-demand for quiet operation.
	5. All requirements should be met in a platform independent manner.
Device Power Requirements	6. Devices, especially for portable systems, must be extremely power conscious.
	7. Devices should be aggressive in power savings:
	a. Should provide just-in-time capabilities.
	b. Low transition latency to higher states.
	c. When possible, the device logic should be partitioned into separate power buses so that portions of a device can be turned off as needed.
	d. Should support connected standby as appropriate for quick connection.
Windows Hardware Certification Requirements	8. The Windows HCK tests require that all devices must support S3 and S4 without refusing system sleep request.
	9. Standby and connected standby must last for days.
	10. Device must queue up and not lose the I/O request while in D1-D3 states.

Power Policy

Power policies (also known as *power plans* or *power schemes*) are preferences defined by the operating system for the choice of system and BIOS settings that affect energy consumption. For each power policy, two different settings are usually set by the operating system by default, one with battery power supply, the other with AC power supply. In order to preserve battery as much as possible, the settings with the battery power supply are geared toward saving power more aggressively. Windows defines three power policies, by default:

- **Performance mode:** In performance mode, the system attempts to deliver maximum performance without regard to power consumption.

- **Balanced mode:** In this mode, the operating system attempts to reach a balance between performance and power.

- **Power saver mode:** In this mode, the operating system attempts to save maximum power in order to preserve battery life, even sacrificing some performance.

Users may create or modify the default plans but the power policies are protected by the access control list. Systems administrators may override a user's selection of power policies. On Windows, a user may use an applet called "powercfg.cpl" to view and edit a power policy. A console version of the applet, called "powercfg.exe," is also available, which permits changing the access control list permissions.

Application software can obtain a notification of the power policy by registering for the power plan and can use power policies in various ways:

- Tune the application behavior based on the user's current power policy.

- Modify the application behavior in response to a change in power policy.

- Move to a different power policy as required by the application.

The Windows Driver Model

In the Windows Drive Model (WDM), the operating system sends *requests* to the drivers to order the devices to a higher or lower power state. Upon receiving such requests, a driver only saves or restores the state, keeping track of the current and next power states of the device, while a structure called Physical Device Object (PDO) performs the work of actually increasing or lowering the power to the device.

However, this arrangement is problematic, as the model requires the drivers to implement a state machine to handle the power IRPs (I/O request packets), and may result in unwanted complexity due to the time needed to perform a power state transition. For example, for a *power down* request, the driver saves the state of the device in memory, and then passes the request down to the PDO, which subsequently removes power from the device; only then can it mark the request as *completed*. However, during a *power up* request that may follow, the driver must first pass the request to the PDO, which then restores the power *before* restoring the device state, and informs the driver to mark the request as *completed*. To overcome this difficulty, the Windows driver framework (WDF) was proposed.

The Windows Driver Framework

In order to simplify power management within a driver, Windows introduced the concept of *events* in the latest Windows Driver Framework (WDF) driver model. In this model, there are optional *event handler* functions in the driver, whereby the framework calls the event handlers at the appropriate time to handle a power transition, thereby eliminating the need for a complex state machine.

Windows 8 offers a new, more granular way to address the power needs of functions on multifunction devices in the form of a *power framework* called *PoFx*. Additionally, it introduces the concept of *connected standby*, allowing a powered-off device to occasionally connect to outside world and refresh state or data for various applications. The primary benefit is a quick recovery from standby state to ON state, as if the system had been awake the whole time. At the same time, the power cost is low enough to allow the system to be in standby state for days.

Device Power Management in Windows 8

In Windows 8, in response to a query from the plug-and-play (PnP) manager, the device drivers announce their device's power capabilities. A data structure called DEVICE_CAPABILITIES is programmed by the driver, indicating the information as shown in Table 6-6.

Table 6-6. *Device Capabilities Structure*

Field	Function
Device D1 and D2	Indicates whether the device supports D1, or D2, or both.
Wake from Dx	Indicates whether the device supports waking from a Dx state.
Device state	Defines the Dx state corresponding to each Sx state.
DxLatency	Nominal transition time to D0.

There is latency for the devices when moving from Dx to D0, as the devices require a small period of time before they can become operational again. The latency is longer for the higher Dx states, so that a transition from D3 to D0 would take the longest time. Furthermore, a device needs to be operational before it can respond to new requests—for example, a hard disk must spin up before it can be slowed down again. The latency is announced by the driver via the DEVICE_CAPABILITIES data structure.

■ **Note** A device state transition may not be worthwhile if sufficient time is not spent in the lower power state, as it is possible that the transition itself would consume more power than when the device had been left in a particular state.

In order to manage the device power, the Windows Power Manager needs to know the transition latency of each device, which can vary for different invocations even for the same device. Therefore, only a nominal value is indicated by the driver. The Windows OS controls the time gap between a query and a set power request, during which time the device sleeps. A high value for this time gap would increase the sleep time, while a low value would cause the OS to give up on powering down the device.

Dealing with Power Requests

There are kernel mode data structures called *I/O request packets* (IRPs) that are used by the Windows Driver Model (WDM) and the Windows NT device drivers to communicate with the operating system and between each other. IRPs are typically created by the I/O Manager in response to I/O requests from the user mode. In Windows 2000, two new managers were added: the plug-and-play (PnP) and Power manager, which also create IRPs. Furthermore, IRPs can be created by drivers and then passed to other drivers.

In Windows 8, the Power Manager sends *requests*--that is, the power IRPs--to the device drivers, ordering them to change the power state of the relevant devices. Power IRPs use the major IRP data structure IRP_MJ_POWER, with the following four possible minor codes:

- IRP_MN_QUERY_POWER: A query to determine the capability of the device to safely enter a new requested Dx or Sx state, or a shutdown or restart of the device. If the device is capable of the transition at a given time, the driver should queue any further request that is contrary to the transition before announcing the capability, as a SET request typically follows a QUERY request.

- IRP_MN_SET_POWER: An order to move the device to a new Dx state or respond to a new Sx state. Generally, device drivers carry out a SET request without fail; the exception is bus drivers such as USB drivers, which may return a failure if the device is in the process of being removed. Drivers serve a SET request by requesting appropriate change to the device power state, saving context when moving to a lower power state, and restoring context when transitioning to a higher power state.

- IRP_MN_WAIT_WAKE: A request to the device driver to enable the device hardware so that an external wake event can awaken the entire system. One such request may be kept in a pending state at any given time until the external event occurs; upon occurrence of the event, the driver returns a success. If the device can no longer wake the system, the driver returns a failure and the Power Manager cancels the request.

- IRP_MN_POWER_SEQUENCE: A query for the D1-D3 counters--that is, the number of times the device has actually been in a lower power state. The difference between the count before and the count after a sleep request would tell the Power Manager whether the device did get a chance to go to a lower power state, or if it was prohibited by a long latency, so that the Power Manager can take appropriate action and possibly not issue a sleep request for the device.

For driver developers, one of the difficulties in calling the various power IRPs is determining when to call them. The Kernel Mode Driver Framework (KMDF) in Windows 8 implements numerous state machines and event handlers, including those for power management. It simplifies the task of power management by calling the event handlers at the appropriate time. Typical power event handlers include: D0 entry, D0 exit, device power state change, device arm wake from S0/Sx, and device power policy state change.

Power Management by the Processor

For fine-grained power management, modern Intel processors support several partitions of *voltage islands* created through on-die power switches. The Intel Smart Power Technology (Intel SPT) and Intel Smart Idle Technology (Intel SIT) software determine the most power efficient state for the platform, and provide guidance to turn ON or OFF different voltage islands on the processor at any given time. Upon receiving a direction to go into a lower power state, the processor waits for all partitions with shared voltage to reach a safe point before making the requested state change.

CPU States (*C*-states)

The CPU is not always active. Some applications need inputs from the system or the user, during which the CPU gets an opportunity to wait and become idle. While the CPU is idle or running low-intensity applications, it is not necessary to keep all the cores of the CPU powered up. The CPU operating states (*C*-states) are the capability of an idle processor to turn off unused components to save power.

For multi-core processors, the *C*-states can be applied at a package level or at a core level. For example, when a single threaded application is run on a quad-core processor, only one core is busy and the other three cores can be in low-power, deeper *C*-states. When the task is completed, no core is busy and the entire package can enter a low-power state.

The ACPI specification defines several low-power idle states for the processor core. When a processor runs in the *C0* state, it is working. A processor running in any other *C*-state is idle. Higher *C*-state numbers represent deeper CPU sleep states. At numerically higher *C*-states, more power-saving actions, such as stopping the processor clock, stopping interrupts, and so on, are taken. However, higher *C*-states also have the disadvantage of longer exit and entry latencies, resulting in slower wakeup times. For a deeper understanding, see the brief descriptions of various *C*-states of an Intel Atom processor, given in Table 6-7.

Table 6-7. *Cx State Definitions, Intel Atom Processor Z2760*

State	Function	Description
C0	Full ON	This is the only state that runs software. All clocks are running and the processor core is active. The processor can service *snoops* and maintain cache coherency in this state. All power management for interfaces, clock gating, etc., are controlled at the unit level.
C1	Auto Halt	The first level of power reduction occurs when the core processor executes an Auto-Halt instruction. This stops the execution of the instruction stream and greatly reduces the core processor's power consumption. The core processor can service snoops and maintain cache coherency in this state. The processor's North Complex logic does not explicitly distinguish *C1* from *C0*.

(continued)

Table 6-7. (*continued*)

State	Function	Description
C2	Stop Grant	The next level of power reduction occurs when the core processor is placed into the *Stop Grant* state. The core processor can service *snoops* and maintain cache coherency in this state. The North Complex only supports receiving a single *Stop Grant*.
		Entry into the *C2* state will occur after the core processor requests *C2* (or deeper). Upon detection of a break event, *C2* state will be exited, entering the *C0* state. Processor must ensure that the PLLs are awake and the memory will be out of self-refresh at this point.
C4	Deeper Sleep	In this state, the core processor shuts down its PLL and cannot handle *snoop* requests. The core processor voltage regulator is also told to reduce the processor's voltage. During the *C4* state, the North Complex continues to handle traffic to memory so long as this traffic does not require a *snoop* (i.e., no coherent traffic requests are serviced).
		The *C4* state is entered by receiving a *C4* request from the core processor/OS. The exit from *C4* occurs when the North Complex detects a *snoop*-able event or a break event, which would cause it to wake up the core processor and initiate the sequence to return to the *C0* state.
C6	Deep Power Down	Prior to entering the *C6* state, the core processor flushes its cache and saves its core context to a special on-die *SRAM* on a different power plane. Once the *C6* entry sequence has completed, the core processor's voltage can be completely shut off.
		The key difference for the North Complex logic between the *C4* state and the *C6* state is that since the core processor's cache is empty, there is no need to perform *snoops* on the internal front side bus (FSB). This means that bus master events (which would cause a popup from the *C4* state to the *C2* state) can be allowed to flow unhindered during the *C6* state. However, the core processor must still be returned to the *C0* state to service interrupts.
		A residency counter is read by the core processor to enable an intelligent promotion/demotion based on energy awareness of transitions and history of residencies/transitions.

Source: Data Sheet, Intel Corporation, October 2012. *www.intel.com/content/dam/www/public/us/en/documents/product-briefs/atom-z2760-datasheet.pdf.*

Performance States (*P*-states)

Processor designers have realized that running the CPU at a fixed frequency and voltage setting is not efficient for all applications; in fact, some applications do not need to run at the operating point defined by the highest rated frequency and voltage settings. For such applications there is a power-saving opportunity by moving to a lower operating point.

Processor performance states (*P*-states) are the capability of a processor to switch between different supported operating frequencies and voltages to modulate power consumption. The ACPI defines several processor-specific *P*-states for power management, in order to configure the system to react to system workloads in a power-efficient manner. Numerically higher *P*-states represent slower processor speeds, as well as lower power consumption. For example, a processor in *P3* state will run more slowly and use less power than a processor running at *P1* state.

While a device or processor is in operation and not idling (D0 and C0, respectively), it can be in one of several *P*-states. *P0* is always the highest-performance state, while *P1* to *Pn* are successively lower-performance states, where *n* can be up to 16 depending on implementation.

P-states are used on the principles of dynamic voltage or frequency scaling, and are available in the market as the SpeedStep for Intel processors, the PowerNow! for AMD processors, and the PowerSaver for VIA processors.

P-states differ from *C*-states in that *C*-states are idle states, while *P*-states are operational states. This means that with the exception of *C0*, where the CPU is active and busy doing something, a *C*-state is an idle state; and shutting down the CPU in the higher *C*-states makes sense, since the CPU is not doing anything. On the other hand, *P*-states are operational states, meaning that the CPU can be doing useful work in any *P*-state. *C*-states and *P*-states are also orthogonal—that is, each state can vary independently of the other. When the system resumes *C0*, it returns to the operating frequency and voltage defined by that *P*-state.

Although *P*-states are related to the CPU *clock frequency*, they are not the same. The CPU clock frequency is a measure of how fast the CPU's main clock signal goes up or down, which may be a measure of performance, as higher performance would generally mean using a higher clock frequency (except for memory-bound tasks). However, this is backward looking, as you can measure the average clock frequency only after some clock cycles have passed. On the other hand, *P*-state is a performance state the OS would like to see on a certain CPU, so *P*-states are forward looking. Generally, higher clock frequency results in higher power consumption.

When a CPU is idle (i.e., at higher *C*-states), the frequency should be zero (or very low) regardless of the *P*-state the OS requests. However, note that all the cores on a current-generation CPU package share the same voltage, for practical reasons. Since it is not efficient to run different cores at different frequencies while maintaining the same voltage, all active cores will share the same clock frequency at any given time. However, some of the cores may be idle and should have zero frequency. As the OS requests a certain *P*-state for a logical processor, it is only possible to keep a core at zero frequency when none of the cores is busy and all cores are kept at the same zero frequency.

While using a global voltage supply for all the cores leads to the situation where certain *P*-state requests cannot be met, separate local voltage supplies for each core would be cost-prohibitive. A tradeoff concerning global- and local-voltage platforms is to adopt multi-core architecture with different *voltage islands*, in which several cores on a voltage island share the same but adjustable supply voltage. An example of this is available in Intel's many-core platform called the *single-chip cloud computer*. Another example is the use of a *fully integrated voltage regulator* (FIVR), available in some processors, that allows per core P-states so that cores can be operated at frequencies independent of each other.

Just as the integrated GPUs behave similarly to the CPU, C-states and P-states have been defined for the GPU like the CPU C-states and P-states. These are known as *render C-states* (RC-states) and *render P-states* (RP-states).

Turbo

It is possible to *over-clock* a CPU, which means running a core at a higher frequency than that specified by the manufacturer. In practice, this behavior is possible due to the existence of multiple cores and the so-called *turbo* feature resulting from the power budget. Turbo is a unique case in which the frequency and voltage are increased, so the turbo state functions as an opposite to the various P-states, where voltage and frequency are decreased. The CPU enters turbo state while operating below certain parameter specifications, which allows it some headroom to boost the performance without infringing on other design specifications. The parameters include the number of active cores, processor temperature, and estimated current and power consumption. If a processor operates below the limit of such parameters and the workload demands additional performance, the processor frequency will automatically dynamically increase until an upper limit is reached. In turbo state, complex algorithms concurrently manage the current, the power, and the temperature with a view toward maximize the performance and the power efficiency.

To better understand a turbo scenario, let us consider an example. Suppose a quad-core CPU has a TDP limit of 35W, so that each core has a maximum of 8.75W power budget. If three of the four cores are idle, the only operational core can utilize the whole 35W of the power budget and can "turbo" up to a much higher frequency than would be possible with a less than 9W power budget. Similarly, if two cores are active, they use the same higher frequency and share the power budget while the idle cores do not consume any energy.

A maximum turbo frequency is the highest possible frequency achievable when conditions allow the processor to enter turbo mode. The frequency of Intel Turbo Boost Technology varies depending on workload, hardware, software, and overall system configuration. A request for a P-state corresponding to a *turbo-range* frequency may or may not be possible to satisfy, owing to varying power characteristics. So a compromise is made because of many factors, such as what the other cores and the GPU are doing, what thermal state the processor is at, and so on. This behavior also varies over time. Therefore, as the operating frequency is time-varying and dependent on C-state selection policy and graphics subsystem, selecting a P-state value does not guarantee a particular performance state.

In general, the *P1* state corresponds to the highest guaranteed performance state that can be requested by an OS. However, the OS can request an opportunistic state, namely *P0*, with higher performance. When the power and thermal budget are available, this state allows the processor configuration where one or more cores can operate at a higher frequency than the guaranteed *P1* frequency. A processor can include multiple so-called turbo mode frequencies above the *P1* frequency.

Some processors expose a large turbo range and typically grant all cores the maximum possible turbo frequency when the cores seek to turbo. However, not all applications can effectively use increased core frequency to the same extent. Differences may arise from varying memory access patterns, from possible cache contention, or similar sources. So allowing all cores to be at a highest level of turbo mode can unnecessarily consume power. In order to combat such inefficiencies, techniques have been proposed to efficiently enable one or more cores to independently operate at a selected turbo mode frequency.[7] One of the techniques periodically analyzes all cores granted turbo mode to determine whether their frequency should be increased, decreased, or left unchanged based on whether the core has been classified as stalled or not over the observation interval.

Thermal States (T-States)

In order to prevent potential damage from overheating, processors usually have thermal protection mechanisms, where, in an effort to decrease the energy dissipation, the processor throttles by turning the processor clocks off and then back on according to a pre-determined duty cycle. The thermal states, or T-states, are defined to control such throttling in order to reduce power, and they can be applied to individual processor cores. T-states may ignore performance impacts, as their primary reason is to reduce power for thermal reasons. There are eight T-states, from 0 to 7, while the active state is T_0. These states are not commonly used for power management.

The Voltage-Frequency Curve

It is important to understand the relationship between voltage and frequency for processors, as both CPU and integrated GPU follow such relationship in order to scale. A P-state requested by the operating system is in fact a particular operating point on the V-F curve.

As can be seen in Figure 6-4, the voltage vs. frequency curve tends to have an inflection point, at which the voltage starts to scale with the frequency.

---

[7]M. K. Bhandaru and E. J. Dehaemer, U. S. Patent No. US 20130346774, A1, 2013. *Providing energy efficient turbo operation of a processor.* Available at www.google.com/patents/WO2013137859A1?cl=en.

Voltage-Frequency Relationship

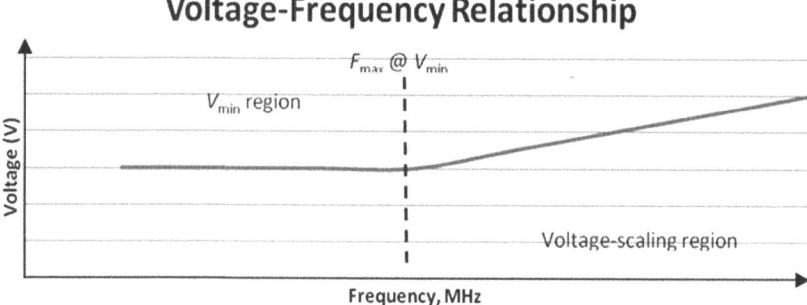

Figure 6-4. *Voltage-frequency relationship of a typical processor*

Up to this point, a minimum voltage, V_{min}, is required to make the circuit operational regardless of the frequency change. The maximum frequency F_{max} at the minimum voltage V_{min} is the highest frequency at which the processor part can operate at V_{min}. This is the point where power efficiency is the best, as we see from the power-frequency relationship shown in Figure 6-5. Increasing the frequency beyond F_{max} requires increased voltage supply for the circuit to be operational. At the voltage-scaling region, the required voltage scales are reasonably linear with frequency. This region offers power-reduction opportunities, as discussed later.

Power-Frequency Relationship

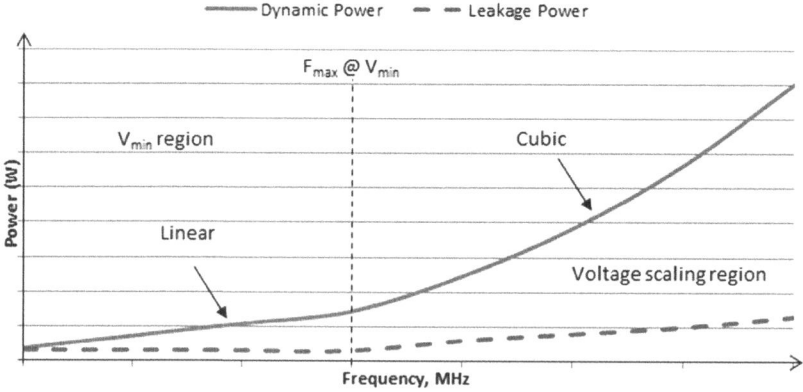

Figure 6-5. *Power-frequency relationship and optimization opportunities*

Figure 6-5 shows the power vs. frequency relationship and power optimization opportunities in a typical processor.

In the V_{min} region, power does not fall as fast as frequency; as dynamic power falls with frequency, the leakage power stays constant. On the other hand, in the voltage-scaling region, power increases much faster than frequency, as voltage scales are roughly linear with frequency; dynamic power goes up as V^2f and leakage goes up roughly as V^3. The leakage power depends only on the small amount of leakage current, and it follows almost the same pattern as the voltage curve with respect to frequency.

Power Optimizations

Let's recall the power equation:

$$Total\ power = leakage\ power + A\ C_{dyn}\ V^2\ f, \hspace{2cm} \text{(Equation 5-2)}$$

Here, A is the activity, C_{dyn} is the dynamic capacitance, V is the voltage, and f is the operating frequency. From this equation it is easy to see that, in order to reduce power, the following approaches can be taken:

- Reduce voltage and frequency

- Reduce activity and C_{dyn}

As voltage increases approximately linearly with frequency in the voltage-scaling region, the term V^2f implies a cubic relationship for the power with respect to the frequency (see Figure 6-5). Therefore, reducing the voltage and/or the frequency results in a dramatic reduction in the power. The power that is conserved in such a manner can be given to other parts of the system so that the overall system operation can benefit.

However, the frequency reduction cannot help below F_{max} at V_{min} (below which voltage cannot be reduced, as there is a minimum voltage necessary for operation of the circuit). In the V_{min} region, the voltage stays constant, so reduction in frequency can yield very little power savings. At this point, only activity and C_{dyn} reduction can provide further power optimization. This calls for more efficient algorithms and micro-architecture design, as well as dynamically turning off unused portions of the circuit.

The above power-reduction considerations have given birth to new ideas and approaches of power optimization, including various *gating* optimizations and use of special-purpose heterogeneous hardware components, such as the integration of a GPU capable of multimedia processing, camera image processing, and so on. Overall, it is not a hardware-only problem; it requires careful consideration at the software micro-architecture level as well.

In general, good power optimization requires incorporation of many approaches, all working in harmony toward the goal of saving power and increasing battery life. From a systems engineering point of view, optimizations can be made in various individual power domains within the system by selectively turning off power to certain idle parts of the system.

Power optimizations can be done at various levels of the system. Typically, power optimization is done at the following levels:

- Architectural optimization

- Algorithmic optimization

- System integration optimization

- Application level optimization

In general, power optimization combines all of these approaches. Architectural optimizations deal with the optimization opportunities at the processor hardware level and try to obtain a suitable hardware-software partitioning. Algorithmic optimizations look for power-saving opportunities in system and application algorithms. For example, above the hardware and hardware abstraction layer, the graphics execution stack includes hierarchical layers of the application, the middleware, the operating system, and the graphics driver. Opportunities to save power exist within each layer and are exploited using algorithmic optimization.

Inter-layer optimization opportunities, however, are more complex and addresses inefficiencies by employing optimization at the system integration level. For example, efficiency can be improved by choosing to use fewer layers and by redefining the boundaries of the layers in order to find the most power-efficient places for the optimization. Furthermore, at the application level, load sharing between the CPU and the integrated GPU may be considered for reuse of power in one unit that is saved from the other, by running a task on the most power-efficient device. Discussions of these optimization techniques are follows in more detail.

Architectural Optimization

The techniques for optimizing power efficiency at the processor architecture level include:

- Hardware-software partitioning

- Dynamic voltage and frequency scaling

- Power gating

- Clock gating

- Slice gating

- Use of low-level cache

Hardware-Software Partitioning

There has been a paradigm shift in the approach to optimizing hardware and software interaction. The earlier philosophy was to obtain performance by removing execution bottlenecks. For example, if the CPU was the bottleneck in a graphics application, then the main power and performance tuning approach was to use a better CPU or to tune the CPU code to maximize graphics performance. However, processor architects soon realized that removing execution bottlenecks alone is not sufficient; it is also prohibitive from a power-consumption perspective to run all subparts of the system simultaneously at maximum performance, as various components compete for their share of the power envelope.

This realization opened two optimization opportunities: (a) power saved in one subpart can be applied to another; and (b) unused power can be applied to turbo behavior. Accordingly, considerations of power management for the overall system and shifting power between the CPU, graphics, and other subsystem are taken into account. As such, the new philosophy of hardware-software interaction aims not only to eliminate performance bottlenecks but also to continue tuning to increase efficiency and save power as well. For example, focus is now given to design goals including:

- Reducing CPU processing

- Optimizing driver codes to use the fewest CPU instructions to accomplish a task

- Simplifying the device driver interface to match the hardware interface to minimize the command transformation costs

- Using special-purpose hardware for some tasks with a balanced approach for task execution

Fixed-purpose hardware is often implemented with a minimum number of gates that switch states or toggle between states to perform certain specific tasks. As dynamic power consumption is a function of the number of gates that are switching, and as less switching means less dynamic power consumption, it is beneficial to perform the same task on the special-purpose fixed function hardware, as opposed to general-purpose hardware that may not use the optimum number of switching gates for that particular task. Obviously, if the nature of the task changes, the special-purpose hardware cannot be used, as it is often not flexible enough to accommodate changes in how it is used. In this case, power saving may be achieved by sacrificing flexibility of tasks, and often by migrating workloads from general-purpose hardware to fixed-function hardware. Careful design of hardware-software partitioning is necessary to save power in this manner, and non-programmable tasks may be migrated from general-purpose execution units to fixed-purpose hardware. For example, video processing algorithms that are run using GPU hardware designed explicitly for that task typically consume less power than running those same algorithms as software running on the CPU.

Dynamic Voltage and Frequency Scaling

To decrease power consumption, the CPU core voltage, the clock rate, or both can be altered, at the price of potentially lower performance, using dynamic voltage and/or frequency scaling. Alternatively, higher performance can be achieved at the expense of higher power consumption. However, as mentioned in the P-state discussion earlier, with the advancement of generations of CPU technology, this process is becoming increasingly complex, and there are many contributing factors, such as the load balancing among the multiple CPU cores and the GPU, thermal states, and so on. On the other hand, new techniques beyond dynamic voltage and frequency scaling are emerging to combat the challenges.

Power Gating

Processors can selectively power off internal circuitry by not supplying current to the parts of the circuitry that are not in use, and thereby reduce power consumption. This can be accomplished either by hardware or software. Examples of this technique include Intel Core and AMD CoolCore, where in a multi-processor environment only certain core processors (or part of the circuit in those processors) are active at a given time.

Power gating generally affects the design more than clock gating, and may introduce longer entry and exit latency from a gated state. Architectural tradeoffs are generally considered between the amount of power saved and the latency involved. Another important consideration is the area used for power gating circuitry if implemented in hardware. For example, in fine-grained power gating, switching transistors may be incorporated into the standard cell logic, but it still has a large area penalty and difficult independent voltage control per cell. On the other hand, in coarse-grained power gating, grid style sleep transistors drive cells locally through shared virtual power networks, and save area at the expense of sensitivity.

For quick wakeup from a power gated state, sometimes *retention* registers may be used for critical applications. These registers are always powered up, but they have special low-leakage circuits as they hold data of the main register of the power gated block, enabling quick reactivation.

Clock Gating

Clock gating is a popular technique for reducing dynamic power dissipation by using less switching logic and by turning off unnecessary clock circuitry, thereby saving power needed to switch states that are not useful at a given time. Clock gating can be implemented in RTL code or can be manually inserted into the design.

There are several forms of clock gating, ranging from manual to fully automated, that may be applied together or separately, depending on the optimization. On the one hand, there is the manual clock gating performed by driver software, where a driver manages and enables the various clocks used by an idle controller as needed. On the other hand, in automatic clock gating, the hardware may detect idle or no-workload states, and turn off a given clock if it is not needed. For example, on a particular board, an internal bus might use automatic clock gating so that it is temporarily gated off until the processor or a DMA engine needs to use it, while other peripherals on that bus might be permanently gated off if they are unused or unsupported on that board.

Slice Gating

Current Intel processors such as the fourth-generation core processor architecture or later have integrated graphics processing units that have arrays of programmable execution units (EUs) in addition to fixed-function hardware for specific tasks. The EUs, along with media samplers, are further arranged in slices. For example, some fourth-generation core SKUs have 40 EUs distributed between two equivalent slices, each containing 20 EUs and located in two different power domains. Figure 6-6 shows the slice structure of typical Intel fourth-generation core processor graphics execution units.

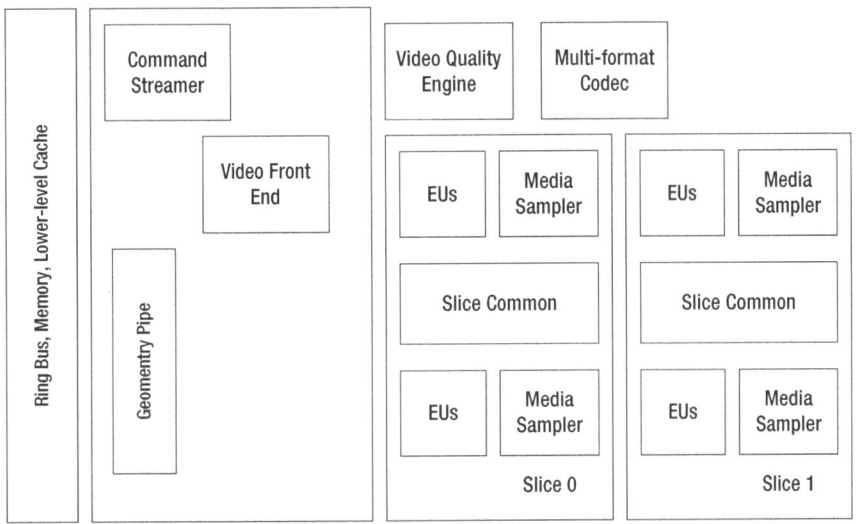

Figure 6-6. *Intel fourth-generation core processor graphics execution unit slice structure*

As hardware-accelerated multimedia tasks require many different assets in the graphics hardware, certain media tasks may place different demands on the media assets inside the slices, such as the EUs or the media samplers, and on the assets that are outside the slices, such as the Video Front End or Video Quality Engine. For some media workloads that require relatively little work from slice-based assets, the processor can shut down one slice to save leakage power. For example, for some media workloads, fewer than 20 EUs are needed, whereupon the driver software may power off one slice without affecting the performance. This is slice gating, also known as slice shutdown. The advantage of slice gating is that it maximizes power efficiency across a broad range of tasks.

Use of Low-level Cache

Memory power can be significantly reduced by using low-level caches and by designing algorithms to utilize these caches in an efficient manner. Video applications are typically compute-bound and not memory-bound, unless a memory-restricted system is used. Therefore, algorithms can take advantage of memory bandwidth reduction approaches, and thereby lower power consumption. For example, in the Intel core architecture, the cache is arranged in hierarchical levels, where both a low-level cache and a level-three cache are used. This enables power optimization, owing to the lower cost of memory access.

Algorithmic Optimization

The goal of algorithmic optimization is to reduce execution time by running the tasks fast and turning off processing units whenever they are not necessary. This can be achieved in many ways, including:

- As power consumption is proportional to execution residency, running less code in the CPU translates to less power consumption. So, performing code optimization of key software modules contributes to algorithmic optimization.

- Processing tasks can be offloaded to dedicated power-efficient fixed-function media hardware blocks as supported by the platform.

- In order to perform various stages in a pipeline of tasks for a given usage, it is generally necessary to expand the data into some intermediate representation within a stage. Storing such data requires a much larger bandwidth to memory and caches. The cost of memory transactions in terms of power consumption can be reduced by minimizing the memory bandwidth. Bandwidth reduction techniques are, therefore, important considerations for algorithmic optimization.

- The concurrency available among various stages or substages of the pipeline may be explored and appropriate parallelization approaches may be made to reduce the execution time.

- The I/O operations can be optimized by appropriate buffering to enable the packing of larger amounts of data followed by longer idle periods, as frequent short transfers do not give the modules a chance to power down for idle periods. Also, disk access latency and fragmentation in files should be taken into account for I/O optimization, as they may have significant impact in power consumption.

- Appropriate scheduling and coalescing of interrupts provide the opportunity to maximize idle time.

- All active tasks can be overlapped in all parts of the platform—for example, the CPU, the GPU, the I/O communication, and the storage.

Algorithmic optimization should be made with the power, performance, and quality tradeoffs in mind. Depending on the requirements of an application, while attempting to save power, attention should be paid to maintaining the performance and/or visual quality. A few common algorithmic optimization techniques are described in the following sections.

Computational Complexity Reduction

A computing device or system consumes very little power when it is not actively computing, as only the display engine needs to be awake; other compute engines may be temporarily in a sleeping state. The idea behind reducing computational complexity is to keep the system in a high power or busy state only as long as necessary, and to allow the system to return to idle state as often as possible. Improving the performance of an application can easily achieve power savings, as it allows the system to go back to idle state earlier because the work is done faster.

There are several approaches to computational complexity reduction, including algorithmic efficiency, active-duty cycle reduction, minimizing overheads such as busy-wait locks and synchronization, reducing the time spent in privileged mode, and improving the efficiency of I/O processing. We discuss some of these approaches next, but for a thorough treatment of them, see *Energy Aware Computing*.[8]

Selecting Efficient Data types

It is possible to optimize an algorithm that is heavy in floating point calculations by using integer arithmetic instead. For example, the calculation of discrete wavelet transforms using the lifting scheme usually involves a number of floating point operations. But the lifting coefficients can be implemented by rational numbers that are powers of 2, so that the floating point units in the data path can be replaced by integer arithmetic units.[9] This leads to power savings, as the hardware complexity is reduced.

Similarly, rearranging the code in a way suitable to take advantage of compiler optimization, or in a way where certain data dependency allows a computation to be done before entering a loop instead of inside the loop, can yield significant performance gain and thereby power savings. In an audio application example,[10] show some sine and cosine functions being repeatedly called on fixed values inside a busy loop; as the values are fixed, the computation can be made before entering the loop. This optimization yields about a 30 percent performance gain and also saves power.

In another example, motion vector and discrete cosine transform calculations were done on a vector of pixels instead of using each pixel separately,[11] which not only gives a 5 percent overall performance improvement in a software-only H.263 video encoder, but also provides power saving in two ways: by doing the computation faster, and by using improved memory access and cache coherency.

[8] B. Steigerwald, C. D. Lucero, C. Akella, and A. R. Agrawal, *Energy Aware Computing* (Intel Press, 2012).

[9] P. P. Dang and P. M. Chau, "Design of Low-Power Lifting Based Co-processor for Mobile Multimedia Applications," *Proceedings of SPIE* 5022 (2003): 733–44.

[10] Steigerwald et al., *Energy Aware Computing*.

[11] S. M. Akramullah, I. Ahmad, and M. L. Liou, "Optimization of H.263 Video Encoding Using a Single Processor Computer: Performance Tradeoffs and Benchmarking," *IEEE Transactions on Circuits and Systems for Video Technology* 11, no. 8 (August 2001): 901–15.

Code Parallelization and Optimization

Removing run-time inefficiency is also the goal of code parallelization and optimization. Multithreading, pipelining, vectorization, reducing the time spent in a privileged mode, and avoiding polling constructs are common techniques for code parallelization and optimization.

A properly threaded application that uses all available resources usually completes earlier than a single-threaded counterpart, and it is more likely to provide performance and power benefits. In this context, selecting the right synchronization primitives is also very important. Some applications, especially media applications, are particularly amenable to improvement using multithreading in a multi-core or multi-processor platform. In a multithreaded media playback example mentioned by Steigerwald et al.,[12] while almost linear performance scaling was achieved, the power consumption was also halved on a four-core processor at the same time, as all the four cores were busy running a balanced workload.

Similarly, the same operation on different data can be efficiently performed by using vector operations such as single-instruction multiple-data (SIMD) in the same clock cycle on a vector of data. Most modern processors support SIMD operations. The Intel Automatic Vectorizing Extensions (AVX) support eight 32-bit floating-point simultaneous operations in a single processor clock cycle. As Steigerwald et al.[13] claims, for media playback applications, use of such SIMD operations can result in approximately 30 percent less power consumption.

In Listing 6-1, note the following Direct3D query structure and the polling construct that only burns CPU cycles, resulting in wasted power.

Listing 6-1. Polling Example with Direct3D Query Structure (Power Inefficient)

```
        while ( S_OK != pDeviceContext->GetData( pQuery, &queryData,
sizeof(UINT64), 0 ) )
        {
                sleep (0); // wait until data is available
        }
```

It is better to use blocking constructs to suspend the CPU thread. However, Windows 7 DirectX is nonblocking. Although a blocking solution using the OS primitives would avoid the busy-wait loop, this approach would also add latency and performance penalty, and may not be appropriate for some applications. Instead, a software work-around may be used, where a heuristic algorithm detects the GetData() call in a loop. In an example of such work around,[14] up to 3.7W power was reduced without performance degradation. Listing 6-2 shows the concept of the workaround:

---

[12] Steigerwald et al., *Energy Aware Computing.*
[13] Ibid.
[14] D. Blythe, "Technology Insight: Building Power Efficient Graphics Software," Intel Developer Forum, 2012.

Listing 6-2. *Example of an Alternative to the Polling Construct*

```
INT32 numClocksBetweenCalls = 0;
INT32 averageClocks = 0;
INT32 count = 0;

// Begin Detect Application Spin-Loop
// ... ...
UINT64 clocksBefore = GetClocks();
if ( S_OK != pDeviceContext->GetData( pQuery, &queryData, sizeof(UINT64),
0 ) ) {
        numClocksBetweenCalls = GetClocks() - clocksBefore;
        averageClocks += numClocksBetweenCalls;
        count++;

        if ( numClocksBetweenCalls < CLOCK_THRESHOLD )
        {
                averageClocks /=count;
                if ( averageClocks < AVERAGE_THRESHOLD )
                {
                        WaitOnDMAEvent( pQuery, &queryData, sizeof(UINT64) );
                        return queryData;
                }
                else
                {
                        return queryBusy;
                }
        }
        else
        {
                return queryBusy;
        }
}
else
{
        return queryData;
}
// End Detect Application Spin-Loop
```

Memory Transfer Reduction

Limiting data movement and efficient data processing lead to better performance and lower power consumption. In this connection, it is more efficient to keep data as close to processing elements as possible by using the memory and cache hierarchy, and to minimize data transfer from main memory.

Reduction of memory transfer can curtail the power consumption, owing to the reduced number of memory accesses, even at the expense of a moderate increase in computational complexity.[15] Bourge and Jung proposed to reduce memory transfer by using embedded compression for the predictive pictures in the encoding feedback loop. It is possible to use an embedded coding scheme that would keep the reference frame in the frame memory in compressed format so as to use about a third of the memory compared to regular uncompressed coding method. If a lossless compression is used, then the required memory would be halved instead.

However, by using block-based memory access and by carefully managing the computational complexity of the embedded coding scheme, Bourge and Jung show that an overall power saving is possible.[16] They achieve this by imposing some restrictions on the coding scheme, which is a lossy scheme and is capable of obtaining better compression ratio and corresponding power saving than a lossless scheme. The restrictions include coding each block independently, fixing the compression ratio for each block, and jointly storing the luminance and chrominance blocks in memory. The end result is that even with an increase in computational complexity, the memory transfer, which dominates power consumption, is saved by 55 percent.

Although this particular scheme resulted in visual quality degradation at higher bitrates, using an appropriate lossless scheme may bring about overall power savings due to less memory transfer. Most important, owing to such reduction in memory transfer, a smaller memory embedded closer to the CPU can be used, leading to less cable dissipation during access. In some hardware implementations, it is possible to use low-cost on-chip memory instead of off-chip SDRAM.

System Integration Optimization

The interaction between various layers in the software stack can be optimized during system integration to yield a more power-efficient solution. The operating system, the graphics drivers, the middleware such as Intel media software development kit (SDK), and the applications can cooperate in such optimization. As these layers are typically developed by different companies, it is natural to expect inefficiencies resulting from such interactions. To improve the inter-layer efficiency, the following approaches to system integration optimization may be considered:

- Reducing the number of layers.

- Improving the understanding of the authors of various layers regarding each other's capabilities and limitations.

- Redefining the boundaries of the layers.

However, lacking such radical approaches, and until these become available, system integration optimization can still be done at various levels, some of which are as follows.

---

[15]A. Bourge and J. Jung, "Low-Power H.264 Video Decoder with Graceful Degradation," *Proceedings of SPIE* 5308 (2004): 372–83.

[16]Ibid.

System Operating Point on the P-F Curve

Figure 6-7 shows typical system operating points on the power-frequency curve, compared to the minimum operating point (F_{max} at V_{min}) and the maximum operating point (running at turbo frequency).

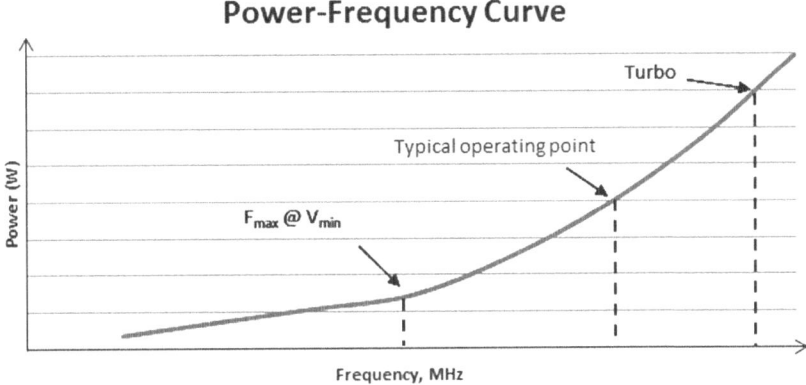

Figure 6-7. *Typical system operating point*

    As seen in Figure 6-7, in the voltage-scaling region of the power curve, tuning the system's operating frequency is important for power saving. It is possible to occasionally run the system at a lower frequency and save power as long as performance requirements are met. From the power consumption point of view, the best operating point is F_{max} at V_{min}; however, this frequency may not be sufficient for some applications. On the other hand, from the performance point of view, the best operating point is in the turbo frequency region. Based on the resource utilization profile, it is possible for power-aware graphics drivers to determine how to tune the frequency of the processor, and it is possible to dynamically move between turbo and regular operating frequency.

    As the operating system manages power, some systems offer various power policies ranging from low-power usage with low performance to high-power usage with high performance. In addition, the BIOS provide some flexibility to set the system frequency. End-users may take advantage of these power policies to adjust the system operating point to appropriate levels; for example, using the *power-saver* policy can lower the operating frequency and thereby save power.

Intelligent Scheduling

The levels of hardware-software partitioning are generally in the scope of architectural optimization. However, system-level optimization should also carefully consider the power-saving opportunities that are not covered by architectural design alone. For example, scheduling and migrating tasks between a software layer and special-purpose hardware units is a way such power-saving opportunities may be made available.

The operating system performs the scheduling of tasks for the CPU, while graphics drivers can schedule and manage the tasks for the GPU. Intelligent scheduling and load sharing between the CPU and the GPU is an active area of research, for which the middleware and the application layer may also make significant contributions. It is important, then, to find the most efficient place to do the processing; for instance, it may not be sufficient to simply multithread a CPU work, and it may be less efficient in terms of Joules per operation than operations per second.

Accomplishing migration of such a task from the CPU to a more power-efficient dedicated hardware module requires cooperation from all layers of the execution stack. To facilitate the scheduling, sometimes it is necessary to partition a piece of the system into several smaller chunks. For example, a shared user mode driver (UMD) that would interact with three run-time environments, such as OpenGL run-time, Direct3D 11 run-time, and Direct3D 9 run-time, may be redefined and divided into three components: OpenGL UMD, D3D 11 UMD, and D3D 9 UMD. This would facilitate both specific hardware access and interaction with the run-time environments; and it would make the system more amenable to power gating.

Similarly, some fixed work repeatedly done by the kernel mode driver for every invocation may be moved to the hardware itself. Examples of such system-level optimization can be found in the Intel fourth-generation core processor architecture, where using such system-level optimizations achieves a 2.25W decrease in CPU power for a popular 3D game application.[17]

Duty Cycle Reduction

By parallelizing the essential active tasks in a system—for example, tasks in the CPU, the GPU, the memory, and the I/O subsystem—the overall *uncore duty cycle* can be minimized. This would keep the related power subdomains active only for the required operations as needed and only for the minimum period of time, turning them off otherwise. The power subdomains include the various sensors, the PLLs, the memory interface interconnect buses, and so on, which can be separately controlled to minimize power consumption.

Furthermore, in order to run at a more efficient operating point, the duty cycle of the processor can be reduced by moving along the voltage-frequency curve, and using a higher frequency and higher power consumption for a shorter period of time, before going to an idle state for a relatively longer period of time. For the overall duration, this would typically result in lower power consumption. Conversely, for the same frequency, power can be saved with a lower voltage setting, as power is proportional to the square of the voltage. Duty cycle reduction is usually done at the system integration optimization level by the graphics kernel mode driver.

Figure 6-8 depicts the effect of a duty cycle reduction algorithm that focuses on using a higher frequency for a shorter period to accomplish the task of a video application, while the CPU is idle for longer period of time. In this example, the CPU utilization is reduced by approximately 20 percent.

---

[17]Blythe, "Technology Insight."

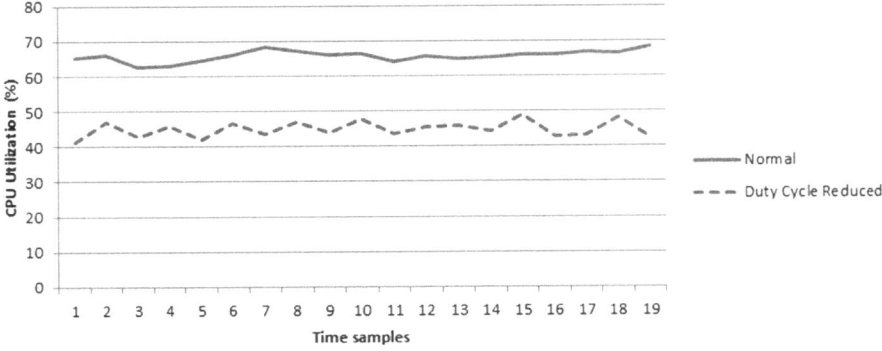

Figure 6-8. *Effect of duty cycle reduction on CPU utilization*

Application-Level Optimization

With the desire to support a plethora of functionalities in mobile computing devices comes the use of multiple sensors. A contemporary platform therefore includes light sensors, gyroscopes, accelerometers, GPS receivers, and near-field communications. By becoming aware of the available system resources and the user environment where multiple sensors may be active at a given time, applications can help avoid power misuse and can help users determine the priority of the sensors and features for a power-starving scenario.

Context Awareness by the Application

It is possible for a badly written application to burn power unnecessarily that could otherwise be saved. On the other hand, if an application is aware of the system resources that it runs on, and can sense a change in the system resource availability, it is possible for that application to react in a friendly manner to overall power consumption. For example, upon detecting low battery and subsequently notifying the user, an application may wait for intervention from the user before going to a lower power state. Alternatively, in a more active response, it may dim the display by default after sensing a darker ambient light condition.

It is the duty of the operating system to allocate system resources for each application, as requested by the application. The application's registering for power-related events allows the operating system to notify the application of a power event so as to enable the application to make an appropriate response. The application can also query for system state information using the APIs (application programming interfaces) provided by the operating system. For example, depending on whether the system is powered by a battery or connected to AC wall power, applications can make various power-saving decisions:

- Instead of a full system scan as done while on AC power, a virus checker may start a partial scan of the system on battery power.

- A media player may decide to trade off video quality to achieve longer playback of a Blu-ray movie.

- A gaming application may choose to sacrifice some special effects to accommodate more sections of the game.

In Windows, applications can query the operating system using a unique GUID (globally unique identifier) called GUID_ACDC_POWER_SOURCE to obtain the power setting information, and use this knowledge when a power event occurs. Similarly, to determine the remaining battery capacity, the GUID_BATTERY_CAPACITY_REMAINING can be used. And to learn about the current power policy, the GUID_POWERSCHEME_PERSONALITY can be used. It is also possible to use the GUID_BACKGROUND_TASK_NOTIFICATION to determine whether it is suitable to run a background task at the current state or it is better to wait for the *active* state so as not to perturb an idle state. In Linux, similar approaches also exist, where CCBatteryInfo structure can be used to determine the battery state. Furthermore, if the application switches contexts, it is possible to lower the power for the application's context that is no longer running.

Applications Seeking User Intervention

An application may invite favorable user intervention to save power. For example:

- An application can monitor battery capacity, and when the battery charge drops to a certain fraction of its capacity--say, 50 or 25 percent--the application may indicate a warning to the user interface to alert the user of the remaining battery capacity.

- An application can respond to a power source change from AC to DC by notifying the user of the change and providing an option to dim the display.

- An application can respond to ambient light level and request the user to adjust the display brightness.

Some of these actions can also be automatically taken by the system, but depending on the application, some may require user intervention. In general, user-configurable options allow the user to personalize the system, the application, and the experience. System and application designers may need to consider various tradeoffs when deciding which choices to give to the user and which to implement by default. For example, Windows provides the user with three power policies to choose from, or to define one's own settings. These options and settings drive the system-level behaviors that significantly impact the power efficiency of the platform.

Power Measurement

Now that we have covered different areas of power optimization, let us consider how to actually measure the power. In this section, we present the measurement methodology and various power-measurement considerations.

The ability to measure and account for power at various levels of the system allows system designers or users to understand existing power-management policies or to deploy optimized power-management policies as needed. Measuring power can uncover power-related problems that result in higher cost for the system. The major motivations for measuring power include:

- Understanding the impact of an application on power consumption by the system, and potentially finding optimization opportunities by tuning the application.

- Determining the effect of software changes at the user level, at the driver level, or at the kernel level; and understanding whether there is any performance or power regression owing to code changes.

- Verifying that a debug code was removed from the software.

- Determining the amount of power savings from power-management features, and verifying that such features are turned on.

- Determining the *performance per watt* in order to drive performance and power tuning, thereby obtaining the best tradeoff in practical thermally constrained environments.

However, few tools and instructions are available to measure the power consumed in a platform. Also, depending on the need for accuracy, different power-measurement methods can be used, ranging from simple and inexpensive devices to specialized data acquisition systems (DAQs). We present various approaches to power measurement.

Methodology

Within a computing system, power is measured at various system levels and at the motherboard level. In particular, this applies to the CPU package power, memory power, and display power measurement. Depending on the type of power supply, such measurement is of two types: AC power and DC power.

AC Power Measurement

For the system AC power or wall-power measurement, generally an AC power meter is connected between the power source and the system under measurement. The price for this measurement equipment may vary, depending on the accuracy and precision requirements. Simple, low-cost equipment typically has several drawbacks, including small ranges, low and imprecise sampling rates, inability to be used with other devices such as AC to DC converters or data acquisition systems, low resolution, and incongruity for measuring small power changes. On the other hand, they are easy to use and require little or no setup time.

For purposes of system-level and motherboard measurement, AC power measurement is not suitable, as these methods cannot provide insight into the system's power consumption.

DC Power Measurement

Although DC power can be measured using scopes and multi-meters, the easiest, most accurate, and most precise way of measuring DC power is by using automated *data acquisition systems* (DAQs). DAQs take analog signals as inputs and convert them to digital data sequence for further processing and analysis, using specialized software. Typically, DAQs can support several input channels, and can interface with the data-analyzing computer via standard serial or USB ports. They are capable of handling very high data rates and can measure tiny voltage differences, making them ideal for automated power measurements.

The power dissipated across a resistor can be expressed as follows:

$$P = V^2/R,$$ (Equation 5-3)

where V is the voltage in volts, R is the resistance in ohms, and P is the power in watts.

The current through the circuit is determined by the ratio of V to R. To measure the power of a black box circuit, it is a common practice to add a very small *sense resistor* with a low resistance, r, in series with the black box, which has a larger resistance, R, so the total resistance of the circuit is approximately equal to R. In this case, the power needed for the black box can be approximated in a modified version of Equation 5-3:

$$\cdot P = \frac{v \times \Delta v}{R},$$ (Equation 5-4)

where ΔV is the voltage drop across the sense resistor and V is the potential of a channel input with respect to ground.

Since voltage is the potential difference between two points, for each voltage measurement two inputs are required: one to represent the ground, or reference potential, and the other to represent the non-zero voltage. In a single-ended measurement, the reference is provided by the DAQ's own ground and only the non-zero voltage is measured for an input channel voltage. Compared to using separate grounds for each channel, single-ended measurements may be less accurate, but they have the advantage of using faster sampling or more input channels.

DAQs can take as inputs the general-purpose analog signals in the form of voltage. The analog signals may have originally been captured using a sensor before being converted to the voltage form, or they may already exist in a voltage form. In the latter case, a simple low-resistance *sense resistor* can act as a sensor.

In order to measure the power of a certain system or motherboard component, typically the appropriate power rails are instrumented so that a sense resistor is connected in series on the rail. As current flows through the sense resistor, a voltage drop ΔV is created, which can be measured by the DAQ, as shown in Figure 6-9, where a very small sense resistor (e.g., 2 milliohm resistance) is used.

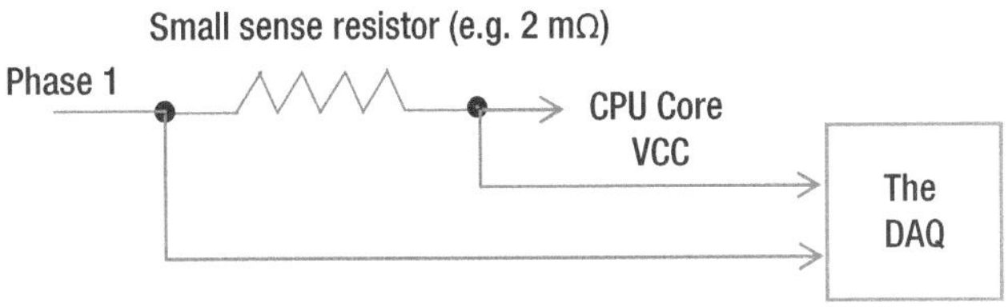

Figure 6-9. *Power measurement setup in a power rail*

The data can be analyzed and aggregated to give the measured power over a period of time, using special software accompanying the DAQ, such as the National Instrument LabView.

Considerations in Power Measurement

The following factors are generally taken into account while measuring power:

- The TDP of the processor part under measurement.

- The accuracy and precision of the data acquisition system; The ability of the DAQ and associated software for real-time conversion of analog voltage signals to digital data sequence, and for subsequent processing and analysis.

- Ambient temperature, heat dissipation, and cooling variations from one set of measurements to another; to hedge against run-to-run variation from environmental factors, a three-run set of measurements is usually taken and the median measured value is considered.

- Separate annotation of appropriate power rails for associated power savings, while recording the power consumption on all power rails at a typical sampling rate of 1 kHz (i.e., one sample every one millisecond), with a thermally relevant measurement window between one and five seconds as the moving average.

- Recognition of operating system background tasks and power policy; for example, when no media workload is running and the processor is apparently idle, the CPU may still be busy running background tasks; in addition, the power-saving policy of the operating system may have adjusted the high-frequency limit of the CPU, which needs to be carefully considered.

- Consideration of average power over a period of time in order to eliminate the sudden spikes in power transients, and consideration only of steady-state power consumption behavior.

- Benchmarks included for both synthetic settings and common usage scenarios; appropriate workloads considered for high-end usages so that various parts of the system get a chance to reach their potential limits.

- Consideration of using the latest available graphics driver and media SDK versions, as there may be power optimizations available in driver and middleware level; also, there is a risk of power or performance regression with a new graphics driver such as potential changes to the GPU core, memory, PLL, voltage regulator settings, and over-clock (turbo) settings.

Tools and Applications

Power-measurement tools include both specialized and accurate measurement systems such as DAQs, as well as less accurate software-based tools and applications. We consider a specialized DAQ system and introduce several software tools with varying capabilities.

An Example DC Power-Measurement System

An example DC power measurement system is based on the National Instruments* PXIe 6363 PCI-express based DAQ and the associated LabView Signal Express software application for signal analysis. The PXIe 6363 has a signal capture bandwidth of 1.25 million samples per second and an A/D conversion resolution of 16 bits on every voltage input channel. This input voltage is programmable down to ±1V, so that it is easy to zoom into the low-voltage signals. Similarly, for today's low-power devices, newer versions of PCIe DAQs with higher-precision input voltages are also available.

Typically a 2 milli-ohm current sense resistor is used in series with all power rails of interest—for example, the CPU package, the memory DIMMs, and the display, for which the peak, the average, and the minimum DC power consumption are measured. Also, the run-time CPU and GPU frequencies are monitored to determine proper turbo operation. The power setup is calibrated automatically on each run for sense resistor and test harness variations that may occur due to ambient temperature.

To capture and compute power in watts, it is necessary to measure both voltage and current for each power rail. This is accomplished by using current sense resistors in series with the incoming power supply on each voltage rail. The voltage drop across the current sense resistor is a small amplitude signal that directly correlates to the amount of current flowing through the sense resistor. The voltage for each power rail (positive and negative wire), and the output of the current sense resistor (positive and negative wire), connects directly to the PXIe 6363 via the removable TB-2706 terminal block analog input modules.

The measured power is logged and plotted using the LabView Signal Express to produce a detailed and comprehensive power-performance profile. This application captures and processes the voltage and current measurements from the PXIe 6363 DAQ modules and computes the power in watts simply by multiplying the measured voltage and sense current.

This application also supports various statistical measurements, such as moving average, peak, average, and minimum power used for detailed signal analysis. Figure 6-10 depicts a sample of a LabView configuration interface for a power measurement system. In this interface, selections can be made for the voltage channels of interest. Figure 6-11 then shows an example of the LabView interface when a power measurement is in progress. The top and bottom windows show voltage, current, or power signals from all input channels and a single channel (Channel 0), respectively.

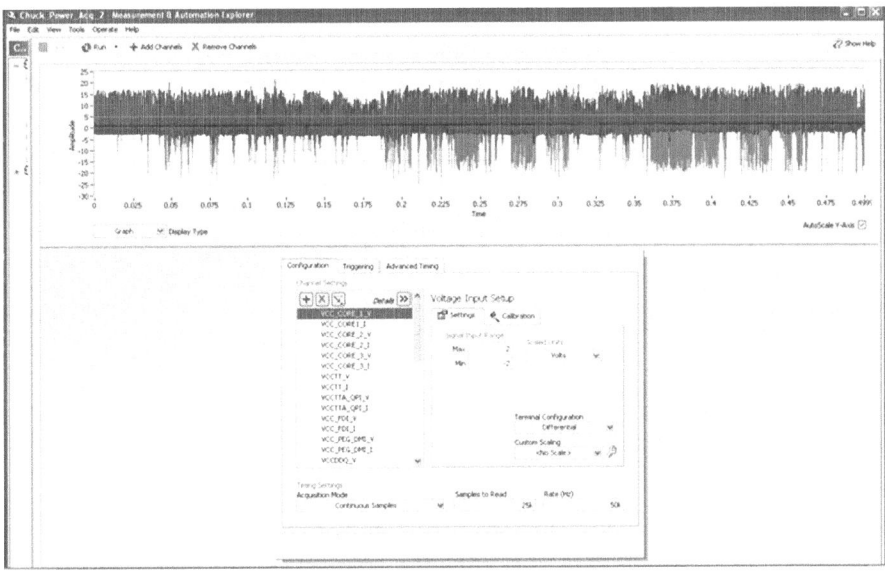

Figure 6-10. *LabView data acquisition setup for power measurement*

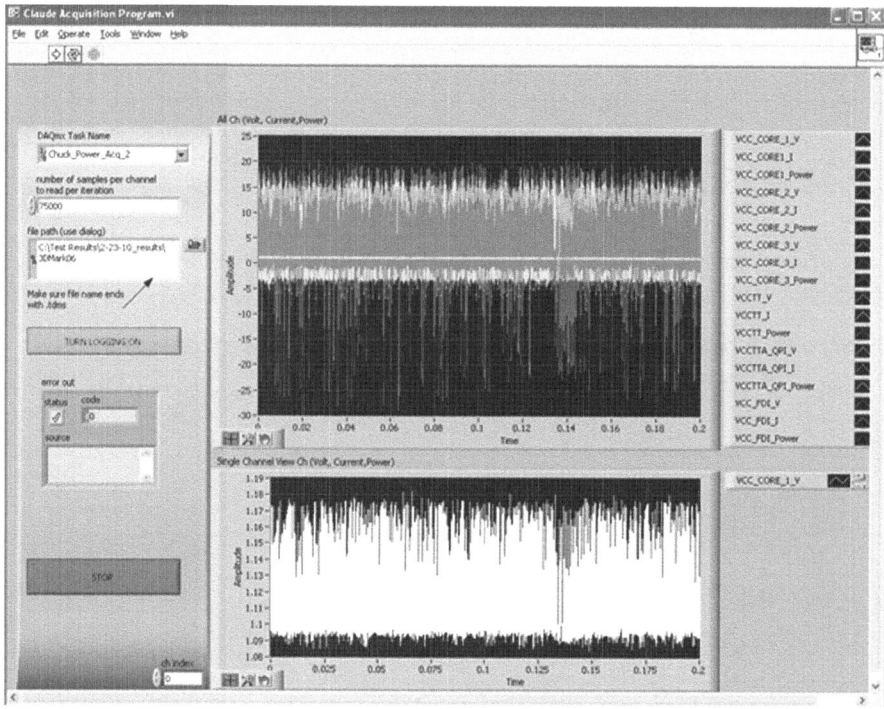

Figure 6-11. *Power data acquisition in progress*

Software Tools and Applications

To get the best and most accurate data on how much energy a computer platform is using during operation, a hardware power meter is needed. The Networked Data Acquisition Unit (NetDAQ) from Fluke, the National Instrument DAQ, and the Yokogawa WT210 are examples of such acquisition systems. However, these are expensive and the cost may not be justifiable to a regular consumer or an application developer who is only interested in a one-time or so power measurement. For these users it makes more sense to select a software tool or application that measures power consumption.

The tool and applications are primarily used to identify power issues, with and without workloads running, in order to optimize the system's power consumption. The issues typically encountered include:

- **CPU/Chipset Power:** Such problems are identified by examining the CPU C-state residency to determine whether the CPU and the chipset power are optimally managed, and to get some insight into what is causing any increase in platform power consumption. For example, high residency at deep C-states such as $C3$ may indicate frequent C-state transition due to device interrupt or software activity.

- **CPU Utilization:** CPU utilization samples are commonly taken at every timer tick interrupt--i.e., every 15.6 millisecond for most media applications and some background applications. However, the timer resolution can be shortened from the default 15.6 millisecond in an attempt to capture activities within shorter periods. For multi-core CPUs, CPU utilization and power consumption depend on the active duration, while each core may only be active for a partial segment of the total duration for which the platform is active. Therefore, CPU core utilization and platform utilization should be counted separately. Logically, when the activities of two cores overlap, the CPU utilization is shown as the sum of two utilizations by most power measurement tools. Only few tools, the Intel Battery Life Analyzer among them, can actually use fine-grain process information to determine the total active duration of both the platform package and the logical CPU. By investigating the CPU utilization, inefficient software components and their hotspots can be identified, and the impact of the software component and its hotspots can be determined to find optimization opportunities.

- **CPU Activity Frequency:** Power tools can help identify software components causing frequent transition of CPU states. It is valuable to determine the frequency of the activity of each component and the number of activities that are happening in each tick period. Understanding why the frequent transitions are happening may help point to power-related issues or improvement prospects.

- **GPU Power:** On the modern processors, as most media applications run on the GPU, it is also important to understand the impact of GPU C-state transitions and GPU utilization. GPU utilization largely controls the power consumption of media applications. However, there are only few tools that have the ability to report GPU utilization; the Intel GPA is one such tool.

In general, there are several tools and applications available for the measurement and analysis of power consumption of various components of a computing device. Some are especially relevant for analysis of the idle system behavior, while others are suitable for media applications. In the next section, we discuss some of these tools, starting with the Linux/Android based PowerTop and going into several Windows-based tools. We then discuss specific tools for monitoring and analyzing battery life. The Power Virus is also mentioned, which is mainly used for thermal testing. However, as new tools are constantly being developed, some tools obviously are not covered.

PowerTop

PowerTop is a software utility developed by Intel and released under GPL license that is designed to measure and analyze power consumption by applications, device drivers, and kernels running on Android, Linux, or Solaris operating systems. It is helpful in identifying programs that have power issues and to pinpoint software that results in excessive power use. This is particularly useful for mobile devices as a way to prolong the battery life.

PowerCfg

PowerCfg is a command line tool in Windows that allows users control the power-management settings of the system and to view or modify the power policy. It is typically used to detect common issues in power efficiency, processor utilization, timer resolution, USB device selective suspend, power requests, and battery capacity.

PwrTest

PwrTest is a power management test tool available in the Windows Driver Kit that enables application developers and system integrators to obtain power-management information such as the various sleep states information (e.g., C-state and P-state information) and battery information from the system and record over a period of time.

Perfmon and Xperf

The Windows Perfmon provides abilities to monitor the performance counters available in Windows, including C-state and P-state residencies, which are useful in understanding CPU utilization and activity related issues.

The Xperf is a command-line tool that helps developers in system-wide performance analysis by monitoring system and kernel events such as context switches, interrupt service routines, and deferred procedure calls for a period of time and by generating reports for graphical review. It is useful to correlate the events with system status in scenarios where the system is idle, running web browsing, or during media applications. Xperf generates event trace logs that can be viewed using Xperfview; both of these tools are available in the Windows Performance Toolkit.

Joulemeter

Developed by Microsoft Research, Joulemeter is a modeling tool to measure the energy usage of virtual machines (VMs), computers of various form factors and power capacity, and even individual software applications running on a computer. It measures the impact of components such as the CPU, screen, memory, and storage on their total power use. One of its advantages is that it can measure the impact of software components, such as VMs, that do not have a hardware interface and therefore are not amenable to measurement by hardware power meters.

The data obtainable from Joulemeter includes the current energy usage for each component, such as the base or idle energy usage, CPU usage above the baseline idle, monitor, and hard disk. The output data is presented in watts and is updated every second. Details can be found in *Joulemeter: Computational Energy Measurement and Optimization.*[18]

Intel Power Gadget

To assist end-users, independent software vendors, original equipment manufacturers, and the application developers to precisely estimate power consumption without any hardware instrumentation of the system, Intel developed a software tool named Intel Power Gadget, which is enabled for the second- Generation Intel Core processors. Additional functions of the tool include estimation of power on multi-socket systems and externally callable APIs to extract power information within sections of the application code.

The gadget includes a Microsoft Windows sidebar gadget, driver, and libraries to monitor and estimate real-time processor package power information in watts, using the energy counters in the processor. After installation, the gadget can be simply brought up to monitor processor power usage while running a workload or when the system is idle. An "Options" pop-up window allows setting the sampling resolution in milliseconds and the maximum power in watts. The output data, notably the processor package power and frequency, is generated in real time and can be logged in a file with a comma-separated values (CSV) format. The gadget can be downloaded from Intel's website.[19]

Intel Power Checker

The Intel power or energy checker tool determines the power efficiency of a system in terms of useful work done with respect to energy consumed during that work. It is an easy way for media or game application developers to check the power efficiency of their applications on mobile platforms with Intel Core or Atom processors. This tool does not require an external power meter, and it is useful for power analysis of any application compiled for Intel processors or Java framework applications.

By default, this tool checks the system capability to provide power consumption data and whether a particular driver called EzPwr.sys (part of Intel Power Gadget) is installed, which would be necessary if an external power meter device is used. Typically, the tool first measures the baseline power without the target application running, while unnecessary processes such as operating system updates, Windows indexing service, virus scans, Internet browsers, and so on are turned off. In the next step, the target application is run, and power is measured again starting from a desired point of the target application's execution. Finally, it measures power again when the target application is completed and returned to an idle state. The tool provides analysis on elapsed time, energy consumption, and average *C3* state residency, and gives the platform timer duration in milliseconds. This tool is now part of the Intel Software Development Assistant.[20]

---

[18]Available from Microsoft Research at research.microsoft.com/en-us/projects/joulemeter/default.aspx.

[19]Available from software.intel.com/en-us/articles/intel-power-gadget.

[20]Available from software.intel.com/en-us/isda.

Intel Battery Life Analyzer

The Intel Battery Life Analyzer (BLA) is a software tool running on Microsoft Windows that is primarily used to monitor the activities of hardware and software platform components and determine their impact on battery life. It can identify drivers, processes, or hardware components that prevent the platform from entering low-power states. BLA has many modules to support the power analysis, including CPU C-state and software activity analysis.

The more time the system spends in the deep C-state, the less power it consumes. BLA recommends threshold values for C-state residencies, in particular, that the deepest C-state residency at idle should be greater than 95 percent for the processor package (i.e., socket) containing multiple processor cores and 98 percent per core. Also, $C0$ and $C1$ states for the package should be less than 5 percent at idle. There are options in the BLA tool to set the appropriate C-state threshold values. Copies of the BLA tool can be requested via e-mail from Intel.[21]

Intel Graphics Performance Analyzer

The Intel Graphics Performance Analyzers 2013 (Intel GPA) is a suite of three graphics analysis and optimization tools--namely, the system analyzer, the frame analyzer, and the platform analyzer—to help game and media application developers optimize their games and other graphics-intensive applications. Intel GPA supports the latest generations of Intel Core and Intel Atom processor-based platforms running Microsoft Windows 7, 8, 8.1, or the Android operating system. The system analyzer provides the CPU and the GPU performance and power metrics in real time, and allows users to quickly identify whether the workload is CPU- or GPU-bound so the user can concentrate on specific optimization efforts. The frame analyzer provides ability to analyze performance and power down to the frame level. The platform analyzer provides off-line analysis of the CPU and GPU metrics and workloads with a timeline view of tasks, threads, Microsoft DirectX, and GPU-accelerated media applications in context. The tool is also available from Intel.[22]

GPU-Z and HWiNFO

GPU-Z is a lightweight system utility from TechPowerUp, designed to provide vital information about a video card and/or the integrated graphics processor; it supports nVIDIA, ATI, and Intel graphics devices. HWiNFO is free software, available from the Internet, that combines the functionalities of CPU-Z and GPU-Z and provides the CPU, the GPU, and memory usages, along with other system information.

---

[21]Request for BLA tool can be made at batterylifeanalyzer@intel.com.
[22]Available from software.intel.com/en-us/vcsource/tools/intel-gpa.

Power Virus

Power virus executes specific machine code in order to reach the maximum CPU power dissipation limit—that is, the maximum thermal energy output for the CPU. This application is often used to perform integration testing and thermal testing of computer components during the design phase of a product, or for product benchmarking using synthetic benchmarks.

Summary

In modern processors, power considerations go beyond battery life and attempt to dictate performance. We reviewed the power-consumption behavior by media applications running on mainstream computing devices.

First, we discussed the requirements and limits of power consumption of typical systems, the power equation, and aspects of various sources of power supply. Then, we covered how a mobile device is expected to serve as the platform for computing, communication, productivity, navigation, entertainment, and education. We also surveyed three major topics: power management, power optimizations, and power measurement considerations.

Finally, we learned about several power-measurement tools and applications, and their advantages and limitations. In particular, we showed as an example a specific DC power measurement system using a DAQ, and several software-based power measurement tools.

While there is no single tool or application suitable for all types of power-measurement scenarios, some tools are quite capable of providing important insights into the power profiles of video applications, and are useful for this purpose.

CHAPTER 7

Video Application Power Consumption on Low-Power Platforms

Some mobile applications, particularly those on low-power devices, unjustifiably use a lot of energy, causing unnecessary strain on the device battery. Some applications start automatically as soon as the device is powered on or do not go to sleep when the user stops using them; in most cases these applications keep performing noncritical tasks. Other applications may devote only a small fraction of their energy budget to the core task, while spending a larger portion on user tracking, uploading user information, or downloading ads.

To conserve battery power, systems usually go to deep sleep at every opportunity, from which they can be awaken as needed; some applications abuse this feature by regularly waking the device for nonessential tasks, such as checking the server for updates, new content, or mail and messages, or for reporting back on user activity or location. For this reason, judicious use of available resources and capabilities by these applications can save significant energy and thereby improve battery life.

However, most mobile media applications have a different emphasis in their power-saving attempts. They usually don't have sufficient performance headroom for doing auxiliary tasks; so they need to consider the characteristically complex nature of the media usages, as well as the resource constraints of the device.

Viewed primarily from the media application point of view, this chapter begins by extending the power consumption, management, and optimization discussions that were begun in Chapter 6 to the domain of the low-power device. The chapter starts with a discussion of the power-consumption priorities of low-power devices. Then, it presents typical media usages on these low-power devices. Four such common usages—namely video playback, video recording, video conferencing, and video delivery over Miracast

Wireless Display—are analyzed and discussed. Examples of activities and a power profile for each of these usages are also offered. This is followed with a discussion of the challenges and opportunities present in system low-power states, which include finer-grained states than the ACPI power states presented in Chapter 6.

The next section discusses power-management considerations for these low-power platforms, particularly some display power-management techniques for saving power. Then, software and hardware architectural considerations for power optimization are surveyed and various power-optimization approaches are offered. In the final section, we briefly mention low-power measurement considerations and metrics.

The Priorities for Low-Power Devices

In the past several decades, we have experienced exponential growth in terms of computer speed and density, in accordance with the well-known Moore's law. Ultimately, this trend will come to an end, as the growth is restricted by the realism of power dissipation and the limits of device physics. But for now, the power consumption of individual computing elements has yet to level off. Mobile computing devices, such as smartphones, laptops, and tablets, as well as home entertainment electronics like set-top boxes, digital cameras, and broadband modems, essentially follow this same trend, and so there is still room for power optimization.

On the other hand, increased functionalities for the mobile devices are becoming common for many low-power embedded systems, ranging from electronic controllers of household appliances to home energy-management systems, and from in-vehicle infotainment to sophisticated medical equipment. Not only must such devices consume very little power, they must also be "always on, always available."

Thus, fueled by insatiable consumer demand and intense competition, manufacturers of mobile devices continue to offer progressively more functions and faster clocks, but these are required to conform to a lower power budget for extended battery life. Furthermore, the advent of wearable computing devices, such as smart earphones and smartwatches, require extremely low-power processors within amazingly tiny forms. For example, a smartwatch with a surface area as small as 1600 mm^2 generally requires more than a week's battery life, ushering in the heralded *Internet of Things* (IoT).

Designers aim to keep reducing the packaging, manufacturing, operational and reliability costs of these devices while simultaneously supporting increasingly complex designs. Such ambitious goals generally materialize with the introduction of new silicon process technologies of finer geometry roughly every two years. However, every generation of process shrinkage results in higher power leakage, owing to increased gate and junction diode leakage. Although the dynamic power consumption scales down with process geometry, growing wire density tempers such reductions. Therefore, a simple scaling down of dynamic power alone is often insufficient for next-generation applications. The increased insistence on performance obligations for evermore complex applications imposes aggressive battery-life requirements, so this calls for more aggressive management of leakage and active power.

In general, the priorities for low-power devices moving from one process geometry to the next, smaller geometry include:

- *Decreasing the dynamic power.* This is possible as the dynamic capacitance and the voltage are both generally reduced for a smaller geometry.

- *Maintaining the total static leakage power.* This is an active area of focus for hardware architects because, due to process technology, the leakage power tends to increase for a smaller geometry. So it is necessary to optimize in this area with a view to maintaining the leakage power.

- *Keep active leakage at a small percentage (e.g., approximately 10-15%) of the dynamic power.* Also, leakage power must not be allowed to dominate the power consumption equation.

Let us recall the power-frequency relationships presented in Chapter 6. Figure 7-1 shows the dynamic and leakage power with respect to frequency.

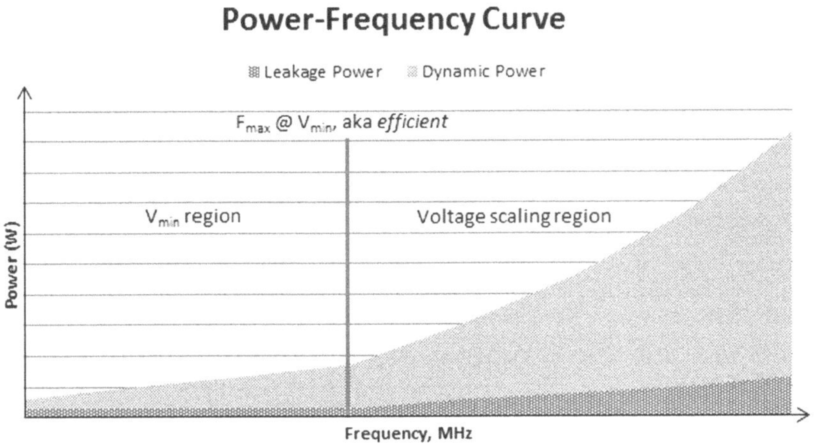

Power-Frequency Curve

Figure 7-1. Power-frequency relationship. For low-power design, the goal is to keep leakage power at ~10-15% of the dynamic power, even as the process shrinks to the next generation

As is evident from Figure 7-1, there is an efficient frequency below which voltage and frequency scaling do not achieve good power reduction. In this V_{min} region, voltage only scales linearly with frequency, so leakage current, and consequently leakage power, becomes a determining factor for further power reduction. It is also necessary to point out that voltage cannot be arbitrarily reduced, as there is a minimum voltage needed to drive the circuit into the active state. Above the efficient frequency point, however, good voltage scaling can be achieved, as the voltage scales with frequency according to a cubic relationship. As a result, in this relatively higher-power region, power reduction can happen with an easy voltage-frequency tradeoff.

An example of mobile low-power platform architecture is shown in Figure 7-2. Power management and optimization are essential for each module of the architecture—namely the system-on-a-chip (SOC), storage module, input and output (I/O) module, sensors and cameras, controllers and communications module. In typical modern SoCs, there are dedicated circuits available for thermal control and power management—for example, the *power management controller* (PMC), which is discussed later.

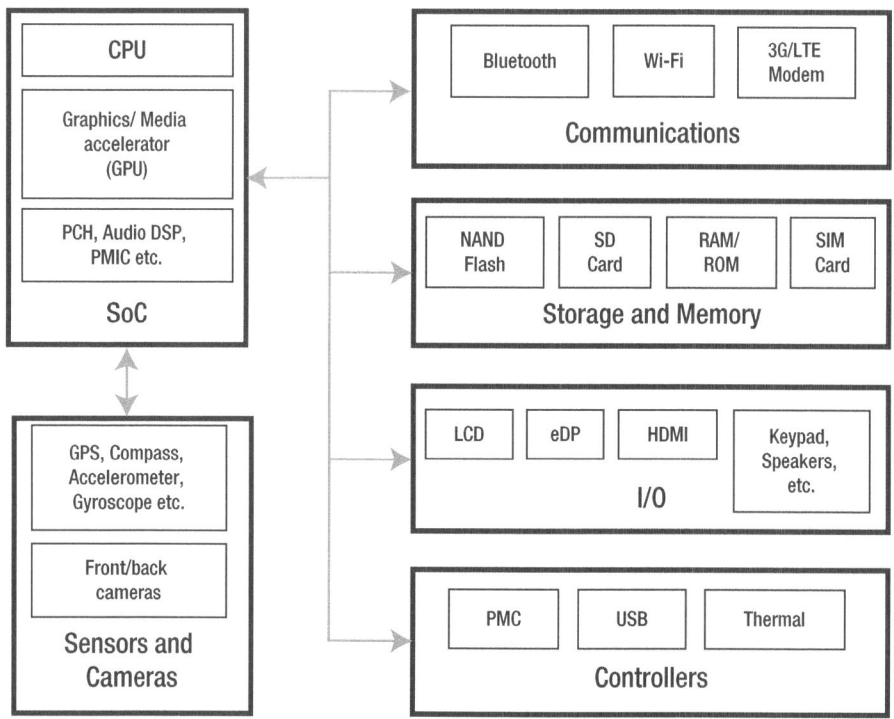

Figure 7-2. *An example of mobile platform architecture*

Typical Media Usage on Low-Power Platforms

Low-power computing platforms span a wide range of devices and applications, all of which need to save power while they are not in use, or only partially in use, regardless of whether media are involved. Table 7-1 lists some examples of low-power platforms.[1]

---

[1]B. Steigerwald, C. D. Lucero, C. Akella, and A. R. Agrawal, *Energy Aware Computing* (Hillsboro, OR : Intel Press, 2011).

Table 7-1. *Examples of Low-Power Platforms*

Category	Example	Special needs and requirements
Computing on the go	Smartphones, tablets, netbooks, wearable devices	Small form factor and limited storage capabilities; minimum heat dissipation and special cooling requirements; reduced memory bandwidth and footprint; multiple concurrent sessions; hardware acceleration for media and graphics; sensors; real-time performance; extended battery life
Medical equipment	Imaging systems, diagnostic devices, point of care terminals and kiosks, patient monitoring system	Amenable to sterilization; secure, stable and safe to health; resistant to vibration or shock; lightweight and portable; high-quality image capture and display capability; fanless cooling support, etc.
Industrial control systems	Operator-controlled centralized controller of field devices	Special I/O requirements, vibration and shock withstanding capabilities, variety of thermal and cooling requirements, ranging from fanless passive heat sink to forced air, etc.
Retail equipment	Point of sale terminals, self-checkout kiosks, automatic teller machines (ATMs)	Ability to withstand extreme ambient temperature, good air-flow design, security from virus attacks, non-susceptibility to environmental conditions (dust, rain etc.)
Home energy management systems	Centralized monitor of a home's usage of utilities such as electricity, gas and water; data mining of energy usage for reporting, analysis and customization (e.g., warning users when washing clothes is attempted during peak electrical rates or when a light is left on without anyone present, etc.)	Internet connectivity; various sensors; ability to control various smartphone apps; wireless push notification capability; fanless operation; ability to wake from a low-power or power-down state by sensors or signals from Internet; low power consumption while working as an energy usage monitor

(*continued*)

Table 7-1. *(continued)*

Category	Example	Special needs and requirements
In-vehicle infotainment systems	Standalone navigation systems, personal video game player, Google maps, real-time traffic reporting, web access, communication between car, home and office	Ability to withstand extreme ambient temperature; special cooling mechanisms in the proximity of heating and air-conditioning system; small form factor to fit behind dashboard; very low power level (below 5W); fast return from deep sleep (S3) state
Digital signage	Flight information display system, outdoor advertising, wayfinding, exhibitions, public installations	Rich multimedia content playback; display of a single static image in low-power state; auto display of selected contents upon sensor feedback (e.g., motion detection); intelligent data gathering; analysis of data such as video analytics; real-time performance
Special equipment for military and aerospace	Air traffic control, special devices for space stations and space missions, wearable military gears	Security; real-time performance; fast response time; extreme altitude and pressure; wearable rugged devices with Internet and wireless connectivity; auto backup and/or continuous operation
Embedded gaming	Video gaming devices, lottery, slot machines	High-end graphics and video playback; keeping attractive image on the screen in low-power state; various human interfaces and sensors; high security requirements; proper ventilation
Satellite and telecommunications	Scalable hardware and software in datacenters and base stations, throughput and power management and control	Compliance with guidelines requirements for environmental condition such as National Equipment Building Systems (NEBS); Continuous low-power monitoring
Internet of Things	Smart appliances, smart watches, smart earphones, smart bowl	Very low power for control and monitoring; Internet connectivity

Although the requirements are manifold from different low-power platforms, practical considerations include tradeoffs among processor area, power, performance, visual quality, and design complexity. It may be necessary to reduce nonessential functions or sacrifice visual quality while maintaining low power and keeping usability and elegance. This approach would favor simplicity over performance, so as to support the energy-efficiency requirements for a specific device or application. Consequently, there are only a few media usages that are overwhelmingly commonplace in low-power platforms. Of particular importance are video playback and browsing over local and remote Wireless Display, video recording, and videoconferencing. Some details of these usages are discussed next.

Video Playback and Browsing

Video playback, either from local storage or streaming from the Internet, is one of the most prevalent and well-known media usage model for low-power devices. Video can be played back using a standalone application or via a browser, while the content is streamed from the Internet. As digital video is usually available in various compressed and coded formats, video playback involves a decompression operation during which the video is decoded before being played back on the display. If the resolution of the video is not the same as the display resolution, the decoded video is resized to match the display resolution.

Video decode and playback are quite complex and time-consuming operations when performed using the CPU only. Often, hardware acceleration is used to obtain higher performance and lower power consumption, as optimized hardware units are dedicated for the operations. In modern processors, hardware-accelerated playback is the norm, and it is essential for low-power platforms.

Figure 7-3 shows the software stack for a video playback usage model based on the Android operating system. The Media Extractor block demultiplexes the video and audio data. That data is fed to the corresponding Open MAX (OMX) decoder, which delivers the video bitstream to the LibMix library to begin decoding. The hardware decoder is used for the actual video decoding, which is driven by the media driver. The decoded video buffers are delivered to the Android audio-video synchronization (AVsync) software module. By comparing the video buffer's *presentation time stamp* (PTS) and the audio clock, AVsync queues the buffers to the Surfaceflinger to be rendered on the display at the appropriate time.

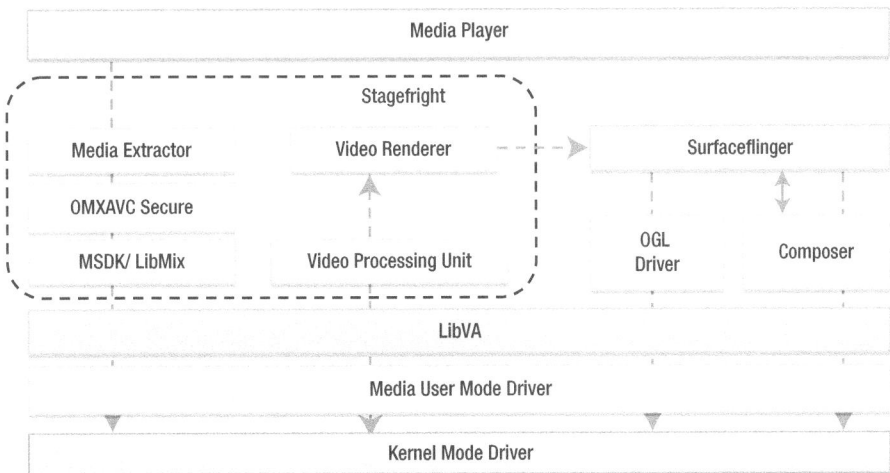

Figure 7-3. *Video playback software stack on Android*

Figure 7-4 shows a block diagram of the decode process in video playback. This process is typically defined by an international standards study group, and all compliant decoders must implement the specified definitions of the codec formats. Careful optimizations at various hardware and software levels are necessary to obtain a certain performance and power profile, especially on low-power platforms. Optimization approaches include implementation of repetitive operations on special-purpose hardware, optimized memory load/store/copy, cache coherency, scheduling, load balancing of tasks in various hardware functional units, optimized post-processing, opportunistically entering power-saving states, and so on.

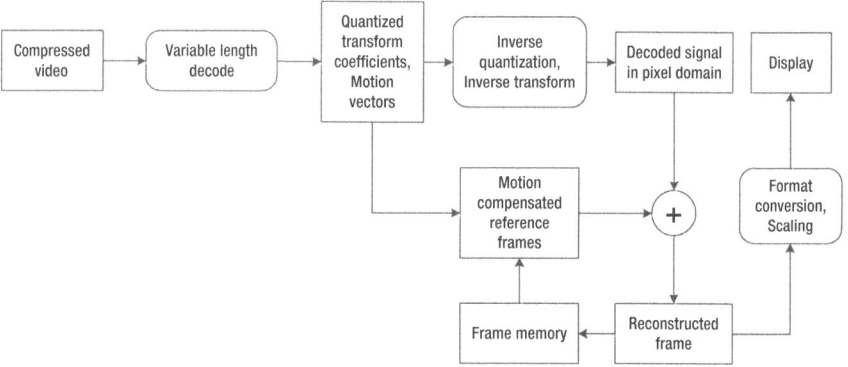

Figure 7-4. *Block diagram of video decode process*

In Figure 7-5, an example activity profile of the video playback usage is presented for AVC 1080p30 and 1080p60 resolutions. In this Intel Architecture-based low-power platform, a 1080p playback generally requires ~10-25% CPU and ~15-30% GPU activity depending on the frame rate. While the display unit consumes about half of the platform power, the SoC consumes ~20-30%. Notable power loss occurs at the voltage regulator (~15%), memory (~5%), and other parts of the platform.

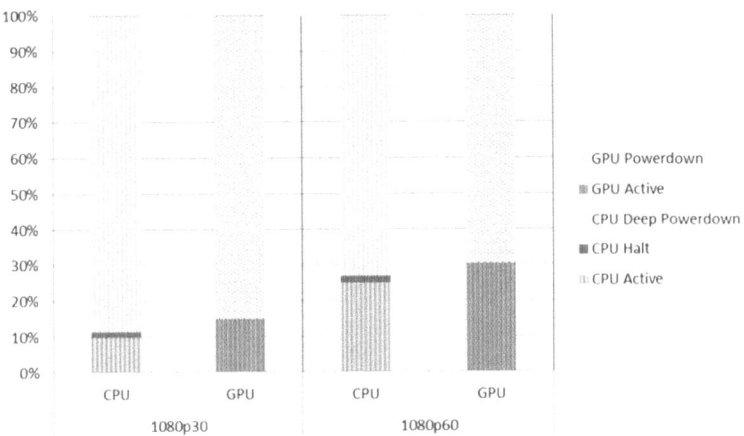

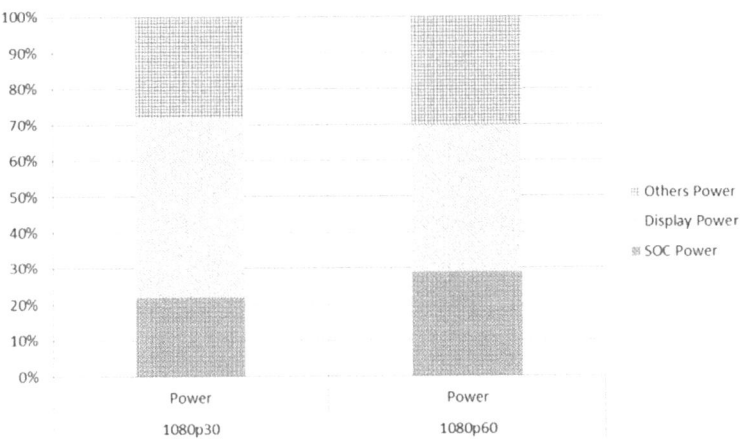

Figure 7-5. *Example of video playback activity and power profile*

Video Recording

In this usage model, the user captures a video of 1080p resolution with the device's main camera while the audio is captured with the integrated microphone. The output of the camera (usually compressed) is decoded and pre-processed with de-noising and scaling as needed. The resulting uncompressed source is encoded using hardware acceleration. Encoded and multiplexed video and audio streams are stored in local files. (While preview mode is fairly common for this usage model, for simplicity it is not considered in the following software stack, as shown in Figure 7-6.)

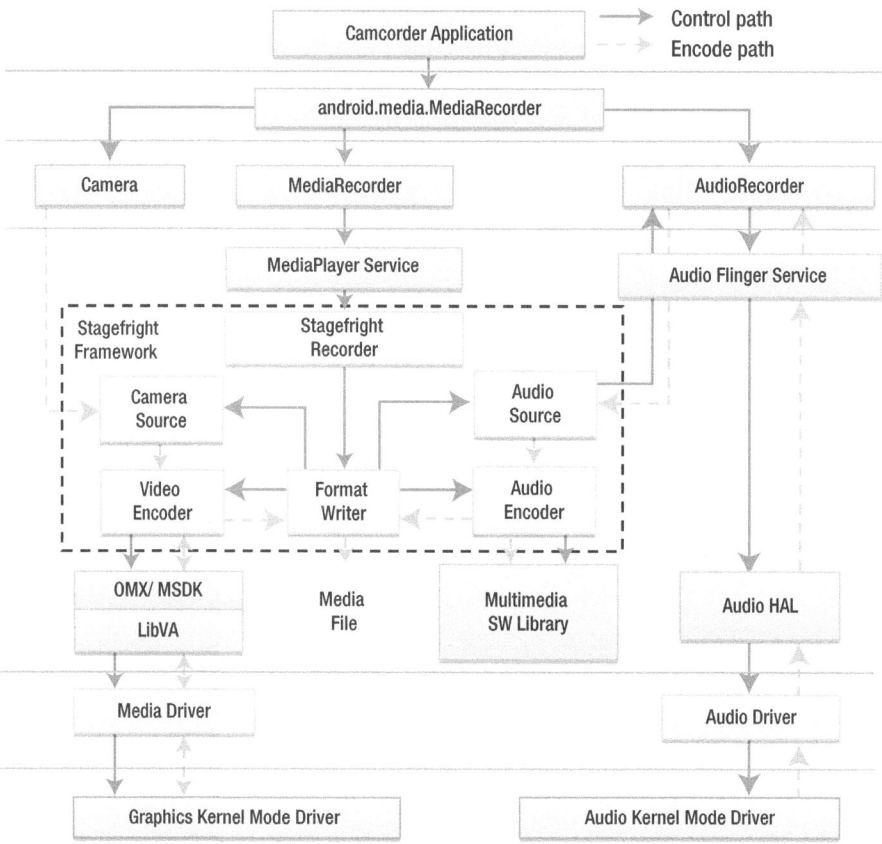

Figure 7-6. *Video recoding software stack on Android*

In Figure 7-7, an example video recording activity profile is presented for AVC 1080p30 and 1080p60 resolutions on an Intel Architecture platform. On this platform, recording a 1080p video generally requires ~40-60% of CPU activity, as well as ~30-60% of GPU activity depending on the frame rate. The SoC and the display unit both consume about a third of the platform power. Similar to the playback usage, significant power dissipation occurs in the voltage regulator (~15%), memory (~7%), and other parts of the platform. Note that compared to the playback usage, GPU activity is higher because of the heavier encode workload.

Video Recording Activity Profile

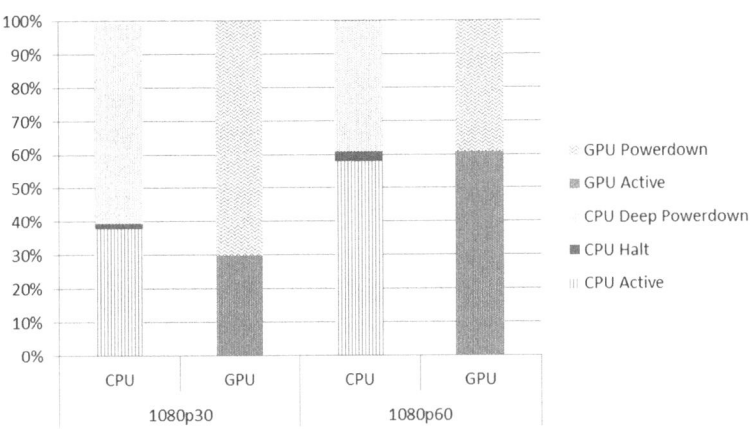

Video Recording Power Profile

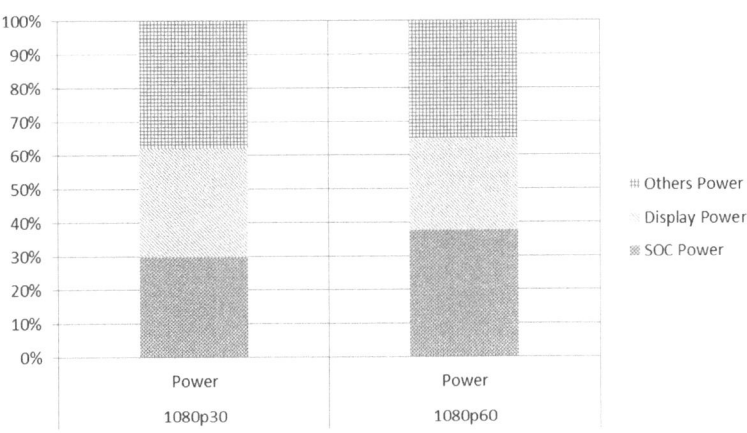

Figure 7-7. *Example video recording activity and power profile*

Video Delivery over Wireless Display and Miracast

In the video delivery over Wireless Display (WiDi) and Miracast usage, the device's Wi-Fi connection is used to stream encoded screen content captured from a local display to a remote HDTV or monitor. WiDi supports wireless transport of the associated audio content as well. In particular, the platform's video and audio encoding hardware acceleration capabilities are exploited to generate an audio-video stream encapsulated in Wi-Fi packets that are streamed over a peer-to-peer connection to a WiDi adaptor device connected to a remote display. Owing to the multi-role support of Wireless Local Area Network (WLAN), multiple connections may simultaneously be available at a wireless access point, serving the WiDi peer-to-peer connection as well as a dedicated connection to the Internet while sharing the same frequency. The multi-role allows, for example, browsing of the Internet in clone or multi-tasking modes, as described below. Playback of video or browsing of the Internet is commonly protected by a digital rights management (DRM) protocol over the WiDi. An overview of a WiDi complete solution and the Miracast industry standard was presented in Chapter 6.

WiDi supports two main usage models:

- Clone mode, where the same content is presented on both a local and a remote display. The local display's resolution may be modified to match the remote display's maximum resolution. Also, if the WiDi performance is insufficient, the frame rate of the remote display may be downgraded.

- Extended mode, where there is remote streaming of a virtual display and the content is not shown on the device's embedded local display. There are two scenarios for the extended display:

  - Extended video mode shows the content on the remote display, while local display displays only the UI controls.

  - Multi-tasking is allowed, where a video is shown on the remote display while an independent application such as a browser may also run and show content on the local display.

Ideally, the platform should be designed so that there is no performance degradation when wireless display is activated.

Figure 7-8 shows the WiDi flow diagram for an Intel Architecture platform. One or more screen content surfaces are captured, composited, scaled, and converted to a format appropriate to the hardware-accelerated video encoder, while the audio is independently captured and encoded as well. The encoded video and audio bitstreams are then encrypted according to the HDCP2 protocol and the encrypted bitstreams are multiplexed and packetized to produce MPEG-2 transport stream packets that are ready to send over the Wi-Fi channel.

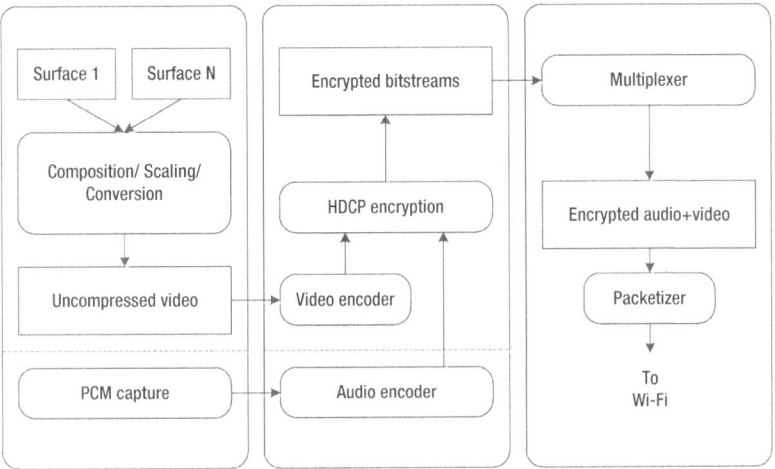

Figure 7-8. *Wireless Display flow diagram on an Intel Architecture platform*

Figure 7-9 shows the activity profile of three WiDi scenarios:

- Clone mode while the device is idle–that is, the local display shows a mostly static screen.

- Video playback in extended mode where the video is played back in the remote display.

- Multi-tasking in extended mode where a browser (or some other window) is independently presented on the local display while a video is played back in the remote display.

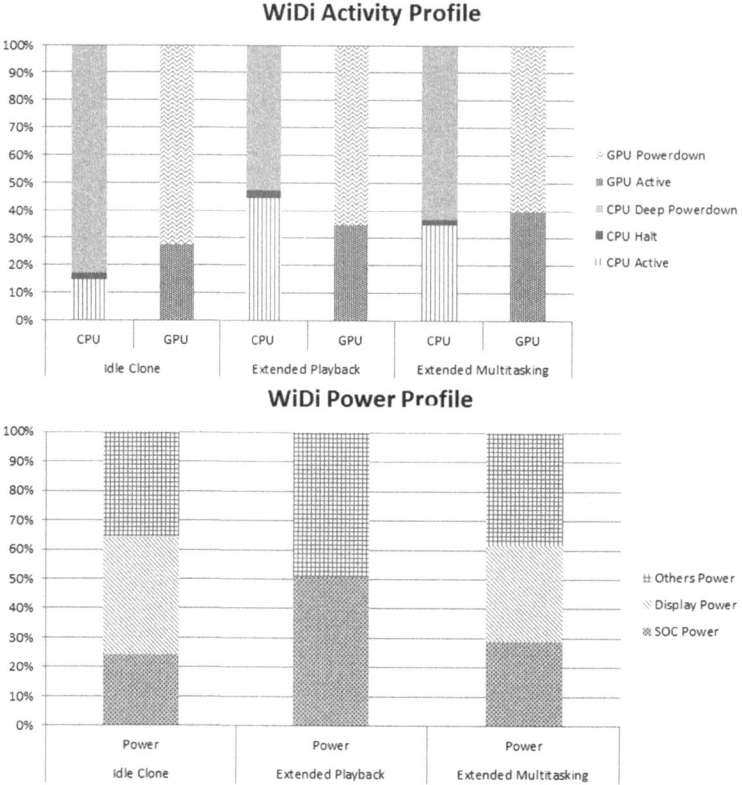

Figure 7-9. *Example WiDi activity profile*

For the idle clone scenario, depending on the display refresh rate, the screen content may be captured and encoded at 720p60 or 1080p30 resolution using hardware acceleration. The encoded video bitstream is typically encrypted, multiplexed with the audio bitstream, and divided into transport packets before transmission using the Wi-Fi protocol.Hardware-accelerated video —pre-processing may also be done before composition, scaling, and conversion to the uncompressed format that is fed to the encoder. When a video is played back in the clone mode, simultaneous decode, capture and encode operations happen using hardware-acceleration. For the extended mode scenarios, the local display does not show the video, the content is only shown in the extended wireless display; no composition, scaling, or format conversion is done and only decoding and encoding are performed, In all cases the audio is typically encoded in the CPU.

On the Intel Architecture platform, WiDi usages generally require ~15-45% of the CPU activity and ~30-40% of GPU activity, depending on the WiDi scenario. While the local display does not consume power in extended video mode, the SoC consumes about half of the platform power, the other half being distributed to the rest of the platform. For other WiDi modes, the platform power is more or less equally divided among the display, the SoC, and the rest of the platform.

Videophone or Videoconferencing

Multi-party videoconferencing is a generalized case of a two-party videophone or video chat, which uses a camera that is integrated with a device for capturing a video, usually in motion JPEG (MJPEG) or other compressed formats. Using the device's hardware-acceleration abilities, the camera output is then decompressed (if the camera output is compressed), processed, and re-encoded to a low-delay format at a dynamically varying bit rate with some error-recovery capabilities. This is basically video playback and video recording usage models concurrently applied, with the following stipulations:

- Both encode and decode operations must happen simultaneously in real time, usually accelerated by hardware.

- The camera output should be input to the encoding unit in real time as well. The camera output frame rate should be in concert with the encoding frame rate.

- The end-to-end delay from camera capture to packetized bitstream output should be constant.

- The video elementary streams for both incoming and outgoing bitstreams should be appropriately synchronized with the corresponding audio elementary streams.

For multi-party videoconferencing, usually in the local display there is one main video window and several thumbnail videos of the multiple parties involved. Figure 7-10 illustrates a flow diagram of typical videoconferencing. Assuming the camera capture format is a compressed one, the captured video is decoded, scaled, and/or pre-processed before encoding to a suitable format such as AVC with appropriate bit rate, frame rate, and other parameters to obtain good tradeoffs among video quality, delay, power consumption, and amount of compression. Also, the incoming video bitstreams are decoded and composed together before display. All these typically are done using hardware acceleration. Multiplexing-demultiplexing and packetization-depacketization are generally done in the CPU, while audio can be processed by special hardware units or audio drivers and kernels.

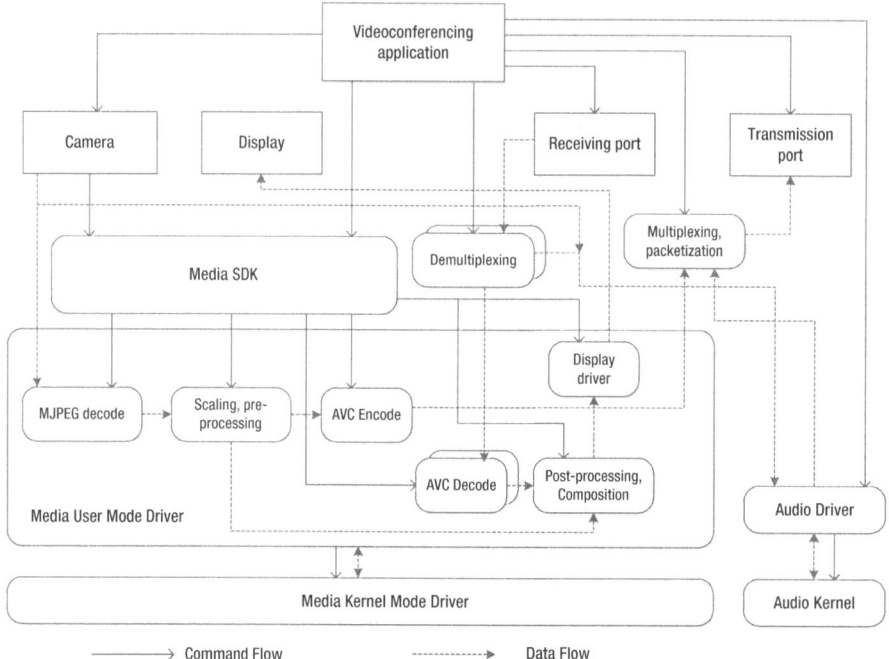

Figure 7-10. *Flow of a typical videoconferencing event*

The activity profile of the videoconferencing usage is similar to the WiDi usage. In both cases, simultaneous encode and multiple decodes are involved; encoding is done with low delay and variable bit rates amenable to changes in network bandwidth, and there are real-time requirements for both cases as well.

System Low-Power States

The goal of low-power platform design, from both hardware and software points of view, is to successfully meet the challenges posed by limited battery life, heat dissipation, clock speed, media quality, user experience, and so on while addressing the ever-increasing need to support complex multi-tasking applications. To this end, as was detailed in Chapter 6, the ACPI defines several processor states to take advantage of reduced power consumption when the processor is not fully active. However, the ACPI model of power management is not sufficient for modern mobile applications, especially with the requirement of "always on" connectivity. To address this issue, new low-power states have been defined, but they present some problems.

Drawbacks of the ACPI Simple Model

The ACPI simple model of power management has a few limitations:

- The model assumes that the operating system or the power manager will manage the power. However, this is not always the case. Some devices, such as disk drives, CPUs, and monitors, manage their own power and implement power policies beyond what is achievable by the ACPI model.

- The four device states are not sufficient. For example, D3 has two subsets. Also, a processor in D0 state has additional CPU (Cx) states and performance (Px) states, as described in the next section. In addition, a device may perform multiple independent functions and may have different power states for each function.

- Recent advances in the operating system impose new requirements on power management. For example, the concept of *connected standby* requires the system to turn off all activities except that of listening for an incoming event, such as a phone call.

Recognition of these drawbacks resulted in special S0iX states within the S0 state in Windows. These are intended for fine-tuning the standby states. A higher integer X represents higher latency but lower power consumption.

Connected Standby and Standby States

Connected Standby (CS) mimics the smartphone power model to provide an instant on/instant off user experience on the PC. Connected Standby is a low-power state in Windows 8 and later versions that uses extremely low power while maintaining connectivity to the Internet. With CS, the system is able to stay connected to an appropriate available network, allowing the applications to remain updated or to obtain notification without user intervention. While traditional Sleep state (S3) has wake latency of more than two seconds and Hibernate (S4) may take indefinitely longer, CS-capable ultra-mobile devices typically resume in less than 500 ms.

When the system goes into CS, the OS powers down most of the hardware, except the bare minimum required to preserve the contents of DRAM. However, the Network Interface Controller (NIC) still gets a trickle of power so it is able to scan incoming packets and match them with a special wake pattern. To preserve the connection, the NIC wakes the OS as needed. The OS also wakes up every few hours to renew its DHCP lease. Thus, the system maintains its layer-2 and layer-3 connectivity, even while the NIC is mostly powered down.

In addition, the system can be wakened in real time. Consider when a Skype call arrives and the system needs to quickly start the ring tone. This works through special wake-pattern packets. The Skype server sends a packet on a long-running TCP socket. The NIC hardware is programmed to match that packet and wake the rest of the OS. The OS receives and recognizes the Skype packet and starts playing the ring tone.

Connected Standby has the following four elements:

- Most of the hardware is in a low-power state.

- The NIC wakes the OS when the NIC needs OS intervention to maintain layer-2 connectivity.

- The OS periodically wakes the NIC to refresh its layer-3 connectivity.

- The NIC wakes the OS when there's a real-time event (e.g., an incoming Skype call).

Connected Standby platforms are usually designed with SoCs and DRAMs that have the following characteristics:

- *Less than 100 ms switch time between idle and active modes.* The active mode allows running code on the CPU(s) or the GPU, but may not allow accessing the storage device or other host controllers or peripherals. The idle mode may be a clock-gated or power-gated state, but should be the lowest power consumption state for the SoC and DRAM.

- *Support of self-refresh mode for the DRAM to minimize power consumption.* Typically mobile DRAM (LP-DDR) or low-voltage PC DRAM (PC-DDR3L, PC-DDR3L-RS) is used.

- *Support of a lightweight driver called Power Engine Plug-in (PEP) that abstracts SoC-specific power dependencies and coordinates device state and processor idle state dependencies.* All CS-capable platforms must include a PEP that minimally communicates to the operating system when the SoC is ready for the lowest power idle mode.

The process of preparing the hardware for low-power during Connected Standby can be visualized as an upside-down pyramid, as shown in Figure 7-11. The lowest power is achieved when the whole SoC is powered down, but this can occur only when each set of devices above it has been powered down. Therefore, the latency for this state is the highest within Connected Standby.

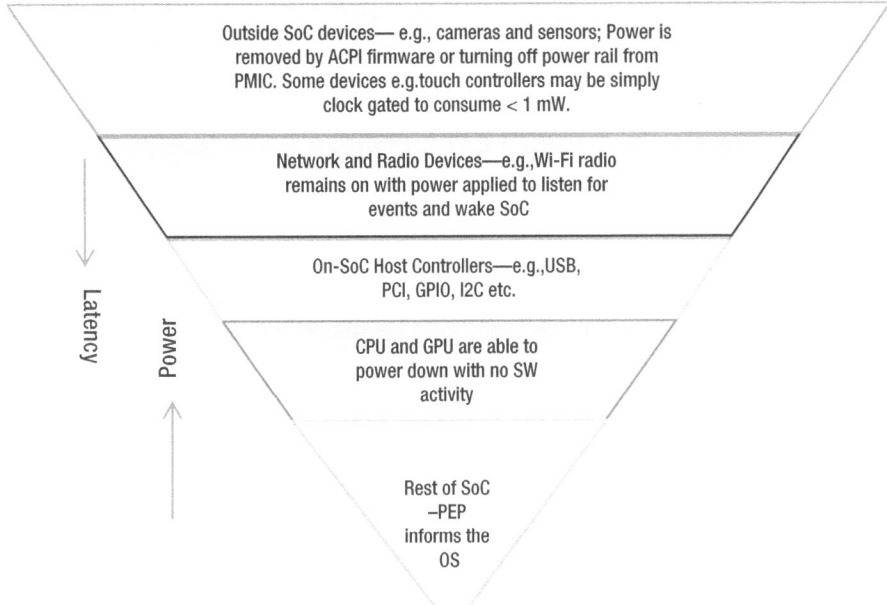

Figure 7-11. *Preparation for transitioning to low-power state during Connected Standby*

The S0iX states are the latest standby states, which include S0i1, S0i1-Low Power Audio, S0i2, and S0i3. These states are available on recent Intel platforms and are enabled for all operating systems. They are special standby states that allow the system to consume a trickle of power, and so they are crucial for Connected Standby to work. A device enters CS when the off/on button is pushed or after an idle time-out; while in CS mode, it consumes very low power. For example, at CS mode, an Intel Z2760-based device consumes <100 mW, and can stay in this mode for more than 15 days without requiring a recharge. Figure 7-12 shows the flow of actions in CS mode.

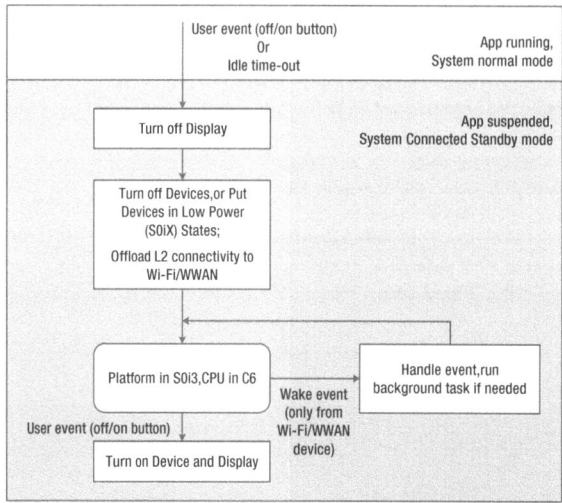

Figure 7-12. *Activity flow for Connected Standby*

Contrast this with ACPI S3 state, where all activities of the processor are paused when the system is sleeping and activities resume only upon signals from keyboard, mouse, touchpad, or other I/O devices. Connected Standby automatically pauses and resumes activity on the system in a tightly controlled manner, using the various S0iX states to help ensure low power and long battery life. A typical system with a 45 Watt-hour battery may achieve 100 hours of battery life in S3 state, while in S0i3 the battery life becomes three times as long.

Figure 7-13 shows a comparison of the S0i3 state with respect to the S0 and the S3 states. The S0i3 essentially gives the best of both power consumption and latency: it is less power consuming than the S0 (Active) state, consuming almost the same power as the S3 (Sleep) state while yielding wake latency in the order of milliseconds, similar to the S0 state.

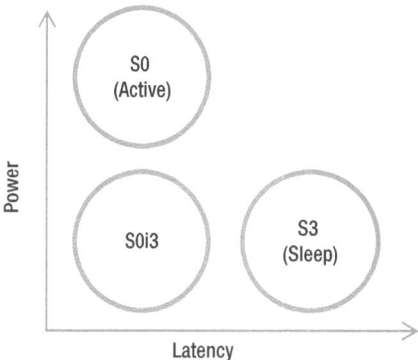

Figure 7-13. *Relative power-latency relationship for S0i3 compared to traditional ACPI Sleep (S3) state*

Table 7-2 provides estimates of the entry and exit latency for each of the low-power standby states on an Intel Architecture platform. Especially on Windows 8 and Windows 8.1 operating systems, these quick latencies translate to an exceptionally snappy wake performance from a user experience point of view, while delivering significant power savings.

Table 7-2. *Estimated Entry and Exit Latency of Standby States of an Intel Architecture Platform*

State	Entry Latency	Exit Latency
S0i1	~200 μsec	~400 μsec
S0i2	~200 μsec	~420 μsec
S0i3	~200 μsec	~3.4 msec

Combination of Low-Power States

Table 7-3 lists the combined system low-power states and ACPI power states for a typical low-power processor. The ACPI power states were described in Chapter 6, so only the effect of S0iX is added here for a clearer understanding.

Table 7-3. *Definitions of System Low-Power States*

State or Substate	Description
G0/S0/PC0	Full on. CPUs are active and are in package C0 state.
G0/S0/PC7	CPUs are in C7 state and are not executing with caches flushed; controllers can continue to access DRAM and generate interrupts; DDR can dynamically enter deep self-refresh with small wake-up latency.
G0/S0	Standby ready. CPU part of the SoC is not accessing DDR or generating interrupts, but is ready to go standby if the PMC wants to start the entry process to S0iX states.
G0/S0i1	Low-latency standby state. All DRAM and IOSF traffic is halted. PLLs are configured to be off.
G0/S0i1 with audio	Allows low-power audio playback using the low-power engine (LPE), but data transfer happens only through specific interfaces. Interrupt or DDR access request goes through the PMC. The micro-architectural state of the processor and the DRAM content are preserved.
G0/S0i2	Extended low-latency standby state. S0i2 is an extension on S0i1—it *parks* the last stages of the crystal oscillator and its clocks. The DRAM content is preserved.

(continued)

Table 7-3. *(continued)*

State or Substate	Description
G0/S0i3	Longer latency standby state. S0i3 is an extension on S0i2—it completely stops the crystal oscillator (which typically generates a 25 MHz clock). The micro-architectural state of the processor and the DRAM content are preserved.
G1/S3	Suspend-to-RAM (STR) state. System context is maintained on the system DRAM. All power is shut to the noncritical circuits. Memory is retained, and external clocks are shut off. However, internal clocks are operating.
G1/S4	Suspend-to-Disk (STD) state. System context is maintained on the disk. All power is shut off, except for the logic required to resume. Appears similar to S5, but may have different wake events.
G2/S5	Soft off. System context is not maintained. All power is shut off, except for the logic required to restart. Full boot is required to restart.
G3	Mechanical off.

Source: Data Sheet, Intel Corporation, April 2014. `www.intel.com/content/dam/www/public/us/en/documents/datasheets/atom-z36xxx-z37xxx-datasheet-vol-1.pdf`.

Power Management on Low-Power Platforms

The main power-management tasks are done by the operating system—for example, taking the idle CPU core offline; migrating and consolidating tasks, threads, and processes to a minimum number of the cores to allow other cores to go idle; limiting frequent back-and-forth switching of tasks between cores for variable workloads; managing tradeoffs between power gain and hardware or software latencies; load balancing between the active cores, and so on. In addition to the power-management features provided by the operating system according to the ACPI standard, there are several hardware and software approaches to power management, all with a singular goal of achieving the most power savings possible. We discuss some of these approaches next.

Special Hardware for Power Management

As power management is crucial for low-power platforms, in most contemporary devices there are special-purpose hardware units dedicated to these power-management tasks. Here are two such units.

Power Management Circuits

With the increase in system complexity comes the need for multiple sources of power at various voltage levels. Such need cannot be fulfilled by voltage regulators alone; power management ICs (PMICs) have been introduced for greater flexibility. Power Management ICs (PMICs) are special-purpose integrated circuits (or a system block in a SoC device) for managing the power requirements of the host system. A PMIC is often included in battery-operated devices such as mobile phones and portable media players. Typical PMIC efficiency is 80 to 85 percent, depending on idle or active states of applications. A PMIC may have one or more of the following functions:

- Battery management
- DC-to-DC conversion
- Voltage regulation
- Power-source selection
- Power sequencing
- Miscellaneous other functions

Power management is typically done with the following assumptions:

- All platform devices are ACPI compliant.
- SOC devices are PCI devices except GPIOs.
- The PCI device drivers directly write to PMCSR register to power down/power up the device; this register triggers an interrupt in the PMC.
- The device drivers use ACPI methods for D0i3/D3 entry/exit flows.

Power-Management Controller

The drivers and the OS alone are not adequate for power-management tasks such as save and restore context or handling special wake events. A special microcontroller, called the Power Management Controller (PMC), is typically used for powering up and powering down the power domains in the processor. It also supports the traditional power-management feature set along with several extended features. For example, in Intel fourth-generation Atom SoCs, the PMC performs various functions, including:

- Power up the processor and restore context.
- Power down the processor and save context.
- Handle wake events for various sleep states.
- Secure the boot.
- Perform low-power audio encode (using LPE) and general-purpose input-output (GPIO) control in order to be able to go to S0i1 when LPE is active.

The PMC is also used in ARM11 processors in power-saving mode,[2] where it determines (for instance, when instructed by the processor) that the processor should be put into dormant or shutdown mode; it asserts and holds the core reset pin; it removes power from the processor core while holding reset; and so on.

Display Power Management

The display is the major power-consuming component in most low-power platforms. Several power-management features have been implemented to manage and optimize the display power in various platforms. For example, in recent Intel platforms, the following are among the many display power-management features:

Panel Self-Refresh

The panel self-refresh (PSR) feature, typically used for *embedded display port* (eDP) displays, allows the SOC to go to lower standby states (S0iX) when the system is idle but the display is on. The PSR achieves this by completely eliminating the display refresh requests to the DDR memory as long as the frame buffer for that display is unchanged. In such cases, the PSR stops the display engine from fetching data from external memory and turns off the display engine.

Display Power-Saving Technology

The display power-saving technology (DPST) is an Intel backlight control technology. In recent versions, the host side display engine can reduce up to 27 percent of the panel backlight power, which linearly affects the energy footprint. The Intel DPST subsystem analyzes the frame to be displayed, and based on the analysis, changes the chroma value of pixels while simultaneously reducing the brightness of backlight such that there is minimum perceived visual degradation. If there is considerable difference between the current frame being displayed and the next frame to be displayed, new chroma values and brightness levels are calculated.

The DPST needs to be enabled by the display driver; the DPST cannot yet work in parallel with the PSR. If a MIPI[3] panel does not have a local frame buffer, then DPST should be enabled and the PSR can be disabled. In general, the DPST should be enabled for additional power savings for bridge chips that do not have local frame buffers.

---

[2]Details available at infocenter.arm.com/help/index.jsp?topic=/com.arm.doc. dai0143c/CHDGDJGJ.html.

[3]The MIPI (Mobile Industry Processor Interface) is a global industrial alliance that develops interface specifications, including signaling characteristics and protocols, for the mobile communication industry. Details are available at www.mipi.org.

Content-Adaptive Backlight Control

A liquid crystal display (LCD) consists of a backlight that shines through a filter. The filter modulates the backlight and is controlled by the pixel values that are to be displayed: the brighter the pixel, the more is the light that passes through. By controlling the array of pixels in an LCD filter, images of different brightness can be shown on the display. Such control is typically based on a power-brightness tradeoff.

Content-adaptive backlight control (CABC) takes advantage of the fact that the perceived brightness is due to the backlight being modulated by the filter. A dark frame appears dark because the filter does not allow much light to shine through. However, the same effect can be achieved by using a dimmer backlight and controlling the filter to allow more light to shine through. In other words, the backlight is dimmed and the brightness of the frame is boosted, resulting in the same perceived brightness.

However, reducing the brightness of the backlight decreases the available dynamic range. The CABC may clip pixels requiring a higher dynamic range, producing a washout effect. Also, it is possible that the target backlight brightness varies greatly from frame to frame; aggressively changing the backlight by such large amounts can result in flickering. Therefore, a tradeoff is necessary between CABC-based power savings and perceived visual quality. To get the best performance, typically separate histogram analysis is done for each color component to determine the desired brightness. Then the maximum of these is chosen as the target backlight. This ensures that all colors are accurately rendered.

The CABC engine resides inside either the display bridge or the panel. With the ability to dynamically control the backlight as the content changes, this feature looks for opportunities to save power by reducing the backlight power. The algorithm processes and analyzes the current frame in order to update the backlight for the next frame. This can work in parallel with the PSR. Power savings from CABC are bigger when playing higher FPS video.

Ambient Light Sensor

The Ambient Light Sensor (ALS) backlight power savings modulates the backlight based on surrounding light. More power savings (e.g., up to 30%) occur when there's less ambient light and there's less power savings (e.g., 20%) when there's more ambient light. The display power management typically uses ALS inside the panel or the bridge.

Low-Power Platform Considerations

It is easy to understand, from the power-frequency relationship (Figure 7-1), that there is an optimal frequency for a given power budget. With this fundamental concept, there are some software and architectural considerations to keep in mind, particularly from the point of view of limited power platforms.

Software Design

Regarding the resource-scarce wearable and ultra-mobile low-power platforms, there are several important ideas to evaluate from the application design point of view. It is a no-brainer that new applications take these considerations into account; however, existing applications can be modified to benefit from them as well.

Given that most low-power devices offer various power-saving opportunities, applications should exploit those prospects appropriately. For example, allowing networking devices to go to a lower power state for longer period may avoid unnecessary network traffic, as will using a buffering mechanism. But here are some additional suggestions for power savings.

Intelligent Power Awareness

Low-power designs typically trade performance for power savings, using techniques such as voltage scaling or frequency scaling, thereby reducing the voltage or the frequency, respectively, to reduce the dynamic power consumption. However, a minimum voltage level must be supplied for the circuit to be operational. Further, reducing the supply voltage increases circuit delay, which restricts the operating frequency of the processor; the delay cannot be too large to satisfy the internal timings at a particular operating frequency. Thus, a careful approach is necessary to determine the appropriate amount of voltage scaling, as demonstrated by the Intel (R) Speed Step (TM) voltage scaling technology.

Unfortunately, system performance requirements in modern processors are too high to make voltage scaling an attractive energy-saving solution. Also, processor frequency is not the only factor that affects the power-performance behavior of complex multimedia applications; memory access latency and memory bandwidth are also important factors. Besides, multimedia applications have real-time deadlines to meet, which further restrict the usefulness of voltage scaling. Therefore, in practical systems, the voltage and frequency need to be dynamically adjusted along the voltage-frequency curve.

Dynamic adjustment of voltage and frequency should normally be the job of the operating system. However, the operating system does not have knowledge of the expected workload, especially when bursty multimedia applications are involved. Also, such workloads are not easy to predict with reasonable accuracy and oftentimes interval-based schedulers are not sufficient. At best, the operating systems may use some heuristics to change the processor frequency and try to control power consumption. But this answer does not work well for bursty workloads such as video decoding, encoding, or processing.

If the application is power-aware, however, the actual workload can be predicted better in a dynamic fashion; and by doing so, a power reduction of over 30 percent is possible for a typical video decoding.[4] The power consumption of applications largely depends on the number of clock cycles needed for instructions and memory references. If applications can provide information about their future demands to the OS, the OS can then perfom more efficient scheduling of tasks and does not need to work with questionable predictions.

---

[4]J. Pouwelse, K. Langendoen, and H. Sips, "Dynamic Voltage Scaling on a Low Power Microprocessor," *Proceedings of the 7ᵗʰ Annual International Conference on Mobile Computing and Networking* (2001): 251–59. Association for Computing Machinery.

This application is particularly useful on resource-scarce mobile devices. Applications must be aware of their processing demands—in particular, the required number of clocks until the next deadline of the application's current task—and must inform the OS of those clock requirements. The OS can then take into account the cycle count along with the power-frequency curve to regulate and achieve the minimum frequency at which the application deadline can be met, thereby achieving optimal power consumption.

For example, for an H.263 decoding, the number of bits used in a coded frame is fairly proportional to the decoding time. Using a hint of the number of bits used in each frame, an application can determine the expected frequency requirement for a video and obtain a power saving of about 25 percent, as reported by Pouwelse et al.[5]

Often there is power-saving firmware logic to control the host system's power draw. However, the firmware does not have knowledge of the work that the application is trying to do, and threfore is unable to optimize the energy investments, as it must rely on measurements and look for opportunities to save the energy. An intelligent power-aware application, on the other hand, can generally use its knowledge of the task and the memory-utilization pattern to send hints to the firmware for further power saving.

In an example of such performance-aware power saving, Steigerwald et al.[6] mentions an online transaction-processing application, where the application monitors the performance counters of the transaction processing to determine the performance impact. This information is provided to the middleware, which learns and updates the optimal power-limiting policy on the system. At the lowest level, a power controller firmware component caps the power consumption of the system.

Quality Requirements

Although it's common to trade performance and visual quality for power saving, users of most modern games and media applications have a high quality expectation that must be met. It is possible to run the processor at a low frequency and lower the power consumption, but that's not acceptable for various games and video codec applications, as the frame-processing deadline would be missed. That would mean the frames would not be displayed in real time, resulting in dropped frames and a poor user experience. However, it is possible to meet the quality requirements while still saving power by carefully considerating both aspects together when designing and optimizing the software.

Hardware-Accelerated Blocks

Recent low-power processors of mobile handheld devices have come with various hardware blocks capable of specific tasks, such as camera image signal processing or video decoding. Often these hardware blocks are optimized for lower power consumption than if the task were performed on a general-purpose CPU. Applications should take advantage of these hardware blocks and offload as much processing as possible in order to lower the computational burden and save power consumption.

---

[5]Ibid.
[6]Steigerwald et al., *Energy Aware Computing*.

Energy-Efficient User Interfaces

From the various usage models described earlier, it is evident that the display is the most power-consuming unit of the platform. However, it is possible to design graphical user interfaces (GUI) in a energy-efficient manner, particularly for liquid crystal displays (LCD) or displays based on organic light-emitting diodes (OLED).[7] For these displays, different colors, color patterns, or color sequences consume different amounts of power, and thus, low-energy color schemes can be used to reduce power. For example, thin film transistor (TFT) LCDs, which are typically utilized in personal digital assistants (PDA), consume less power when black than when white.

Another way toward energy efficient GUI design is to improve the latency caused by interfaces with humans by using fewer and larger buttons for greater user convenience.

Code Density and Memory Footprint

With low-power mobile platforms such as smartphones, poor code density means that the code size of the firmware and driver required to run on the processor is too large. As a result, the device manufacturer has to use more RAM and flash memory to hold the firmware, compared to what optimized firmware would need. Thus, more memory is used resulting in higher cost and reduced battery life.

Since memory cost dominates the cost and power budget of the entire design, firmware optimization and code density can make a big difference. Optimization at the compiler level, and possibly instruction set architecture (ISA), can achieve much denser firmware binaries, thereby helping reduce power consumption.

Another way to shrink the memory footprint of binaries is to perform a lossless compression while writing to memory and the correspoding decompression when reading back from memory. However, the power savings would depend on the nature of the workload. As media workloads have good locality characteristics, it may be beneficial to use memory compression for certain media usages.

Optimization of Data Transfer

Modern mobile applications designed for low-power devices often benefit from available Internet connectivity by processing a large amount of data in the cloud and maintaining only a thin client presence on the device. While such an approach may boost performance, the large number of data transfers means the power consumption is usually high. But by exploiting the pattern of the data to be transferred, it is possible to reduce the data bandwidth usage, thereby also reducing power consumption. Furthermore, applications can take advantage of the new S0iX states for notification purposes, and allow the processor to go to lower power states more frequently.

---

[7]K. S. Vallerio, L. Zhong, and N. K. Jha, "Energy-efficient Graphical User Interface Design," *IEEE Transactions on Mobile Computing* 5, no. 7 (July 2006): 846–59.

Parallel and Batch Processing

Scheduling of tasks is a challenge in multi-tasking environments. Serialization of tasks over a long period of time does not provide the processor with an opportunity to go to lower power states. Parallelization, pipelining, and batch processing of tasks can grant the processor such opportunity and thereby help lower power consumption. If a task does not have specific processing rate, it may be useful to run the processor at the highest frequency to complete the task as fast as possible, and then allow the processor to idle for a long period of time.

Architectural Matters

The architectural considerations for low-power designs are quite similar, even for strikingly different hardware architectures, such as Qualcomm 45nm Serra SoC[8] and Intel 22nm fourth-generation Atom SoC. While it is important to design various system components for lower power consumption, it is also important to optimize the system design and manage it for lower power consumption. In particular, architectural ideas for power saving include the following.

Combined System Components

Integrating the processor and the platform controller hub (PCH) on the same chip can reduce power, footprint, and cost, and also can simplify the firmware interface. Utilizing run-time device power management, as initiated by the kernel and optimizing in the proper context, can significantly contribute to a low-power solution.

Optimized Hardware and Software Interaction

In the era of specialization, code sharing between drivers or hardware versions and operating systems has become less important, in favor of reduced processing and reduced power use. Emphasis has also shifted so that hardware commands are similar to API commands and less command transformation is used. The following ideas embody this concept.

Migrating the Workload from General-Purpose to Fixed-Purpose Hardware

Moving tasks from general-purpose to fixed-function hardware saves power, as a fewer number of gates can be used for a special purpose, thereby reducing the dynamic power consumption that depends on the number of gates that are switching. This technique should especially benefit non-programmable tasks such as setup, clip, coordinate calculation, some media processing, and so on.

---

[8]M. Severson, "Low Power SOC Design and Automation," retrieved July 27, 2009, from http://cseweb.ucsd.edu/classes/wi10/cse241a/slides/Matt.pdf.

Power Sharing Between the CPU and the GPU

To maximize performance within given platform capabilities, technologies such as Intel Burst Technology 2.0 automatically allow processor cores to run faster than the specified base frequency while operating below the specified limits of power, temperature, and current. Such turbo mode and power sharing are supported for both CPU and GPU. The high-performance burst operating points can be adjusted dynamically.

Using Low-Power Core, Uncore, and Graphics

All parts of the processor—the core, uncore, and graphics—should be optimized for low power consumption. The GPU being a significant power contributor, tasks such as composition of the navigation bar, status bar, application UI, video, and so on should go to the display controller as much as possible, so as to facilitate the lowest device activity and memory bandwidth.

Also, as power consumption is highly dependent on the use cases, processor power optimization should be done with practical use cases in mind. That is, a low-power solution is constrained by the worst use case; optimization for and management of both active and leakage power should be considered.

For reducing leakage power, know that although the PMIC can regulate power and can collapse the full chip power during sleep, such power collapsing requires data to be saved to memory and rebooting of all parts of the processor for restoration. The desirable impact, then, is when there is a net savings in power—that is, when the leakage power saved over a period of time is larger than the energy overhead for save and restoration.

Power reduction of RAM/ROM periphery and core array

All memories in the core array need to have leakage control. Power reduction should be considered for peripheral circuits as well, as significant power reduction is achievable from the RAM/ROM periphery.

Reduce V_{DD} during sleep mode to manage leakage

Minimizing V_{DD} of the entire die during sleep is beneficial, especially during short sleep cycles. This has the added advantage of small software overhead and fast restoration, as all memory and register states are retained. Voltage minimization is the most effective way to control leakage without complex management by the software. However, exceptions to this are the domains where voltage cannot be reduced due to the requirements of Connected Standby.

Independent memory bank collapse for large high-density memories

A common power-saving technique is to use power gating on some portions of memory banks that are not accessed by the application. Standby/active leakage reduction can be achieved by gating clock/data access to banks that are not needed. However, this requires proper power management in software and firmware so that the appropriate, potentially idle memory banks are identified and gated correctly. Also, power gating is not possible for all blocks. Careful analysis will compare the savings achieved from dynamic and leakage powers.

Advanced low-power clocking and clock tree optimization

Clock tree power is a major contributor to total active power use. In the case of Qualcomm Serra, clock tree consumes 30 to 40 percent of the total active power. As clock architecture has a high impact on power use, its use and frequency should be carefully planned. In particular, the number of phase-locked loops (PLLs), independent clock domain control, frequency and gating, synchronicity, and so on need to be considered.

Clock domain partitioning

Partitioning the clocks across different clock domains allows better power management for various clocks, as the clocks can be independently turned off. For example, separate clock domains for I/O and the CPU can be controlled individually.

Independent frequency clock domains

Although asynchronous domains are more flexible, there is increased latency across clock domain boundaries. On the other hand, synchronous clock domains can save power in low-performance mode, but they are susceptible to higher power costs for the worst user case.

Fine-grained tuning of clock gating

One of the most effective ways to save active power is to use fine-grained automatic and manual tuning of clock gating. Either the hardware or the software can control the clock gating, but each has a cost. Analysis of clock gating percentage and efficiency of clock gating can reveal further optimization opportunities.

Using Power Domains or Power Islands

The idea of power islands can be illustrated with the example of a house. If there is only one power switch for all the light bulbs in the entire house, more energy must be spent when the switch is turned on, even when there is only one person in the house who needs only one light at any given time. Instead, if there are individual switches to control all the light bulbs, energy savings are achieved when only the needed light bulb is turned on. Similarly, in a typical computing system, 20 to 25 independent power islands are defined to achieve power savings. This allows for the following energy savings.

Independent voltage scaling for active and sleep modes

For different power islands, the voltage scaling can be done independently for both active and sleep modes. This implies both frequency and consequential power savings.

Power gating across power domains

The PMIC can control multiple power domains, yielding better power control and efficiency, as well as independent voltage scaling and power-collapsing abilities. However, the number of power domains is limited by the increased impedance of the power domain network (PDN), increased IR drop due to larger number of power switches, increased bill of materials (BOM), and the fact that level shifters and resynchronization are necessary at the domain boundaries. Dynamically controlling the power collapse of various domains using small, fast, and efficient chip regulators may be the best option.

When a power island is not power gated, all unused peripherals that are parts of the power island should be clock gated. On the other hand, when a power island is power gated, all relevant peripherals are in D0i3 (but the supply voltage may still be on).

Offering Power-Aware Simulation and Verification

To optimize the system for low power consumption, use power-aware simulation tools and verification methodology to check entry to and exit from low-power states, properly model the power collapse, verify the polarity of clamping, verify the power domain crossings, and so on. These methods include the following.

Tradeoffs among power, area, and timing

Custom design flows and circuits can produce more power-efficient results for some tasks; however, custom design requires more time and efforts. A balanced approach makes careful selection of areas for customization and optimization so as to obtain the greatest benefit. Good customization candidates are clock trees, clock dividers, memory and cell intellectual property (IP, a common term for innovative designs), and so on. It is better to move the customization to the IP block and to use automated methods to insert, verify, and optimize that IP block; that way, specific improvements and/or problems can easily be attributed to the appropriate block.

Comparative analysis of low-power solutions

Design decisions are generally complex, especially when many tradeoffs are involved. To judge how good a low-power solution will be, perform a comparative analysis using a similar solution. Usually this method exposes the strengths and weaknesses of each solution.

Power Optimization on Low-Power Platforms

On low-power platforms, several aspects of power optimization typically come into play simultaneously. Here, we use a video playback application to illustrate various power-optimization approaches. Each contributes a different amount to the overall power savings, depending on the application's requirements and the video parameters.[9]

Run Fast and Turn Off

In a hardware-accelerated video playback, the GPU does the heavy lifting by performing hardware-based decode, scaling, and post-processing, such as deblocking. Meanwhile, the CPU cores execute the OS processes, media frameworks, media application, audio processing, and content protection tasks. The I/O subsystem performs disk access and communication, while the uncore runs the memory interface. On a typical platform, these subsystems execute various tasks in sequential order. An obvious optimization is to exploit the overlap of tasks and to parallelize them so that the hardware resources can operate concurrently, making the most of the I/O bursts for a short time before becoming idle. In this case, the optimization focus is on reducing the active residencies and achieving power savings when the resources are idle.

Activity Scheduling

It is important to appropriately schedule the activity of the software and hardware, and make smart use of the memory bandwidth. In video playback, tasks are mainly handled by the CPU and the GPU, and are of three main categories: audio tasks, video tasks, and associated interrupts.

Audio activities are periodic with a 1 to 10 ms cycle, and are handled by a CPU thread that schedules audio decode and post-processing tasks, as well as related audio buffers for DMA operations. In addition to the regular periodic behavior of the audio DMAs, the DMA activity involves Wi-Fi and storage traffic from I/O devices going into memory, which are somewhat bursty in nature.

Video tasks are partially handled by the CPU, which performs the audio-video demultiplexing, while the GPU performs the decoding and post-processing tasks. Parallelizing the CPU and the GPU tasks are obvious scheduling choices to lower the overall power consumption.

As CPU power consumption is influenced by how many times the CPU needs to wake up from a low-power state to an operating state, and the energy required for such state transitions, significant power savings can be achieved with a power-aware application that avoids such transitions. It does this by using interrupts to the maintain execution sequences, instead of timer-based polling constructs, and by appropriately scheduling them. Further optimization can be achieved by scheduling regularly occurring audio DMAs further apart, or even offloading audio to a dedicated audio-processing hardware.

---

[9]A. Agrawal, T. Huff, S. Potluri, W. Cheung, A. Thakur, J. Holland, and V.Degalahal, "Power Efficient Multimedia Playback on Mobile Platform," *Intel Technology Journal* 15, no. 2 (2011): 82–100.

Reducing Wake-ups

Streaming the video playback usually requires communication devices to have smoothing buffers while accessing the memory due to the bursty nature of the incoming network data packets. Smart communication devices can reduce the number of CPU wake-ups by combining the interrupts and by using programmable flush threshold, and by treading the path to memory that is already active.

Burst Mode Processing

Video playback is done on a frame-by-frame basis, while frame processing time is between 15 and 30 ms, depending on the frame rate. The nature of video playback does not offer the CPU much opportunity to go to a low-power state and return to active state within this short time. To overcome this limitation, a pipeline of multiple frames can be created such that the associated audio is decoupled and synchronized properly, thereby allowing the CPU to aggressively utilize its low-power states. Although this approach requires substantial software changes, it should nonetheless be explored for very low-power devices.

Improving CPU and GPU Parallelism

As the decoding and processing tasks are done by the GPU, the main tasks that remain for the CPU are preparing the hardware acceleration requests for decoding, rendering, and presenting the video frames. These tasks are independent and, therefore, can be implemented in separate hardware threads, which can be scheduled and executed in parallel. By parallelizing the execution threads, the CPU package can stay in deep idle states for longer periods, achieving power savings.

GPU Memory Bandwidth Optimization

The GPU performs video decoding and post-processing. These processes require the highest amount of memory bandwidth in the entire video playback application. But memory bandwidth scales with the content and display resolution, as well as with the amount of video processing done on each frame. Displaying a frame with a *blit* method (i.e., drawing the frame directly onto a display window) requires copying the frame twice: once to a composition target surface and once to a render surface. These costly copies can be avoided by using the overlay method, by which the kernel directly renders the frame to the overlay surface.

Display Power Optimization

Display requires more than a third of the device's power for video playback applications. There are many ways to address this problem, most of which were covered earlier. In addition, a media playback application can benefit from frame buffer compression and reduction of refresh rate.

Frame buffer compression does not directly impact the video window; but for full-screen playback when update of the primary plane is not needed, only the overlay plane can be updated, thereby saving memory bandwidth and, consequently, power. Reducing the refresh rate, on the other hand, directly reduces memory bandwidth. For example, reducing from 60 Hz to 40 Hz results in a 33 percent reduction in memory bandwidth. Not only is memory power reduced, owing to less utilization of display circuitry, but the overall display power is also lowered. However, depending on the content, the quality of the video and the user experience may be poorer compared to a video with full refresh rate.

Storage Power Optimization

Frequent access to storage results in excessive power dissipation, while reduction of such access allows the storage device to go to lower power states sooner and contributes to overall power savings. However, unlike the CPU, storage devices typically need to be idle for more than a second before they are transitioned to a lower power state.

As the requirement for media playback storage access is on the order of tens of milliseconds, normally a storage device would not get a chance to sleep during media playback. A power-aware media playback application, however, can pre-buffer about 10 seconds' worth of video data, which will allow a solid-state storage device to enter a low-power state. Storage devices based on a hard drive are slower, however, and require multiple minutes of pre-buffering for any meaningful power savings.

The Measurement of Low Power

Measuring low power is generally done following the same shunt resistor method as described in Chapter 6. However, precise measurement and analysis of several power signals may be necessary to determine the impact of particular power-consuming hardware units.

Processor Signals for Power

In a typical low-power Intel Architecture platform, such as the Intel Atom Z2760,[10] the processor signals that are defined for power interface are as shown in Table 7-4. Power analysis is done by appropriately measuring these signals.

---

[10] "Intel Atom Processor Z2760: Data Sheet," Intel Corporation, October 2012, retrieved from www.intel.com/content/dam/www/public/us/en/documents/product-briefs/atom-z2760-datasheet.pdf.

Table 7-4. Important Processor Signals for Power Interface

Signal	Description
V_{CC}	Processor core supply voltage: power supply is required for processor cycles.
V_{NN}	North Complex logic and graphics supply voltage.
V_{CCP}	Supply voltage for CMOS Direct Media Interface (cDMI), CMOS Digital Video Output (cDVO), legacy interface, JTAG, resistor compensation, and power gating. This is needed for most bus accesses, and cannot be connected to $V_{CCPAOAC}$ during Standby or Self-Refresh states.
V_{CCPDDR}	Double data rate (DDR) DLL and logic supply voltage. This is required for memory bus accesses. It needs a separate rail with noise isolation.
$V_{CCPAOAC}$	JTAG, C6 SRAM supply voltage. The processor needs to be in Active or Standby mode to support always on, always connected (AOAC) state.
LVD_VBG	LVDS band gap supply voltage: needed for Low Voltage Differential Signal (LVDS) display.
V_{CCA}	Host Phase Lock Loop (HPLL), analog PLL, and thermal sensor supply voltage.
V_{CCA180}	LVDS analog supply voltage: needed for LVDS display. Requires a separate rail with noise isolation.
V_{CCD180}	LVDS I/O supply voltage: needed for LVDS display.
$V_{CC180SR}$	Second generation double data rate (DDR2) self-refresh supply voltage. Powered during Active, Standby, and Self-Refresh states.
V_{CC180}	DDR2 I/O supply voltage. This is required for memory bus accesses, and cannot be connected to $V_{CC180SR}$ during Standby or Self-Refresh states.
V_{MM}	I/O supply voltage.
V_{SS}	Ground pin.

Media Power Metrics

Many media applications share characteristics of real-time requirement, bursty data processing, data independency, and parallelizability. So, in media applications, it is important to measure and track several power-related metrics to understand the behavior of the system and to find optimization opportunities. Among these metrics are SoC, display, voltage regulator and memory power as percentages of full platform power, the CPU core and package C-state residencies, the GPU activities and render cache (RC)-state residencies, memory bandwidth, and so on.

As certain media applications may be compute-bound, memory-bound, I/O bound, or have other restrictions such as real-time deadlines, determining the impact of these factors on power consumption provides better awareness of the bottlenecks and tradeoffs.

It is also important to understand the variation of power with respect to video resolution, bit rate, frame rate, and other parameters, as well as with regard to system frequency, thermal design power, memory size, operating system scheduling and power policy, display resolution, and display interface. Analyzing and understanding the available choices may reveal optimization opportunities for power and performance.

Summary

The marriage of low-power devices with increased demand for application performance, and the various challenges for attaining such low-power use, has been described in this chapter. As downscaling of process technology, together with voltage and frequency scaling, provides reductions in power, these techniques fall short of achieving state-of-the-art low-power design targets. Analysis of common low-power scenarios from a media usage standpoint shows that more aggressive power-reduction approaches are necessary while taking the whole system into account.

To this end, various power-management and optimization approaches were discussed. Low-power measurement techinques were also presented. Together, Chapters 6 and 7 provide a good platform for understanding the tradeoffs between increased dynamic ranges for frequency tuning and greater static power consumption—elements that must be carefully balanced. In the next chapter, some of these tradeoffs between power and performance are viewed from the point of view of a media application.

CHAPTER 8

Performance, Power, and Quality Tradeoff Analysis

When you're operating under resource constraints, it is necessary to understand the right tradeoffs to reach your goal. Often one thing must be given up to gain another thing. Depending on the objectives, priorities, and tolerances of a solution, an appropriate balance must be struck between the best use of available resources and the best achievable success measure.

In the case of creating compressed video, measures are used to obtain the best quality and the highest performance at the cost of the least number of bits and the lowest power consumption. For an overall encoding solution, success criteria may also include choices among hardware-based, software-based, and hybrid encoding systems that offer tradeoffs in flexibility, scalability, programmability, ease of use, and price. Users of video coding solutions, therefore, need to be aware of the appropriate choices to be made among cost, adaptability, scalability, coding efficiency, performance, power, and quality so as to achieve a particular video coding solution.

Tradeoff analysis is useful in many real-life situations. Understanding the options, particularly in terms of performance, power, and quality, is a valuable capability for architects, developers, validators, and technical marketers, as much as it is helpful for technical reviewers, procurers, and end-users of encoding solutions. Making informed product decisions by assessing the strengths and weaknesses of an encoder, comparing two encoders in terms of their practical metrics, and tuning the encoding parameters to achieve optimized encoders are among the decision points offered by such analysis. Furthermore, when new features are added to an existing encoder, such analysis can reveal the costs and benefits of those new features in particular measures. This helps users decide whether or not to enable some optional encoding features under various constraints and application requirements.

As coding efficiencies in terms of rate distortion of various algorithms were covered in previous chapters, here we turn the discussion toward an examination of how tradeoff analysis actually works. We focus on three major areas of optimization and the tradeoffs inherent in them—namely, performance, power, and quality. These three areas are of critical importance in present-day video encoding usage models and encoding solutions.

The discussion starts with the common considerations of tradeoff analysis involving these three measures, along with other options that may appear. This is followed by a discussion of the effects of encoding parameter tuning on these three measures. We then briefly discuss a few common optimization strategies and approaches.

With these discussions we present case studies of tradeoff analysis that look at performance power, performance quality, and power quality. These examples view the variables from several different points of view, shedding light on the methodologies commonly used in such analyses.

Considerations in a Tradeoff Analysis

Tradeoff analyses are essentially decision-making exercises. With the wide variability of video complexities and the numerous combinations of tuning parameters available, tradeoffs are commonly made based on the application's requirements for enhanced visual experience. The outcome of a tradeoff—namely, the properties of the encoded bitstream—determine the worthiness of the analysis. It is imperative that an encoded bitstream be syntactically valid, but its merit is typically judged in terms of its properties, including the amount of compression, the perceived quality when decoded, the amount of time it took to generate it, and its power consumption.

An important application of tradeoff analysis is a comparison of two encoding solutions based on both's performance, power, and quality. To make a fair comparison in such cases, various system parameters must be considered and made as equivalent as as possible. Such considerations include:

- The configurable TDP (cTDP) or scenario design power (SDP) settings (in fourth-generation Intel Core or later processors) such as nominal TDP, cTDP up, or cTDP down, which are usually done to accommodate overclocking or available cooling capacities.

- Power mode, whether AC (plugged in) or DC (battery).

- Operating system graphics power settings and power plans, such as maximum battery life and balanced or maximum performance.

- Display panel interface, such as embedded display port (eDP) or high-definition multimedia interface (HDMI).

- Display resolution, refresh rate, rotation and scaling options, color depth, and color enhancement settings.

- Number and type of display units connected to the system (e.g., single or multiple, primary or secondary, local or remote).

- Overhead and optimizations, application's settings for color correction and color enhancement, driver settings, and so on.

- Firmware, middleware, and driver versions.

- Operating system builds and versions.

- • Operating voltage and frequency of the CPU and GPU.

- • Memory configuration and memory speed.

- • Source video content format and characteristics.

When conducting these comparisons, it is also necessary to turn off irrelevant applications or processes, leaving the test workload as the only one running on the platform. This ensures that the available resources are properly allocated to the workload and there is no resource contention or scheduling conflict. Futhermore, it is generally good practice to take the same measurement several times and to use the median result. This reduces any noise that may be present within the measurement tolerance. Keeping the system temperature stable is also necessary to reduce potential noise in the measured data; temperature contollers that automatically activate a dedicated fan when a higher than target temperature is sensed are typically used for this purpose.

Types of Tradeoff Analyses

For mobile devices, saving power is typically the high priority. Therefore, if a decision only moderately impacts visual quality but saves substantial power, that quality tradeoff in favor of power saving is usually preferred for low-power mobile devices. Generally, techniques that provide greater compression while keeping the visual quality level approximately the same are good candidates for tradeoffs. However, these techniques often come with higher complexity, imposing a greater demand on power or performance. For example, HEVC encoding offers improved efficiency at the cost of more complex compression compared to AVC. As such, HEVC encoding in lieu of AVC encoding for an HD video is not an automatic choice on a power-constrained mobile device. Thus, tradeoff analysis must consider the overall benefit or the net gain.

Priorities are driven primarily by the requirements of the usage models and hence they also govern the tradeoffs that are made. For example, consider a videoconferencing application versus a video transcoding application. For the former, the low delay and real-time requirements demand steady power consumption throughout the video session, while for the latter, a run-fast-and-sleep approach is more beneficial. Additional limits may be applied depending on the availablility of resources. For instance, certain techniques that trade visual quality for better performance may not always be feasible, owing to limitations of the system, including the TDP, maximum processor frequency limit, and so on. Similarly, although a low-level cache can increase performance in many video applcations, it may not be available in the system under consideration.

Effects of Parameter Tuning

Various encoding parameters affect the relationship between performance and power consumption, as well as visual quality. The motivation for such tuning parameters is often to expose opportunities for obtaining higher performance, better quality, or power savings. Further, these tuning exercises reveal whether there are inefficiencies in a non-optimized video application, in addition to potential causes for such inefficiencies, all of which can lead to better solutions.

Typically, the impact of such tuning is more easily seen in improved visual quality and performance, rather than in lowered power consumption. As elaborated in Chapters 4 and 5, many parameters have a significant impact on both performance and quality, including the video spatial resolution, frame rate, and bit rate; group of pictures structure; number of reference pictures; R-D optimization in mode decision and determination of motion vectors; adaptive deblocking filter; various levels of independent data units such as macroblocks, slices, frames, or group of pictures; multiple passes of analysis and processing, multiple generations of compression, and pre- and post-processing filters; and special-effects filters.

Some of these parameters have greater effects on the visual quality while others benefit performance and power; your parameter tuning efforts should take these relative benefits into account. For example, using B-pictures significantly affects both visual quality and performance, but using R-D optimization in mode decision and determination of motion vectors slows down the encoding speed more significantly than it improves visual quality. Similarly, using multiple slices slightly reduces visual quality, but it improves parallelizability and scalability and it offers performance and power-saving opportunities.

In addition, it is important to consider just how *much* power savings or improved performance can be achieved while the encoded video retains reasonable visual quality. The nature of the video content and the bit allocation policy are important considerations here.

Optimization Strategies

There are a few optimization strategies with regard to improving performance or saving power, usually without surrendering visual quality. These strategies are typically employed in optimizations of video coding applications and come with appropriate tradeoffs.

- **Reducing scheduling delay:** Batching allows a series of function calls, such as motion estimation calls, for various macroblocks to be done together. Further, it allows macroblock-row-level multithreading for appropriate parallel processing. Typically, all operations within a batch share the same memory surface, thereby improving data fetch and cache hit rate. However, the application must wait for the writes to complete for all macroblocks in a batch before it can read from the memory surface. This introduces a small delay, but that delay is nonetheless suitable for video applications such as video streaming. The performance benefits achieved by batching do not typically sacrifice visual quality, and they give plenty of headroom for other workloads running concurrently.

- **Optimizing slack time:** Proactive energy optimization by workload shaping[1] is another way to obtain power optimization. As opposed to worst-case design philosophy, the video codec implementation framework not only is aware of hardware-specific details but also proactively adapts the implementation strategy to offer the best possible resource utilization. When it's in a traditional reactive energy optimization approach, the system merely adjusts its execution speed to the changing workload by exploiting available slack time. In the proactive scheme, the implementation can alter the shape of the workload at a given time, thereby achieving ~50 to 90 percent more energy savings than traditional implementations. In this case, the slack is accumulated over multiple data units so that the underlying processor can use a more aggressive power-saving approach, such as a deep sleep, for the larger slack period. Further, by reordering the video frames within a tolerable latency increase, additional slack accumulation and consequent power savings are achieved. The proactive workload adaptation is done by using high-level complexity models, while the video codec framework interprets the models at run-time to choose the appropriate frequency and voltage of operation, thereby minimizing energy without loss in quality and without missing any deadlines for a frame.

- **Parallelizing tasks:** The benefits of the parallelization of tasks, data, and instructions have been discussed in Chapter 5. Parallelizing independent tasks and distributing them over multiple processors makes full use of available processing capabilities. This allows the processing to complete quickly and enables the processors to go to deeper sleep states for longer periods of time, thus achieving power savings. Pipelines of tasks also keep the resources busy for as long as necessary and minimize resource conflicts. Further, with appropriate design of parallel applications, bottlenecks can be removed by re-scheduling a task to a different processor when a processor becomes too slow or unresponsive. Prioritization of tasks on various processors also helps overall performance. However, parallelization has its potential disadvantages of added overhead, such as inter-processor communication or synchronization costs.

---

[1]V. Akella, M. van der Shaar, and W. F. Kao, "Proactive Energy Optimization Algorithms for Wavelet-based Video Codecs on Power-aware Processors," in *Proceedings of IEEE International Conference on Multimedia and Expo* (Amsterdam, The Netherlands: IEEE, July 2005), 566–69.

- **Optimizing I/O:** Besides maximizing the use of system resources for the shortest possible time, increasing the data access speed and reducing I/O bottlenecks have prime significance when making power versus processing delay choices. In some off-line video applications, such as cloud-based video distribution, it is possible to obtain better overall power profiles by using dedicated I/O processors while groups of parallel processors are encoding batches of video segments. However, a side effect of this technique is the increased delay; the final video bitstream can only be stitched together when all the encoding processors are done. Therefore, the number of groups served by the I/O processor becomes the parameter of a possible tradeoff. Data prefetching and streaming opportunities should also be exploited as much as possible, noting that video data is particularly amenable to such techniques.

- **Reducing compute operations:** Algorithmic optimization allows techniques such as threshold-based early termination of loops, or exploitation of SIMD-style parallelism. These techniques help reduce the number of compute operations. However, this requires extremely careful analysis to determine and understand the various tradeoffs involved; in some cases, the visual quality may be affected as well. Furthermore, optimizing the code by hand or using the various compiler optimization techniques has direct impact on performance and power consumption by reducing the number of instructions to be executed.

- **Optimizing the cost vs. benefit:** The cost of high performance in terms of implementation complexity, and consequently in terms of power consumption, should be always carefully considered. It may be necessary to redesign the performance optimization approaches to tackle power consumption. In an image-filtering experiment,[2] it was observed that the behavior of this workload is drastically different from, for example, a scalar-vector multiplication-accumulation workload, although both workloads are similarly parallelizable. In the image-filtering case, performance optimizations are possible owing to the available data-reuse opportunities. This is also true for many video coding and processing applications. However, the energy profile of the image-filtering process is irregular, owing to uneven data reuse and resource scheduling. The energy optimal point corresponds to a large unrolling factor, relatively modest array partitioning,

---

[2]B. Reagen, Y. S. Shao, G. Y. Wei, and D. Brooks, "Quantifying Acceleration: Power/Performance Trade-Offs of Application Kernels in Hardware," in *Proceedings of 2013 IEEE International Symposium on Low Power Electronics and Design* (Beijing: IEEE, September 2013), 395–400.

pipelined multipliers, and non-pipelined loops. The large unrolling factor allows for greatest utilization of loaded data for the reasonable bandwidth requirements. The bandwidth needs are usually amortized over multiple cycles, maximizing the reuse and efficiency of given resources. This results in a complex control flow and a non-intuitive energy optimal design solution. For such complex workloads, the cost of additional power consumption by a higher performing design may not always be justified.

The Performance–Power Tradeoff

Recall from Equation 6-1 that power is a linear function of frequency. As improved performance directly depends on increased frequency up to a certain frequency limit, it is desirable to increase the frequency while keeping the power consumption the same. To achieve this, the co-factors of frequency—namely, the voltage, the capacitance, and the activity factor—need to be reduced. Furthermore, leakage current needs to be reduced. The activity factor is typically reduced by using clock gating, while the capacitance is reduced by downsizing the gates. As mentioned in Chapter 6, lowering the voltage can only be done in the voltage scaling region until a minimum voltage is reached, which must be sufficient for transistors to operate. Note from Figure 6-7 that leakage is constant in the low-frequency V_{min} region, while in the voltage-scaling region the leakage keeps increasing at typical operating points. Leakage current is decreased by reducing the transistor width and using lower leakage transistors. However, lower leakage transistors are also slower compared to leaky ones. Therefore, hardware designers need to make the appropriate optimizations to maximize the frequency for a given amount of power consumption.

It is important to note that ~90 percent of all modern mobile platforms are power limited. Therefore, every bit of power savings is considered equivalent to a corresponding gain in frequency. Typically, in the V_{min} region of low-power platforms, ~10 percent power saving translates to some 15 percent gain in frequency, while in the voltage scaling region, ~20 percent power saving corresponds to a mere ~5 percent gain in frequency. Thus the importance of judicious hardware design for optimal voltage, capacitance, activity factor, and leakage cannot be overstated, particularly with regard to obtaining the highest frequency at a given power budget. Operating at higher frequency generally implies better performance for various applications.

For video applications, higher CPU or GPU operating frequencies provide faster encoding speed, but they also consume more energy. A tradeoff between energy consumed and encoding speed is thus necessary at the system-design and hardware-architectural level, particularly for GPU-accelerated encoders. The programmable part of the encoding should also maintain an appropriate balance between performance and power. Usually this is done by parallelizing the encoding tasks, by scheduling appropriate tasks among multiple threads of CPU and GPU, by migrating tasks between the CPU and the GPU on the fly, by adjusting the schedules of the tasks, and/or by optimizing resource utilization for individual tasks, all without significantly affecting visual quality. For example, depending on the complexity of the video content, encoding two slices of a picture in parallel can yield ~10 percent performance gain with negligible quality impact. Tuning of encoding parameters also affects the overall encoding speed and power consumption, as some hardware units, such as the bit-rate control units, scaling units, and so on, may be optionally turned off depending on the parameter setting. Such tuning, however, may influence visual quality.

Let's consider the following case study of a performance–power tradeoff for a video transcoding application. For comparison, the same tests are run on two platforms with different performance-power characteristics. Note that the transcoding comprises decoding of a compressed video into an uncompressed format, which is subsequently encoded using appropriate encoding parameters into the target video in compressed format. The decoding tasks in the transcode operation remain the same for the same source video content, and usually the decoding is much faster than the encoding. Thus, the overall transcode performance can be measured in terms of the encoding speed alone. As such, the terms *performance* and *encoding speed* are used interchangeably.

Case Study

Consider two transcoding workloads, each about 5 minutes long; both of them operate with a spatial resolution of 1920×1080p at 30 fps, and transcode from higher bit-rate H.264 input bitstreams into lower bit-rate bitstreams of the same format. Workload 1 consists of lower complexity frames with infrequent scene changes, fewer details, and slower motion compared to Workload 2.

The transcoding tasks are carried out on two Intel platforms. Table 8-1 shows the platform configurations.

Table 8-1. *Plaform Configuration for Transcode Experiment*

	Platform 1	Platform 2
Processor	4th-gen. Core i7	4th-gen. Core i5
# Cores	4	2
CPU frequency	2 GHz	1.3 GHz
CPU turbo	3.2 GHz	2.6 GHz
TDP	47 W	15 W
Cache size	6 MB	3 MB
Graphics	Intel (R) Iris Pro (TM) 5200	Intel (R) Iris Pro (TM) 5000
Max GPU frequency	1.2 GHz	1 GHz
Embedded DRAM	Yes	No
Memory	4 GB dual channel	4 GB dual channel
Memory speed	1333 MHz	1333 MHz

We also consider three sets of encoding parameters, numbered 1, 2, and 3. In each set, a combination of encoding parameters is used. Table 8-2 shows some of the important distinctions for each parameter set.

Table 8-2. *Important Differences in Settings for Each Parameter Set*

	Param Set 1	Param Set 2	Param Set 3
Configuration	Fast	Faster	Fastest
Motion estimation algorithm	Algorithm 1	Algorithm 1	Algorithm 2
Motion estimation search range	48 pixels	48 pixels	28 pixels
Fractional pixel motion compensation	1/8 pixel	1/8 pixel	1/4 pixel
Adaptive search	Yes	No	No
Multiple reference pictures	Yes	No	No
Multiple predictions	Yes	No	No
Macroblock mode decision	Complex	Complex	Simplified

Figure 8-1 shows transcode performance on the two platforms using the two workloads with various sets of encoding parameters. It is notable that, owing to the different complexities of the two workloads, the tuning of parameters affects them differently. It is also observed that the degree of such impact is different on the two different platforms.

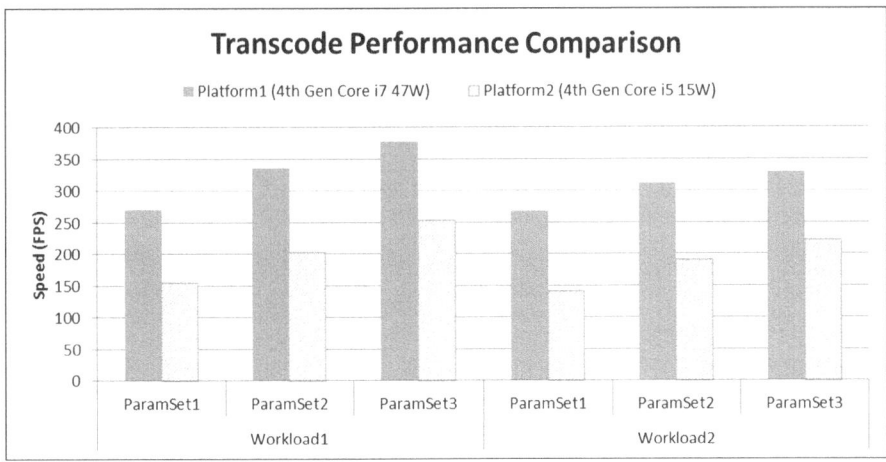

Figure 8-1. *Transcode performance comparison of two platforms*

From Figure 8-1 it can be noted that Platform 1 has an average of ~60 percent better throughput in terms of encoding speed compared to Platform 2, of which the embedded dynamic RAM provides ~10 percent performance throughput difference and the GPU frequency difference accounts for another ~20 percent. The remaining ~30 percent difference can be attributed to a combination of processor graphics hardware optimization, number of GPU execution units, cache size, number of CPU cores, CPU clock speed, turbo capacity, and so on.

While Workload 2 gives consistently increasing performance as the parameters move from fast to fastest cominations, especially on Platform 1, Workload 1 provides a peak performance of over 12-fold faster than real-time speed with parameter set 3. Therefore, it is clear that workload characteristics, along with parameter tuning, greatly influence the transcode performance. Comparing the fastest parameter set (3) for both workloads on Platform 1, it can be observed that Workload 1 provides ~13 percent better performance compared to Workload 2. On Platform 2, a similar trend is observed, where Workload 1 is ~12 percent faster compared to Workload 2.

Note that, owing to the characteristics of Workload 2 and to the constrained resources on Platform 2, parameter set 1 yields significantly lower performance on this platform because this parameter set includes multiple reference pictures, multiple predictions, and elaborate analysis for mode decisions.

Figure 8-2 shows the package power consumptions by the two platforms for the two workloads with the same sets of parameters.

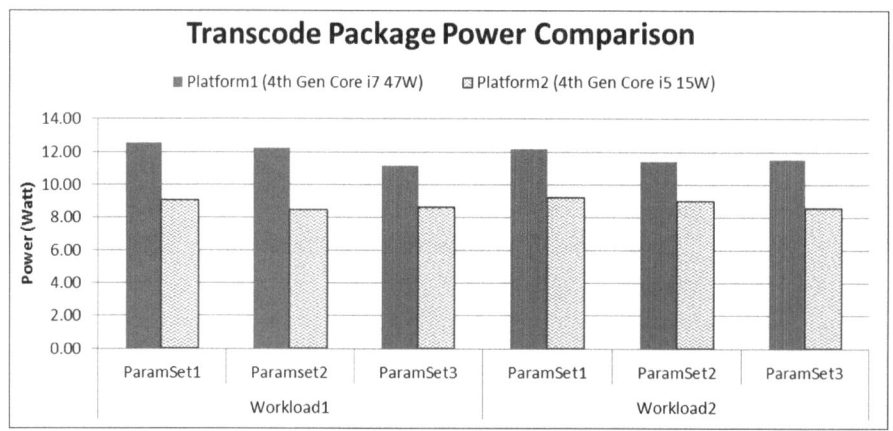

Figure 8-2. *Transcode package power consumption comparison of two platforms*

From Figure 8-2, it is clear that on, average, Platform 1 consumes ~34 percent more package power compared to Platform 2, while neither platform reaches its maximum TDP limit for the workloads under consideration. However, some parameter settings require certain hardware units to turn on and consume power, while others don't. This is evident from the difference in power consumption between the two platforms, ranging from ~14 percent to ~44 percent.

Interesting observations can also be made if the absolute power consumption is considered on each platform. As the parameters are tuned, the power consumption generally decreases, especially on Platform 1. Further, on Platform 1, Workload 1 has a 28 percent dynamic range of power consumption, while Workload 2 has a mere 8 percent dynamic range. On Platform 2, however, these numbers are ~7 and ~10 percent, respectively. This shows that on Platform 1, the power consumption of Workload 1 reacts more quickly to changing parameters compared to Workload 2. However, on Platform 2, these workloads are not compute-bound and therefore do not react to changing parameters. In this case, cache performance and number of GPU execution units become the dominant factors, with little regard to the encoding parameters.

Figure 8-3 shows the platform efficiency in terms of fps per watt for both workloads on the two platforms for each set of parameters. Platform 1 is generally more efficient than Platform 2, with an average of ~23 percent better efficiency, owing to its dedicated embedded dynamic RAM, higher GPU frequency, higher number of GPU execution units, and better cache performance.

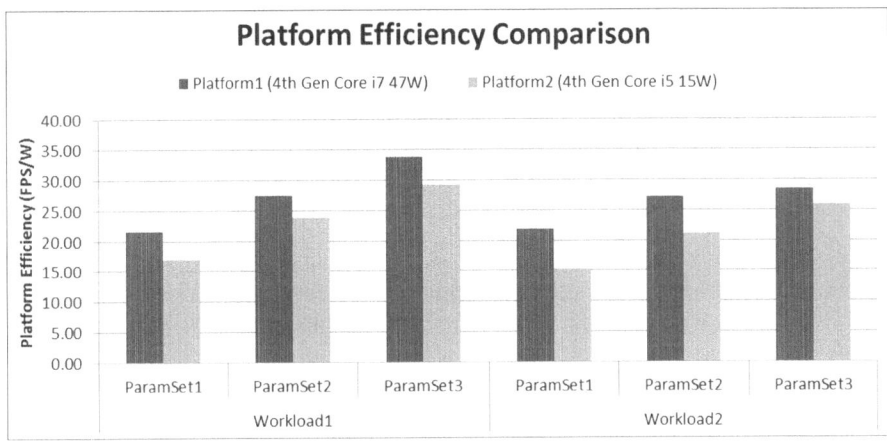

Figure 8-3. *Platform efficiency in terms of fps per watt during a set of transcode experiments*

From Figure 8-3 it is also observed that parameter tuning somewhat similarly impacts Workload 1 on both platforms, but for Workload 2, Platform 2 shows larger variation in terms of platform efficiency. This behavior of platform efficiency is not only due to changing parameters but also to the different characteristics of the workloads.

Figure 8-4 shows another point of view for performance versus power analysis. The two platforms are clearly showing different performance characteristics owing to differences in their available resources. Both workloads are clustered together on the two platforms. Because of the bigger cache size and the presence of an embedded dynamic RAM, Platform 1 generally consumes more power compared to Platform 2, but it provides much higher performance as well.

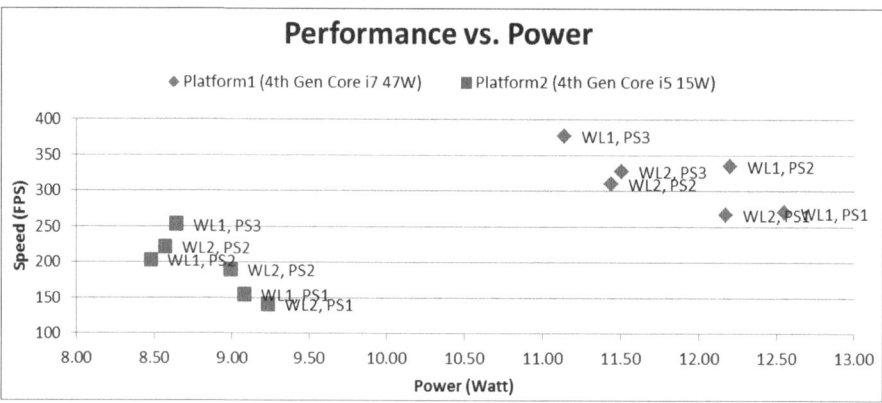

Figure 8-4. *Performance vs. power on the two platforms (WL and PS are abbreviations of workload and parameter set, respectively)*

From Figure 8-4 it can be observed that, on a given platform, appropriate parameter selections can provide good power-saving opportunities. For example, on Platform 2, Workload 1 can provide close to 1 watt of power saving using parameter set 3 compared to parameter set 1.

Note that, while one is performing a performance and power tradeoff analysis, employing parallelization techniques or optimizing resource utilization generally has little impact on visual quality. However, by tuning the encoding parameters, the quality is affected as well. In these cases, the power–performance tradeoff becomes a power–performance–quality three-way tradeoff. If the bit-rate control algorithm tries to maintain the same quality with a variable bit rate, resulting in different bitstream sizes, then the tradeoff becomes a power–performance–encoding efficiency three-way tradeoff. This is a side effect of the power–performance tradeoff.

The Performance–Quality Tradeoff

Higher encoding speed can be obtained by manipulating some video encoding parameters such as the bit rate or quantization parameter. By discarding a large percentage of high-frequency details, there remains less information to be processed, and thus encoding becomes faster. However, this directly affects the visual quality of the resulting video. On the other hand, using B-pictures offers a different performance-quality factor. Although a delay is introduced as the reference frames must be available before a B-picture can be decoded, the use of B-pictures generally improves the visual quality as well as the temporal video smoothness. For example, in a set of experiments with the H.264 encoding, we found that when we used two B-pictures between the reference pictures, the average impact on FPS was ~7 percent, but that some ~0.35 dB better quality in terms of BD-PSNR was obtainable for the same set of HD video sequences.

Similarly, manipulating parameters such as the motion search range, search method, number of reference pictures, two-pass encoding, and so on can impact both performance and quality. Therefore, it is necessary to always look into the potential impact on visual quality of a any performance gain or loss before considering a feature or parameter change in the video encoder. To illustrate the performance–quality tradeoff, we present two case studies and discuss the results obtained.

Case Study I

A 35 Mbps H.264 input bitstream is transcoded into another bitstream of the same format, but with a lower bit rate of 7 Mbps. The original video test clip is about 5 minutes long, with a spatial resolution of 1920×1080p at 30 fps. It comprises several scenes with varying complexities ranging from high spatial details to mostly flat regions, and from high irregular motion to static scenes. The transcoding tasks involve fully decoding the bitstream and re-encoding it with new coding parameters.

The transcoding tasks are carried out on a platform with configurations given in Table 8-3.

Table 8-3. *Platform Configuration for Transcoding in Case Study I*

System Parameter	Configuration
Processor	4th-gen. Core i5
# Cores	4
CPU frequency	2.9 GHz
CPU turbo	3.6 GHz
TDP	65 W
Cache size	6 MB
Graphics	Intel (R) HD Graphics (TM) 4600
Max GPU frequency	1.15 GHz
Embedded DRAM	Yes
Memory	4 GB dual channel
Memory speed	1333 MHz

Two transcoder implementations are used: a software-based transcoder running entirely on the CPU, and a GPU-accelerated transcoder where most of the compute-intensive tasks are done in special-purpose fixed-function hardware units. The two implementations optimize the parameters differently, but both offer three output modes of performance–quality tradeoffs: the best quality mode, the balanced mode, and the best speed mode.

Although the GPU-accelerated implementation provides only a few externally settable parameters, and while there are many choices available for the CPU-only implementation, effort is made to keep these paramteters as close as possible for both implementations. Surely, there are variations in the exact parameters that are tuned for a mode by the two implementations, but there are some commonalities as well. Table 8-4 summarizes the common parameters.

Table 8-4. Common Parameters of the Two Implementations

Parameter	Best Quality Mode	Balanced Mode	Best Speed Mode
Motion estimation and mode decision methods	Algorithm 1	Algorithm 2	Algorithm 3 (with early exits)
Fractional motion compensation	Eighth pixel	Quarter pixel	None
Reference pictures	Many	Few	Single
Adaptive search	Yes	No	No
Motion search range	Large	Medium	Small
Weighted prediction	Yes	Yes	No
Multiple B-pictures	Yes	Yes	No
Sub-macroblock partitions	All	Few	None
Scene change detection	Yes	Yes	No
Look-ahead analysis for bit rate control	Many frames	Few frames	No look-ahead

Note that these parameters are used slightly differently in the two implementations, so the exact same quality is not expected from the two implementations. Also note that the focus of the GPU-accelerated implementation is on achieving higher performance without losing much visual quality, thus only a few parameters are varied from the best quality to the best speed in this implementation. On the other hand, obtaining higher performance is difficult in CPU-only implementation; therefore, the best speed mode in this implementation turns off several features much more aggressively compared to the GPU-accelearted implementation.

The performance is measured in terms of FPS for the three modes of operation for both transcoder implementations. Note that the coding parameters are tuned for each of the three modes to obtain certain performance–quality tradeoffs. Figure 8-5 shows the transcode performance comparison between the CPU-only and the GPU-accelerated implementations. It also shows speedups of the different modes.

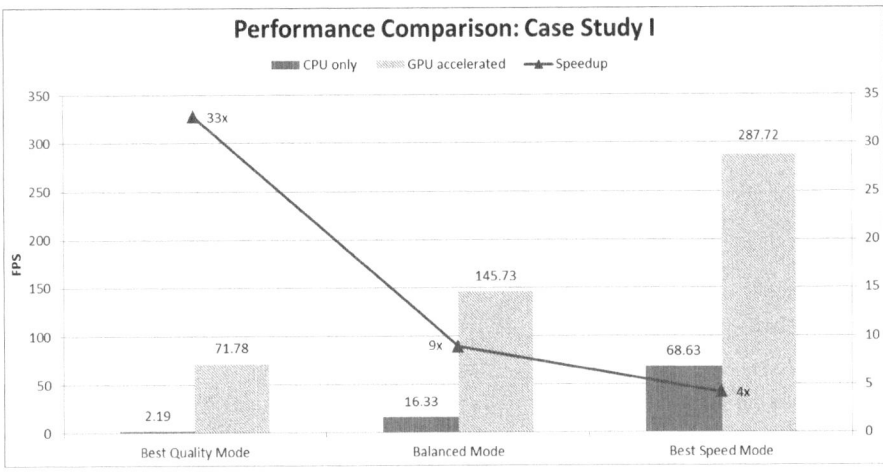

Figure 8-5. *Transcode comparison of various performance modes*

From Figure 8-5, we can see that both implementations scale in terms of speed from the best quality, to the balanced, to the best speed modes. For instance, the GPU-accelerated implementation speeds up the encoding from one mode to the next by a factor of approximately 2. However, with more aggressive tuning of the encoding parameters, the CPU-only implementation scales from the best quality to the balanced mode by performing the optimizations given in Table 8-5 and achieving a 7.45 times speedup. Similarly, from the balanced to the best speed mode, an additional 4.2 times speedup is obtained.

Table 8-5. *Optimizations in Different Modes for the CPU-only Implementation*

Parameters	Best Quality	Balanced	Best Speed
Motion estimation method	Uneven multihexagon search	Hexagonal search with radius 2	Diamond search with radius 1
Maximum motion vector range	24	16	16
Sub-pixel motion estimation	Yes	Yes	No
Partitions	All (p8x8, p4x4, b8x8, i8x8, i4x4)	p8x8, b8x8, i8x8, i4x4	No sub-macroblock partitions
Use trellis for mode decisions	Yes	No	No
Adaptive quantization	Yes, with auto-variance	Yes	No

(continued)

311

Table 8-5. (*continued*)

Parameters	Best Quality	Balanced	Best Speed
R-D mode decision	All picture types	I-picture and P-picture only	None
Max number of reference pictures	16	2	1
Number of references for weighted prediction for P-pictures	2	1	None
Number of frames to look-ahead	60	30	None
Max number of adaptive B-pictures	8	2	No B-pictures
CABAC	Yes	Yes	No
In-loop deblocking	Yes	Yes	No
8×8 DCT	Yes	Yes	No
Scene change detection	Yes	Yes	No

Obviously, these optimizations take a toll on the visual quality, as can be observed from Figure 8-6, which shows the quality comparisons for the two implementations. From the best quality to the best speed, the CPU-only implementation loses on an average of about 5 dB in terms of PSNR, with a tiny reduction of less than 0.1 percent in file size. On the other hand, with the focus on performance improvement while maintaining visual quality, the GPU-accelerated implementation does a good job of losing only an average of about 0.6 dB of PSNR from the best quality to the best speed mode. However, this implementation ends up with a ~1.25 percent larger file size with the best speed mode compared to the best quality mode, thereby trading off the amount of compression achieved.

Figure 8-6. *Quality comparisons of the two implementations*

Another observation can be made from Figures 8-5 and 8-6; in terms of speed for the three performance modes, the GPU-accelerated implementation is faster than the CPU-only implementation by factors of approximately 33, 9, and 4, respectively. This shows the contrast between the two implementations in terms of parameters tuning. While the GPU-accelerated implementation starts with a much better performance in the best quality mode, it has an average of 1.76 dB lower PSNR with a ~1.5 percent larger file size compared to the best quality mode in the CPU-only implementation. Thus, it has already sacrificed significant visual quality in favor of performance. Further, the GPU-accelerated implementation is less flexible in terms of ability to change the algorithms, as some of the algorithms are implemented in the fixed-function hardware units. Nonetheless, in the best speed mode, this implementation shows an average of ~2.8 dB better PSNR, but with a ~2.9 percent larger file size compared to the CPU-only implementation. These results demonstrate the performance–quality tradeoff and the tuning choices inherent in the two implementations.

Figure 8-7 shows the encoded video quality versus the encoding speed for this case study. It is evident that quality and speed scale among the different modes for both CPU-only and GPU-accelerated implementations, although the rate of scaling is different for the two implementations.

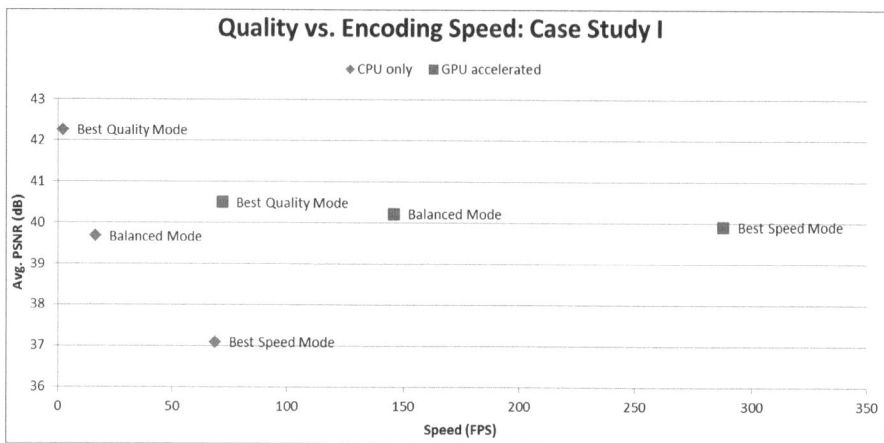

Figure 8-7. *Quality vs.encoding speed for case study I*

Case Study II

This second case shows another sample comparison of two encoding solutions in terms of performance and quality. A set of ten different video contents with varying complexities of motion and details are used. The video resolutions belong to the set {352×288, 720×480, 1280×720, 1920×1080}. Seven sets of video encoding parameters are used, providing a range between best quality and best speed. Encoding tests are carried out using two GPU-accelerated encoder implementations.

In this example, both encoding solutions operate on similar application program interfaces, such that parameter set 1 provides the best quality and parameter set 7 gives the best speed, although there are some differences between a parameter set for Encoder 1 compared to the same level of parameter set for Encoder 2. For example, parameter set 1 for Encoder 1 includes ⅛ pixel precision motion compensation and the use of trellis for mode decision, while Encoder 2 does not include these parameters in its parameter set 1. Some important parameters that are common to both two encoders are shown in Table 8-6.

Table 8-6. *Important Common Parameters between Encoder 1 and Encoder 2*

	PS1	PS2	PS3	PS4	PS5	PS6	PS7
8×8 transform	Yes	Yes	Yes	Yes	Yes	Yes	Yes
¼ pixel prediction	Yes	Yes	Yes	Yes	Yes	Yes	No
Adaptive search	Yes	Yes	Yes	Yes	Yes	No	No
Max references	10	8	6	5	4	2	2
Multiple prediction	Yes	Yes	Yes	Yes	P-picture only	No	No

Figure 8-8 shows the performance comparison between the two encoders in terms of FPS for each of the parameter sets. For both encoders, there are clear trends of improved performance with the progress of the parameter sets. However, the rates of improvement are different for the encoders. While Encoder 2 reaches the best performance of close to nine-fold faster than the real-time performance much more aggressively after parameter set 3, Encoder 1 displays a comparatively gradual rate of rise in performance as it steadily reaches about the same performance by tuning the parameters.

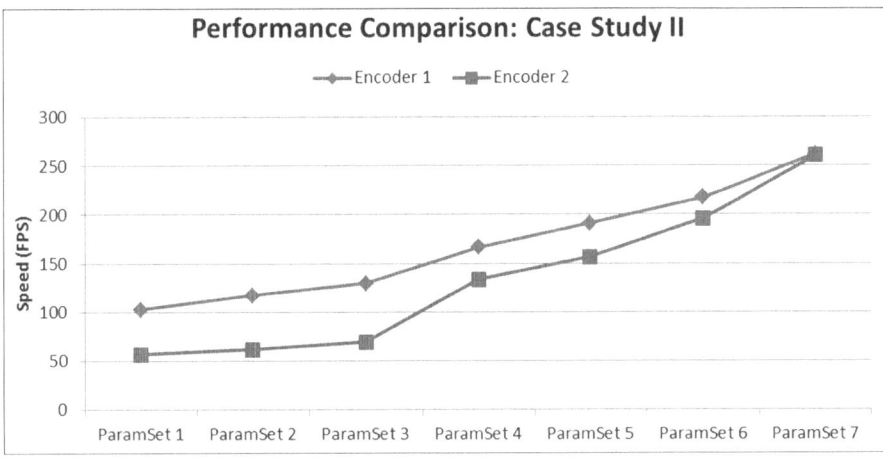

Figure 8-8. *Performance comparison of the two encoders*

Figure 8-9 shows a quality comparison between the two encoders in terms of BD-PSNR with respect to the parameter set 7 of Encoder 2. Again, clear trends of gradually lower quality are observed for both encoders. For Encoder 2, tuning the parameters can yield up to ~0.45 dB gain in BD-PSNR, while Encoder 1 reaches a level of ~0.47 dB quality gain. However, for Encoder 1, the mid-levels of parameter tuning do not show significant quality differences. Noticeable quality improvement for Encoder 1 happens between parameter sets 7 and 6, and between parameter sets 2 and 1.

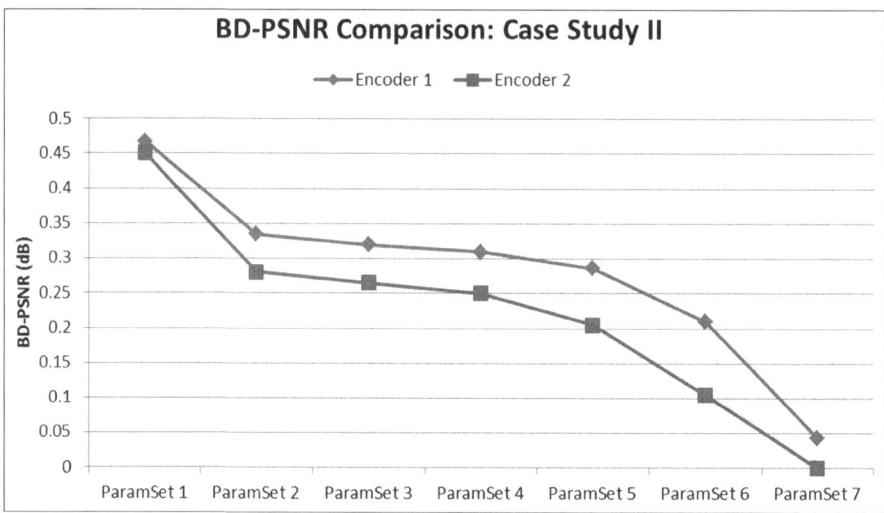

Figure 8-9. *Quality comparison of the two encoders*

Figure 8-10 shows quality versus encoding speed for the second case study. In general, for all sets of parameters, Encoder 1 provides better quality for a given encoding speed compared to Encoder 2. It is also clear that parameter set 1 truly represents the best quality mode, while the set 7 represents the best speed for both encoders. For Encoder 1, parameter set 6 appears to be the most effective, as it provides the largest quality difference (~0.1 dB BD-PSNR), but at the same time it also provides over 11 percent improvement in encoding speed compared to Encoder 2.

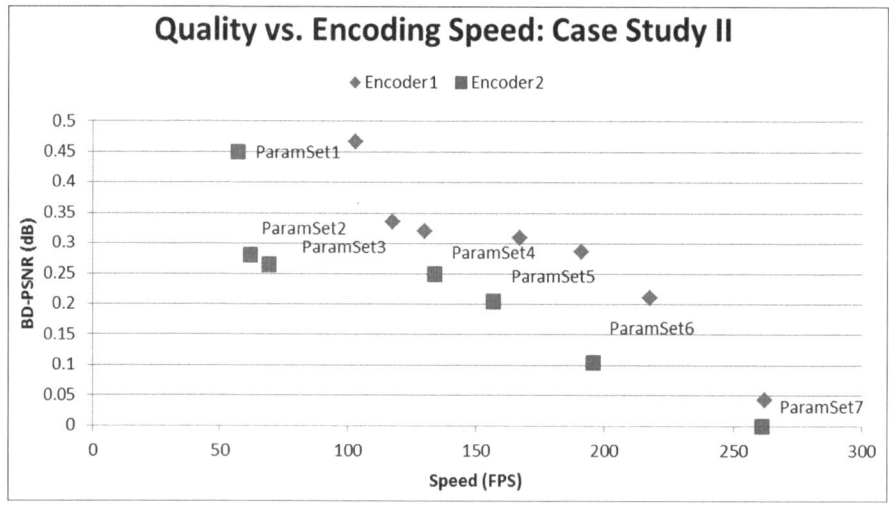

Figure 8-10. *Quality vs. encoding speed for case study II*

The Power–Quality Tradeoff

Noise is one of the most critical problems in digital images and videos, especially in low-light conditions. The relative amount of brightness and color noise varies depending on the exposure settings and on the camera model. In particular, low-light no-flash photo- and videography suffers from severe noise problems. The perceptual quality of video scenes with chroma noise can be improved by performing *chroma noise reduction*, alternatively known as *chroma denoise*. A complete elimination of brightness or luma noise can be unnatural and the full chroma noise removal can introduce false colors, so the denoising algorithms should carefully adapt the filtering strength, depending on the input local characteristics. The algorithm should represent a good tradeoff between reduction of noise and preservation of details. An example of a GPU-accelerated implementation of a chroma denoise filter, as a video processing capability, is typically available as an image-enhancement color processing (IECP) option offered by the Intel processor graphics.

To demonstrate the power–quality tradeoff, we present a case study of chroma denoise filtration. While playing back a video, the chroma denoise filter detects noise in the two chroma planes (U and V) separately and applies a temporal filter. Noise estimates are kept between frames and are blended together, usually at 8-bit precision. As the the GPU-accelerated chroma denoise typically provides sufficient performance for real-time processing, on modern processor platforms the performance is not normally a concern. However, although the visual quality is expected to improve, the additional operations required by the chroma noise reduction filter means that extra power is consumed. Therefore, this case study illustrates a tradeoff between power use and quality.

Case Study

This example uses a third-generation Intel Core i7 system with CPU frequency 2.7 GHz, turbo frequency up to 3.7 GHz, 45 W TDP, and graphics turbo frequency up to 1.25 GHz. The screen resolution is set to be 1920×1080, the same as the resolution of the video content. The balanced OS power policy is used, and the operating temperature is kept at 50°C, which is typical with CPU fans as the cooling system. Two workloads are employed, consisting of playback of an AVC encoded and a VC-1 encoded Blu-ray disc along with chroma denoise filter. Note that the VC-1 encoded content has much higher scene complexity compared to the AVC encoded content.

Figure 8-11 shows effect of the chroma denoise filter on the package power. An average of ~7 percent additional power, up to ~0.48 Watts, is consumed owing use of the chroma denoise filter. This is a significant penalty in terms of power consumption.

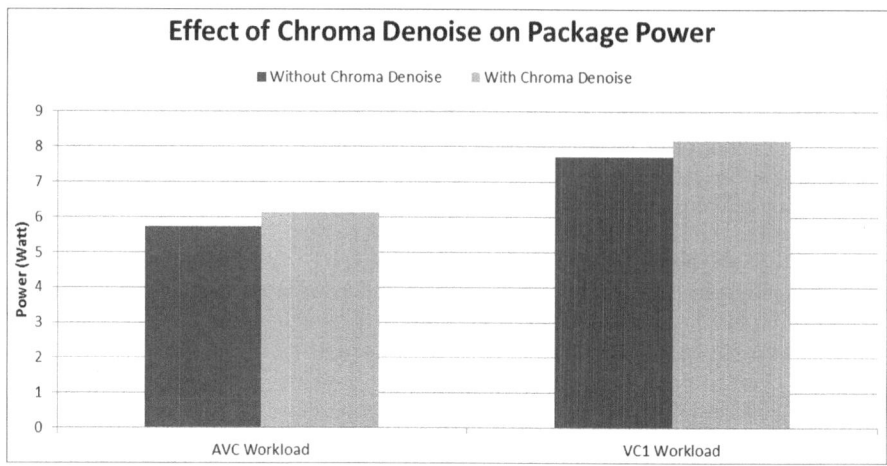

Figure 8-11. *Effect of chroma denoise filter on package power*

Figure 8-12 shows the effect of the chroma denoise filter on the combined CPU and GPU activity. From the activity point of view, there's an average of ~8.5 percent increase owing to the chroma denoise. This corresponds well with the increase in power consumption, and also represents a substantial increase in the time during which the processor is busy.

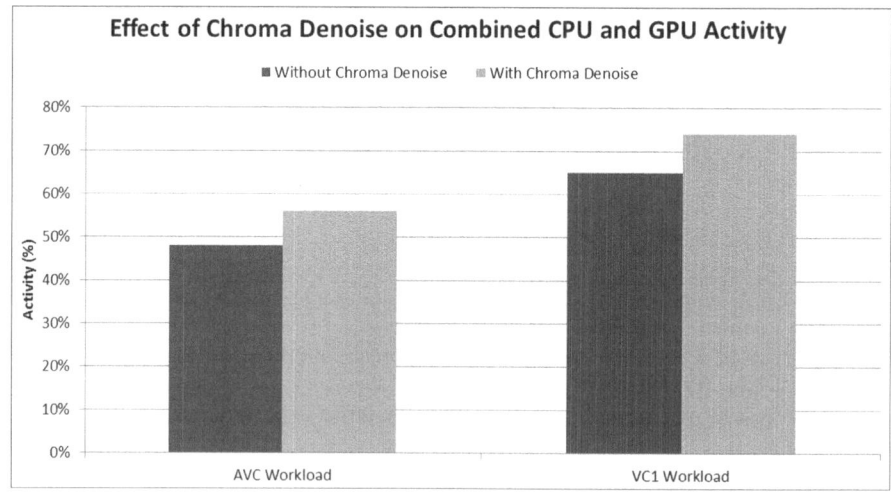

Figure 8-12. *Effect of chroma denoise filter on combined CPU and GPU activity*

Figure 8-13 shows the effects of the chroma denoise filter on visual quality, in terms of PSNR. Although an average of ~0.56 dB improvement is seen in PSNR, the perceived impacts on visual quality for these workloads are small. Note that the absolute PSNR value for the VC1 workload is about 4 dB lower than the AVC workload—this is due to the inherent higher complexity of the VC-1 encoded video content compared to the AVC encoded content.

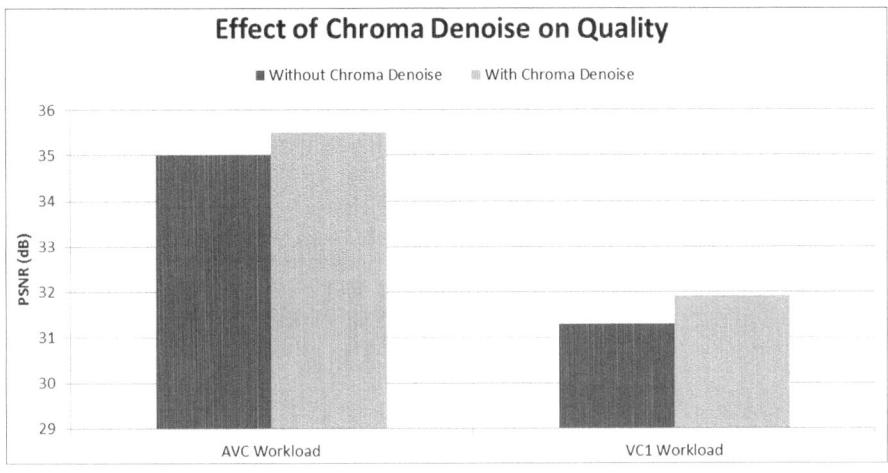

Figure 8-13. *Effect of chroma denoise filter on perceived visual quality*

Figure 8-14 shows the power–quality tradeoff for the chroma denoise filter. Although improved PSNR is observed, such improvement comes at the expense of substantially increased power consumption. Therefore, on power-constrained platforms, this option should be carefully considered. It is possible, for example, that upon detecting a lower level of battery availability, a power-aware playback application would automatically turn off the optional chroma denoise so as to save power.

Figure 8-14. *Power–quality tradeoff for the chroma denoise filter*

Summary

In this chapter we discussed how tradeoff analysis works and we provided four practical examples. While a tradeoff analysis can involve consideration of many different dimensions, we focused on performance, power, and quality, which are the foremost criteria for success in today's low-power computing devices. While power consumption defines the achievable battery life, it joins highest encoding speed and best visual quality as the most desirable features of contemporary video coding solutions.

We discussed encoding parameter tuning and examined some optimization strategies. Further, we offered case studies that embodied performance–power, performance–quality, and power–quality tradeoffs. These case studies reflect several different points of view and help clarify the methodologies commonly used in such analyses.

CHAPTER 9

Conclusion

The increasing demand for compressed digital video and the associated computational complexity, the availability and evolution of video coding standards, the restrictions of low-power computing devices, particularly in mobile environment, the requirements of increased speed and efficient utilization of resources, the desires for the best visual quality possible on any given platform, and above all, the lack of a unified approach to the considerations and analyses of the available tradeoff opportunities—these have been the essential motivating factors for writing this book. In this final chapter, we summarize the book's key points and propose some considerations for future development in the field.

Key Points and Observations

Based on the key points made in this book, the following observations can be made:

- Tradeoffs are possible among the various video measures, including the amount of compression, the visual quality, the compression speed, and the power consumption.

- Tuning of video encoding parameters can reveal these tradeoffs.

- Some coding parameters influence one video measure more than the others; depending on the application, optimization of certain measures may be favored over others.

- Analyzing the impact of various coding parameters on performance, power and quality is part of evaluating the strength of a video coding solution.

- Some video coding solutions are more suitable for certain types of video uses than for others, depending on the optimization performed and the parameters tuned.

- Being able to compare two video coding solutions is not only useful in ranking available solutions but also valuable in making informed choices.

- Sampling and scanning methods, picture rates, color space, chromaticity, and coding formats are among the parameters defined by the ITU-R digital video studio standards in three recommended specifications.

- Visual quality degradation is an option, owing to the natural tolerant characteristics of the human visual system (HVS) and the fact that the HVS is more sensitive to certain types of visual quality loss than others. Many video compression techniques and processes take advantage of this fact and trade quality for compression.

- Chroma subsampling is a common technique to take full advantage of the HVS sensitivity to color information. Many video usages make use of 4:2:0 chroma subsampling.

- Various techniques are available for digital video compression. Most international standards adopt transform-based spatial redundancy reduction, block-matching motion compensation-based temporal-redundancy reduction, and variable-length code-based spectral redundancy-reduction approaches for lossy predictive coding.

- International standards define a range of video applications in the domains of practical compression, communication, storage, broadcast, gaming, and so on. Standard video formats are essential for exchanging digital video among products and applications. The algorithms defined by the standards are implementable in practical hardware and software systems, and are common across multiple industries.

- Video compression is influenced by many factors, including noise present at the input, dynamic range of input pictures, picture resolution, artifacts, requirements for bit rate, frame rate, error resiliency, quality settings (constant or variable from picture to picture), algorithm complexity, platform capabilities, and so on.

- Lossy compression introduces some visual quality impairment, but owing to HVS limitations, a small amount of quality loss is not too objectionable. Common compression artifacts include quantization noise, blurring, blocking, ringing, aliasing, flickering, and so on. Quality is also affected by sensor noise at the video capture device, video characteristics such as spatial and temporal activities, amount and method of compression, number of passes or generations of compression, errors during transmission, and artistic visual effects introduced during post-production.

- The opinions of human viewers are the most important criteria in judging the visual quality of compressed videos, but the opinions are subjective, variable and cannot be repeated reliably.

- Objective measures such as PSNR and SSIM are widey used in the evaluation of video quality. Although they do not correlate perfectly with human experience, they provide a good estimate of visual quality. However, when judging the output of an encoder, such objective measures alone are not sufficient; the cost in terms of bits spent must also be considered.

- Many video quality-evaluation methods and metrics are offered in the literature, with varying levels of complexity. This is an active area of academic research as well as emerging ITU standards.

- Several encoding parameters can be tuned to trade video quality for performance and power consumption. Important parameters here include the bit rate, frame rate, and latentcy requirements; bit-rate control type, available buffer size, picture structure, and picture groups; motion parameters, number of reference pictures, motion vector precision, and motion vector search and interpolation methods; entropy coding type; number of encoding passes; and so on.

- Coding efficiency is determined in terms of the quality achieved in regard to the number of bits used. In the literature, *coding efficiency* is often used to mean performance. However, in this book, *performance* refers to the coding speed.

- Encoding speed is determined by several factors, including the platform and the video characteristics. Platform characteristics include the CPU and GPU frequencies, operating voltages, configurable TDP state, operating system power policy, memory bandwidth and speed, cache policy, disk access speed, I/O throughput, system clock resolution, graphics driver settings, and so on. Video characteristics include formats, resolutions, bit rate, frame rate, group of picture structure, and other parameters. Video scene characteristics include the amount of motion, details, brightness, and so on.

- Various parallelization opportunities can be exploited to increase encoding speed. However, costs of task scheduling and interprocess communication should be carefully considered. Parallelization approaches include data partitioning, task parallelization, pipelining, data parallelization, instruction parallelization, multithreading, and vectorization.

- Faster than real-time encoding is useful in applications such as video editing, archiving, recording, transcoding, and format conversion.

- Visual communication and applications such as screen cast require low-delay real-time encoding, typically on resource-constrained client platforms.

- Performance optimization implies maximal utilization of available system resources. However, power-aware optimization approaches maximize the resource utilization for the shortest possible duration and allow the system to go into deeper sleep states for as long as possible. This is in constrast to the traditional approach of only minimizing the idle time for a given resource.

- Algorithmic optimization, code and compiler optimization, and redundancy removal are noteworthy among the various performance-optimization approaches.

- There are several power-management points in the system, including the BIOS, the CPU, the graphics controller, the hard disk drive, the network, and the display. Memory power management is also possible, but is done infrequently.

- Typically, hardware-based power management involves the various CPU C-states and the render C-states. Software-based power management in the operating system or in the driver includes CPU core offline, CPU core shielding, CPU load balancing, interrupt load balancing, CPU and GPU frequency governing, and so on.

- On low-power platforms, special hardware units are typically needed for power management. Multiple points of power at various voltage levels constitute a complex system, for which fast and precise management of power requirements is handled by these special-purpose units.

- The goal of power management is to allow the processor to go into various sleep states for as long as possible, thereby saving power consumption.

- Total power consumption includes dynamic power and static leakage power; dynamic power depends on the operating voltage and frequency, while static power depends on the leakage current.

- A minimum voltage is required for the circuit to be operational, regardless of frequency change. The maximum frequency at which the processor can operate at minimum voltage ($F_{max}@V_{min}$) is the most power-efficient operating point. Increasing the frequency from this point also increases the dynamic power at a cubic rate and the static power at a linear rate. At this relatively high power, a power reduction can happen with an easy voltage–frequency tradeoff. Reducing the frequency below the most efficient point—that is, into the V_{min} region—reduces the dynamic power linearly while the static power remains constant, drawing a constant leakage current.

- Power optimization can be done at the architecture level, at the algorithm level, at the system integration level, and at the application level.

- On low-power platforms, some practical tradeoffs are possible among processor area, power, performance, visual quality, amount of compression, and design complexity. It may be necessary to sacrifice visual quality in favor of power savings on these platforms.

- The display consumes a considerable portion of the system power for video applications—in many cases, about a third of the system power. There are several display power-management techniques, including panel self-refresh, backlight control using the Intel display power-saving technology, ambient light sensors, and content adaptivity.

- Low-power software design considerations include intelligent power awareness, quality requirements, availability of hardware-acceleration capabilities, energy-efficient UI, code density and memory footprint, optimization of data transfer and cache utilization, parallel and batch processing, and so on.

- Low-power architectural considerations include combining system components on the same chip, optimized hardware-software interaction, workload migration from general-purpose to fixed-function hardware, CPU-GPU power sharing, reduced power core, uncore and graphics units, use of power islands, power-aware simulation and verification, and so on.

- Power-optimization approaches include running fast and turning off the processor, scheduling of tasks and activities, reducing wakeups, burst-mode processing, reducing CPU-GPU dependency and increasing parallelism, GPU memory bandwidth optimization, and power optimization for the display and the storage units.

- To measure power and performance for tradeoff analysis, you calibrate the system and select appropriate settings for operating temperature, voltage and frequency, cTDP, AC or DC power mode, OS power policy, display settings, driver settings, application settings, encoding parameters, and so on.

- The tradeoff analysis discussed in Chapter 8 attempts to fill a void that presently exists in comprehensive analysis methods. Particularly, it is important to examine the impact of tuning various parameters to obtain a better understanding of the costs and benefits of different video measures.

- Understanding the tradeoffs among performance, power, and quality is as valuable to architects, developers, validators, and technical marketers as it is to technical reviewers, procurers, and end-users of encoding solutions.

Considerations for the Future

The topics covered in this book will, I hope, inspire discussions that will take us into the future of video coding and related analysis. Some of the areas where future analysis is likely to extend are the following.

Enhanced Tools and Metrics for Analysis

Although it is possible to look into the details of performance, power, and quality in a given encoding test run, and to understand the relationships between them, it is not easy to determine why there is an increase or decrease for a given metric for that encoding run. This is especially difficult when comparing the results of two tests, likely generated by two different encoding solutions with different capabilities. Similarly, when comparing two metrics for the same run, it is not always obvious why there is an increase or decrease relative to each other. The complexity arises from the presence of many variables that react non-deterministically to changes in the system or video coding parameters, and that affect one another. Also, those influences are different for different video contents, applications, and usage scenarios. There needs to be study, as well as careful and time-consuming debugging, so we can understand these complex relationships.

Researchers are trying to come up with better video metrics, indices, and scores, particularly for visual quality, compression, performance, and power consumption. The analysis techniques are expected to adapt to more comprehensive future metrics. Eventually, there will be a single measure for all the benefits to weigh against a single measure for all the costs for video coding, and that this measure will be universally accepted for evaluation and ranking purposes. With the availability of the new metrics, enhanced benchmarking tools that consider all aspects of video coding are also expected.

Improved Quality and Performance

Techniques to improve visual quality with the same amount of compression will follow a path of continuous improvement. In the past couple of decades, this trend was evident in algorithms from MPEG-2 to AVC, and from AVC to HEVC. Similarly, optimization techniques for performance and power are improving at a rate even faster than that for quality improvement. Every generation of Intel processors is producing roughly 30 to 200 percent performance for the same power profile as compared to the previous generation for GPU-accelerated video coding and processing. Even low-power processors today are capable of supporting video applications that were only matters of dreams a decade ago. It is not far fetched to think that, with appropriate tradeoffs and optimizations, everyday video applications will have better visual quality despite platform limitations.

Emerging Uses and Applications

Wearables pose unique challenges when it comes to power consumption and performance, yet new uses on these emerging computing platforms are appearing every day. The role of video here is an open area of research. It embraces the notions of how to determine measures of goodness in video coding for these uses, how to quantify them, and which metrics to use.

The capabilities, uses, and requirements of video coding in driverless cars and radio-controlled drones are being assessed and developed. With their increasing processing abilities operating on resource-constrained systems, tradeoff analysis and optimization will play major roles in design and application. However, the methodologies and metrics for these uses are still open to definition.

Telemedicine, too, is in its infancy. Compression and communication technologies for high-resolution video are maturing to eventually reach flawless execution on handheld devices that can be used in remote surgical operations. Performance, power, and quality will be factors requiring tradeoffs in these scenarios as well.

Beyond Vision to the Other Senses

Of the five human senses, vision is considered the most important, but human experience is not complete with vision alone. Consequently, video is not the only data type for digital multimedia applications. Typically, audio and video are experienced together; touch and gestures are also rapidly evolving. So their measurement, understanding, and tuning will include audio, touch, and gesture. The relationships among these sense-based data types are complex and will require deep analysis, detailed study, and—ultimately—tradeoffs. This remains another active area of research.

As the challenges of the future are resolved, we will experience the true, full potential of the human senses.

APPENDIX A

Appendix

To the best of our knowledge, there is no benchmark available in the industry that is suitable for comparison of video encoding solutions in terms of performance, power, quality, and amount of compression. However, there is a well-known academic effort carried out by Moscow State University (MSU) to compare available codecs. This academic analysis is able to rank various software-based and/or hardware-accelerated encoder implementations in terms of objective quality measures. Obviously, it is possible to tune the parameters of an encoder to achieve higher coding efficiency, higher performance, or lower power use, resulting in a different ranking.

The discussion of this Moscow effort is followed by short descriptions of common industry benchmarks, which are generally limited to power and performance evaluations and do not consider other aspects of video coding. However, it is possible that new benchmarks will be suitable for a wider ranking of video encoding. Also included in this appendix is a brief list of suggested reading materials. Although existing references do not cover tradeoff analysis methods and metrics, they have in-depth discussions of certain topics only briefly mentioned in this book.

MSU Codec Comparison

A codec comparison project supported by the Computer Graphics and Multimedia Laboratory at Moscow State University compares the coding efficiency of various codecs.[1] The goal of this project is to determine the quality of various H.264 codecs using objective measures of assessment. The annual project reports are available from 2003 to 2012.

---

[1] D. Vatolin et al., *MSU Video Codec Comparison*, http://compression.ru/video/codec_comparison/index_en.html.

In the most recent comparison, done in 2012, the following H.264 encoders were compared:

- DivX H.264 software

- Elecard H.264 software

- Intel QuickSync H.264 encoder using Intel third-generation Core processor graphics

- MainConcept H.264 software

- MainConcept CUDA based H.264 encoder

- XviD MPEG-4 Advanced Simple Profile software

- DiscretePhoton software

- x264 software

The contents of various complexities with resolutions ranging from 352×288 to 1920×1080 were used, including 10 standard-definition, 16 high-definition (HDTV), and five video-conferencing sequences. The PSNR, SSIM, and MS-SSIM were used as the comparison objective metrics on all the color planes Y, U, and V for all frames in the video sequences. In making the comparisons and ranking the encoders, the following facts were recognized:

- For an encoder, the output visual quality is not the same for different frames of the same video sequence. Thus, a fair comparison would consider whether the same frames are being compressed by the various encoders. Frame mismatch can easily make a difference in quality.

- Different encoders are tuned to different content types. In particular, the default settings of an encoder may be best suited for a certain content type or video usage model. Therefore, comparing encoders with default settings may not necessarily be fair.

- Compression quality considerably depends on coding parameters. Setting appropriate coding parameters based on practical usage models is important in obtaining a realistic evaluation of various encoders.

To make a fair comparison, codec settings provided by the developers of each codec were used. The target application was video transcoding, mainly for personal use. The fast presets were taken to be analogous to real-time encoding on a typical home-use personal computer.

The 2012 report ranked the eight codecs by considering the overall average achieved bit rates for approximately the same quality, and presented the following ranking based on this measure alone, without regard to encoding speed. Table A-1 shows the ranking:

Table A-1. *MSU Codec Ranking*

Rank	Codec	Overall Average Achieved Bit Rate for the Same Quality (in percentage of XviD bit rate, lower is better)
1	x264	51
2	MainConcept H.264 Software	62
3	DivX H.264	69
4	Elecard H.264	71
5	Intel QuickSync (3rd -gen. Core)	93
6	XviD	100
7	DiscretePhoton	121
8	MainConcept CUDA	137

While this comparison is useful to some extent, note that only the quality aspects are considered here, regardless of performance and power consumption tradeoffs. This is a weakness of this comparison methodology; choosing different parameters for an encoder could easily provide different coding efficiency than is used for the ranking.

The tradeoffs and methodologies discussed in this book are important for getting an understanding of the big picture. Comparison of encoders should always acknowledge and take into account the various options considered by the encoders for different usage models. An encoder implementation with default settings may work better than one for video conferencing, but may not be as good for transcoding applications. However, the encoding parameters exposed by an implementation may be tuned to obtain better quality or performance. Note that different encoders give different controls to the end-users. Knowledge of parameters for an encoder is necessary to achieve best results for particular scenarios.

Industry Benchmarks

Some common benchmarks in the industry are occasionally used by enthusiasts to compare processors and their graphics and video coding capabilities. Although these benchmarks may include some video playback tests, they are not generally suitable for comparing video encoders, owing to their limited focus. Nevertheless, a few such benchmarks are briefly described below. It is hoped that points made in this book will inspire establishment of benchmarks that overcome this shortcoming and eventually reflect a higher state of the art.

MobileMark 2012

MobileMark 2012 from BAPCo is an application-based benchmark that reflects patterns of business use in the areas of office productivity, media creation, and media consumption. In addition to battery life, MobileMark 2012 simultaneously measures performance, showing how well a system design addresses the inherent performance and power management.

Unlike synthetic benchmarks, which artificially drive components to peak capacity or deduce performance using a static simulation of application behavior, MobileMark 2012 uses real applications, user workloads, and datasets in an effort to reflect the battery life a user might experience when performing similar workloads. MobileMark is commonly used by PC OEMs, hardware and software developers, IT departments, system integrators, publishers, and testing labs, as well as information technologists and computer industry analysts.

However, MobileMark is targeted to run business applications such as Microsoft Office, and uses Adobe Premiere Pro CS5 and Adobe Flash Player 11 to perform the video processing tasks. While this provides an indication of system design and CPU capabilities, it does not take advantage of the GPU-acceleration opportunities available in modern systems. Furthermore, it does not take into account the visual quality of compressed video. Therefore, this benchmark is very limited in its scope.

PCMark and PowerMark

These benchmarks were developed by FutureMark. PCMark is a standard performance benchmarking tool for personal computers of various form factors. With five separate benchmark tests and battery life testing, it can distinguish the devices based on efficiency and performance. It allows measurement and comparison of PC performance using real-world tasks and applications. The applications are grouped into scenarios that reflect typical PC use in the home and office environments.

PowerMark is a battery life and power consumption benchmark designed for professional testing labs. It delivers accurate results from realistic productivity and entertainment scenarios.

Both PCMark and PowerMark have limitations similar to those of MobileMark, as they consider performance or power alone and do not incorporate appropriate tradeoffs. Therefore, using only these benchmarks for ranking video encoders is not sufficient.

GFXBench

GFXBench, previously known as GLBenchmark and DXBenchmark, is a unified 3D graphics performance benchmark suite developed by Kishonti Ltd., who also developed CompuBench (formerly CLBenchmark) for CPUs. It allows cross-platform and cross-API comparison of GPUs in smartphones, tablets, and laptops. GFXBench 3.0 is an OpenGL ES 3 benchmark designed for measuring graphics performance, render quality, and power consumption in a single application. It utilizes OpenGL ES 3 capabilities, such as multiple render targets for deferred rendering, geometry instancing, transform feedback, and so on. It generates relevant workloads and measurement targets on different graphic performance aspects.

However, GFXBench does not deal with natural or synthetic video playback, recording, transcoding, video coferencing, screencast, or similar workloads. Furthermore, 3D graphics such as video games are primarily concerned with real-time performance and good graphics render quality, while video encoding and transcoding online and off-line applications may benefit from faster than real-time performance and an acceptable level of playback quality. Since GFXBench does not consider compressed video or bit-rate variations in quality measurements, it is difficult to ascertain the actual cost of quality. In addition, GFXBench does not report the package power, leaving open the possibility of large variations in power consumption from use of peripheral devices, while the processor package power may have been quite stable.

Therefore, the current version of GFXBench is not sufficient for measuring video applications in terms of power, performance, and quality. Yet, it is encouraging to see some commercial tool developers starting to think in terms of performance, power, and quality; perhaps future versions of GFXBench will fill the gaps that exist today in tools and benchmarking areas.

Suggested Reading

Here are a couple of academic research efforts that may be of interest.

- H. R. Wu and K. R. Rao, eds., *Digital Video Image Quality and Perceptual Coding* (Boca Raton, FL: CRC Press, 2005).

Perceptual coding techniques discard superfluous data that humans cannot process or detect. As maintaining image quality, even in bandwidth- and memory-restricted environments, is very important, many research efforts are available in the perceptual coding field. This collection of research, edited by H. R. Wu and K. R. Rao, surveys the topic from a HVS-based approach. It outlines the principles, metrics, and standards associated with perceptual coding, as well as the latest techniques and applications.

The collection is divided broadly into three parts. First, it introduces the basics of compression, HVS modeling, and coding artifacts associated with current well-known techniques. The next part focuses on picture-quality assessment criteria; subjective and objective methods and metrics, including vision model-based digital video impairment metrics; testing procedures; and international standards regarding image quality. In the final part, practical applications come into focus, including digital image and video coder designs based on the HVS, as well as post-filtering, restoration, error correction, and concealment techniques.

This collection covers the basic issues and concepts along with various compression algorithms and techniques, reviews recent research in HVS-based video and image coding, and discusses subjective and objective assessment methods, quantitative quality metrics, test criteria, and procedures; however, it does not touch on performance, power, or tradeoff analysis.

- Ahmad and S. Ranka, eds., *Handbook of Energy-Aware and Green Computing* (Boca Raton, FL: CRC Press, 2012).

Some power-efficient techniques from various systems points of view, including circuit and component design, software, operating systems, networking, and so on, are presented in this book by Ahmad and Ranka. It is not specific to video applications; however, this two-volume handbook explores state-of-the-art research into various aspects of power-aware computing. Although one paper in the handbook discusses about a particular approach to mobile multimedia computing, future researchers may find some of the other optimization aspects and techniques useful in the general area of video encoding as well.

Index

A

Absolute category rating (ACR) method, 117
Advanced Configuration and Power
       Interface (ACPI) specification, 217
  compliance, 218
  power draw states
    device states, 219
    global states, 218
Advanced video coding (AVC) standard
  coding concepts, 72
  deblocking filter, 82
  encoder and decoder blocks, 76
  entropy coding, 81
  error resilience, 83
  flexible macroblock ordering, 73
  inter coding, 76
  interlaced coding, 82
  inter prediction modes, 78
  intra coding, 76
  intra prediction modes, 77
  NAL abstracts, 73
  picture structure, 75
  profiles, 73
  tools, 73
  transform and quantization, 79
  VCL unit, 73
Algorithmic optimization
  computational complexity reduction
    code parallelization and
      optimization, 240
    data type selection, 239
    memory transfer reduction, 242
  fast entropy coding, 188
  fast intra prediction, 186
  fast mode decision, 188
  fast motion estimation, 186
  fast transforms, 184–185
  goals, 238
  parallelization approaches
    data parallelization, 192–193
    data partitioning, 189
    instruction parallelization, 193
    multithreading, 195
    pipelines, 191–192
    task parallelization, 190–191
    vectorization, 197
Ambient light sensor (ALS), 283
Application-level optimization
  context awareness, 245
  GUID, 246
  user intervention, 246
Application-specific integrated circuits
    (ASICs), 6
Architectural optimization
  clock gating, 236
  dynamic voltage/frequency
    scaling, 235
  hardware-software partitioning, 234
  low-level caches, 237
  power gating, 235
  slice gating, 236
Arithmetic operations optimization, 200

B

Benchmarks, 331
  GFXBench, 332
  MobileMark, 332
  PCMark, 332
  PowerMark, 332
Bjøntegaard delta PSNR (BD-PSNR), 144
  advantages, 145
  limitations, 145

W, X, Y, Z